HETEROSIS IN PLANT BREEDING

HETEROSIS IN PLANT BREEDING

Proceedings of the Seventh Congress of

EUCARPIA

The European Association for Research in Plant Breeding

L'Association Européenne pour l'Amélioration des Plantes

Die Europäische Gesellschaft für Züchtungsforschung

Budapest 24-29 June 1974

Editors: A. Jánossy and F. G. H. Lupton

ELSEVIER SCIENTIFIC PUBLISHING COMPANY
AMSTERDAM – OXFORD – NEW YORK

The distribution of this book is being handled by the following publishers:

for the U.S.A. and Canada

American Elsevier Publishing Company, Inc.
52 Vanderbilt Avenue
New York, New York 10017

for East Europe, China, Democratic People's Republic of Korea, Cuba, the Democratic Republic of Vietnam and Mongolia

Akadémiai Kiadó, The Publishing House
of the Hungarian Academy of Sciences, Budapest (Hungary)

for all remaining areas

Elsevier Scientific Publishing Company
335 Jan van Galenstraat
P.O. Box 211, Amsterdam, The Netherlands

ISBN 0-444-99835-7

Joint edition published by
Elsevier Scientific Publishing Company, Amsterdam, The Netherlands
and Akadémiai Kiadó,
The Publishing House of the Hungarian Academy of Sciences, Budapest, Hungary

Printed in Hungary

CONTENTS

Session IV. HETEROSIS IN OUT-POLLINATING SPECIES

Chairmen: Professor G. Röbbelen and Professor E. Kurnik

Session V. HETEROSIS IN SELF-POLLINATING SPECIES

Chairmen: Professor H. Hänsel and Dr. S. Rajki

ABSTRACTS OF PAPERS SUBMITTED TO THE CONGRESS

GENERAL ASSEMBLY

ANNEX

EUCARPIA

Past Presidents

J. C. Dorst	(1956–1959)
W. Rudorf	(1959–1961)
J. Bustarret	(1962–1965)
E. Akerberg	(1965–1968)
G. Haussmann	(1968–1971)
F. G. H. Lupton	(1971–1974)

President

A. Jánossy

Vice-President

E. Sanchez-Monge

The Board

J. S. Bojanowski
S. Borojević
K. Z. Budin
C. Daskalov
C. Dorsman
F. Garcia-Olmedo
A. Hagberg
H. Holland
A. Jánossy
H. W. Kehoe
R. A. F. Laeremans
F. G. H. Lupton
M. Mesken
M. Mota
A. Munk
D. Palaversić
W. K. F. Plarre
G. Popow
E. Sanchez-Monge
G. T. Scarascia Mugnozza
J. Sneep
O. Steineck
E. Strand
P. Tigerstedt
G. Wricke

Chairmen of Sessions

1. *Biometrics in Plant Breeding*	Professor Dr. G. Wricke	Institut für Angewandte Genetik, Herrenhäuser Str. 2, Hannover, FRG
2. *Cereals*	Dr. H. Lamberts	Institute de Haaff, P. O. Box 117, Wageningen, The Netherlands
3. *Forage Crops*	Dr. G. Julen	Swedish Seed Association, S-26800 Svalöv, Sweden
4. *Fruit Plants*	Dr. C. Dorsman	Institute for Horticultural Plant Breeding, Mansholtlaan 15, Wageningen, The Netherlands
5. *Maize and Sorghum*	Dr. D. Palaversić	Institute for Breeding and Production of Field Crops, Zagreb, Yugoslavia
6. *Mutation and Polyploidy*	Professor Mrs. K. Borojević	Department of Genetics and Plant Breeding, Faculty of Agriculture, 21000 Novi Sad, Yugoslavia
7. *Ornamentals*	Dr. L. D. Sparnaaij	Institute for Horticultural Plant Breeding, Mansholtlaan 15, Wageningen, The Netherlands
8. *Physiology*	Dr. Z. Martinić	Department of Genetics and Plant Breeding, Faculty of Agriculture, Zagreb University, P. O. Box 1009, 41001 Zagreb, Yugoslavia
9. *Potatoes*	Dr. J. G. Th. Hermsen	Department of Plant Breeding, Agricultural University, Wageningen, The Netherlands
10. *Vegetables*	Dr. C. North	Scottish Horticultural Research Institute, Invergowrie, Dundee DD2 5DA, UK
11. *Wild Species and Primitive Forms*	Professor J. G. Hawkes	University of Birmingham, Department of Botany, P. O. Box 363, Birmingham B15 2TT, UK

INTRODUCTION

It was agreed at the VIth Congress held in Cambridge in 1971 that the VIIth Congress should be held in Budapest.

The Vice-President, Prof. A. Jánossy, undertook the responsibility of organizing the Congress. A Secretariat and a broader Organizing Committee were set up to arrange for the scientific programme, post-congress excursions and accommodation of the participants. The following served on the Organizing Committee:

A. Jánossy, *Chairman*
Mrs. A. Aradszky, *Secretary*
W. Jakabffy, *Secretary*

Members

E. Almási	I. Biacsi
G. Auguszta	S. Rajki
A. Bálint	I. Tamássy

The Congress took place from 24 to 29 June, 1974. The scientific meetings and General Assembly were held in the Congress Hall of the University of Horticulture. The Congress was attended by 312 guests from 31 countries, and was pleased to welcome Prof. T. Hosoda and H. Yamaguchi of S. A. B. R. A. O., Dr. C. F. Konzac of F. A. O., Dr. A. Micke of I. A. E. A. and – for the first time – eight delegates from the GDR.

The formal business of the meeting was introduced with a speech of welcome by Prof. A. Somos, Rector of the University of Horticulture. After some introductory words from Prof. A. Jánossy on behalf of the Organizing Committee, the President, Dr. F. G. H. Lupton, opened the Congress with a short opening address.

The scientific programme of the Congress was devoted to the general theme of "*Heterosis in Plant Breeding*" as a subject of paramount importance in up-to-date breeding work.

During the three symposium days 36 papers were delivered in the framework of five Sessions.

An informal reception by the Hungarian Minister of Agriculture and Food was held in the Grand Hotel at Margareth Island and a Conference Dinner took place in the Hotel Gellért.

The Organizing Committee takes this occasion to thank the Rector and the staff of the University of Horticulture for their attention to all the technical work involved with the Congress, and the Directors of the following Institutes for welcoming the participants of post-congress excursions: the Research Institute of Vegetable Crops, Kecskemét, the Research Institute of Horticulture, Budapest-Budatétény, the Experimental Station and Botanical Garden, Budapest-Soroksár, the Agricultural Research Institute, Martonvásár, the Research Institute of Crop Husbandry and Soil Conservation, Kompolt, the Research Institute of Cereal Crops, Szeged, the Institute of Agrobotany, Tápiószele, the Research Institute of Forage Crops, Iregszemcse, the Research Institute of Beet Crops, Sopronhorpács, Biological Centre of the Hungarian Academy of Sciences, Szeged.

The Committee also wishes to thank the following institutions for their generous financial assistance:

Head Centre of State Farms
Hungarian Tobacco Industry
National Brewery Company
Company for Seed Production and Seed Trade
Hungarian Sugar Industry
Cereal Trust
Trust of the Canning Industry
National Distillery Industry
Enterprise for Vegetable Oils and Detergents
Hungarian Refrigeration Industry

OPENING ADDRESS BY THE PRESIDENT, F. G. H. LUPTON

Professor Jánossy, Professor Somos, Ladies and Gentlemen, friends and colleagues, it gives me great pleasure to welcome you to this VIIth EUCARPIA Congress. This is the first occasion at which a General Congress of our Association has met in one of the Eastern countries of Europe, and we are all most grateful for all the careful preparations which have obviously been made, not only to make this an interesting and instructive Congress, but also to look after our physical needs and to ensure that we see something of this historic city and of the beautiful country in which it stands. The flags outside this Congress Hall indicate the worldwide interest which this meeting has attracted, and it is very pleasant to see so many visitors from countries beyond Europe. It also gives me great pleasure to welcome our colleagues from the GDR, who are attending a EUCARPIA Congress for the first time.

It is, I think, appropriate that the theme of this Congress should be "Heterosis in Plant Breeding". From the list of speakers and the range of subjects with which they will deal, we are obviously going to have an interesting and stimulating time. I feel, however, that this Congress, sited at the centre of gravity of Europe and attended by workers from East and West, has the possibility of stimulating a greater heterosis of scientific progress arising from mutual trust and an understanding not only of each other's problems, in feeding the world's increasing population, but of the efforts we are making to solve them. The British novelist Jonathan Swift once wrote that the man who could make two blades of grass grow where one grew before was worth more than all the politicians of the world together. That would be heterosis indeed, and sets a standard to which we all can strive to work. In declaring this Congress open, I would like to express the wish that the heterosis it stimulates may be deep seated and long lasting.

PROFESSOR ANDOR JÁNOSSY, 1908–1975

It is with great regret that we must include in the Proceedings of this VIIth Congress of Eucarpia an obituary notice of our late President, Professor Andor Jánossy, who was responsible for all the arrangements for the Congress and was elected our President at the General Assembly with which it closed. Those who worked closely with him throughout this period appreciate the meticulous care which he took to ensure that this, our first conference in an Eastern European country, should be a success both scientifically and in its attention to our comfort and enjoyment. But Professor Jánossy was more than this – throughout his term of office on our Board he made it his duty to extend the activities of EUCARPIA across the boundaries between East and West, making several journeys with officers of the Association to make the personal contacts upon which such links essentially depend. The wide distribution of the meeting places in the current programme indicates the success which he achieved. He was also greatly concerned with the problems of genetic conservation and did much not only to encourage the exploration and preservation of Hungarian varieties and ecotypes, but also to gain recognition of the importance of this problem on a worldwide scale, and to use the resources of EUCARPIA to help in the coordination of gene banks and collections in Europe and throughout the world.

Professor Jánossy was born in 1908 and graduated from the University of Budapest in 1930. Since this time he has been active in the field of plant breeding and seed production. His work has led to the production of three varieties of red clover, including the tetraploid variety HUNGAROPOLY, the first tetraploid variety to be released in Hungary and now widely grown in Europe and Canada. He was responsible for establishing the system and the institute for plant variety testing in Hungary, and founded in 1959 the National Institute of Agrobotany at Tápiószele, continuing as Director until his death on 4 May 1975. He leaves a wife and three sons, all of whom we met during the Congress; we extend them all our sincere sympathy.

Session I

GENETIC PRINCIPLES OF HETEROSIS

Chairmen:

PROFESSOR G. WRICKE AND DR. I. BÓCSA

Genetic and evolutionary principles of heterosis

J. MAC KEY

Department of Genetics and Plant Breeding, Agricultural College of Sweden, Uppsala, Sweden

SUMMARY

According to the broad, modern concept, there exists positive or negative heterosis, luxuriant, adaptive, selective or reproductive heterosis, and labile or fixed heterosis. They may all be governed by allelic, nonallelic, plasmatic and/or maternal interaction which in function can be described as being either sheltering, inhibiting, stimulating, complementing and/or dosage adjusting. Allelic heterosis depends on dominance or overdominance, nonallelic heterosis has transgressive, recombinative and epistatic components. The relative prevalence of the different mechanisms is dependent on the reproductive system and the stage of evolutionary development and direction. This may be pertinent, if the adjustment is not optimal, since dominance and nonallelic types of heterosis are to a lesser extent than overdominance to be taken as end stations in evolution or breeding.

In a population, heterosis creates a genetic élite paid for by a segregational genetic load. From this consideration of loss in total productivity, heterosis may indicate a less efficient system preserved mainly as a tribute to flexibility. Where a compromise is allowed, the heterotic expression tends to be preserved through either some kind of heterozygous or homozygous balance or by mere disomic propagation. Balanced heterozygosity works with smaller or larger gene blocks, balanced homozygosity with duplications preceded or followed by allelic diversification. Both methods function up to the complete preservation of a genotype. Homoeologous heterosis has a very efficient system in autogamous disomic polyploidy.

In modern agriculture and horticulture, where uniformity rather than flexibility is favoured, all more or less elaborate methods to preserve heterosis should be carefully considered. Recurrent selection and rotational crossbreeding are important initiatives by promoting permanent recombination under maintained variation.

DEFINITION OF HETEROSIS

The definition of heterosis is one of the more delicate and controversal affairs in genetic terminology, and the development of the concept over time is not without interest. Since the congress has chosen heterosis as its theme, it appears appropriate to start with an attempt to define what we will be going to talk about.

Shull (1914) introduced the term heterosis for the special stimulus of heterozygosis upon "cell-division, growth, and other physiological activities of an organism" and suggested later (1948) that the term "should be *free from every hypothesis*" and that "any subsequent student was free to propose his own explanation without thereby being obliged to abandon the word 'heterosis' ".

This generous invitation was vividly accepted with the effect that the original concept became considerably changed. His definition that "heterosis is the *increase* of size, yield, vigor, etc." was enlarged by evolutionists to cover also heterosis in survival, i.e. adaptive, selective, and reproductive advantage. Dobzhansky (1950) even suggested substitution of the term euheterosis for these added qualifications. As Hagberg (1952) remarks, it is hardly fair to use a term, once defined by another scientist, to cover a different concept, give a new definition and then criticize its use as originally defined. Shull (1948) apparently ignored this manoeuvre but expressed his dislike, when Powers (1944) and Stern (1948) extended the concept to cover inferiority as well. Powers' *(l. c.)* example on locule number in tomato may suffice to illustrate the understanding of negative heterosis (Table 1).

Table 1. A case of negative heterosis in tomato (Powers, 1944)

Parental line or hybrid	Mean number of locules per fruit
P_1: 4110	9.6 ± 0.328
F_1: 4110 × 4101	7.3 ± 0.249
P_2: 4101	11.7 ± 0.499

From the evolutionary point of view, negative heterosis is expected as the converse of positive heterosis. This is hardly a reason for trying to find an entirely new term, since the same hybrid may show relative superiority in one situation and relative inferiority in another, or a certain heterotic expression may be evaluated in two entirely opposite ways.

A few examples illustrating this point may be appropriate. Emsweller *et al.* (1962) observed that some hybrid seedlings from the lily cross *Lilium speciosum* × *L. auratum* contained higher concentrations of ferulic acid than the two parental species, a case of positive heterosis. Normally, this phenolic compound is detoxified by glycosidation. This process must at least partly have broken down in the hybrids, which gave rise to malignancy and a pronounced negative heterosis.

When the platyfish, *Xiphophorus (Platypoecilus) maculatus,* is crossed with the swordtail, *X. helleri,* some hybrid individuals also develop tumours, an invasive melanoma (Gordon & Gordon, 1957). This may appear to demonstrate only negative heterosis. Fishes with rapidly growing melanomas are, however, more resistant to certain undesirable environmental conditions, *e.g.* osmotic stress, than individuals without tumours (Anders *et al.,* 1962).

A third example may be taken to illustrate the lability of heterozygous advantage. In the Fayoumi breed of chicken, a monofactorial heterozygote for plumage colour and pattern was found to be socially inferior in 'peck order', if the individuals were raised together but socially superior at maturity if raised apart (Crawford & Smyth, 1964).

Oscillation from positive to negative heterosis is considered to function as a device for cyclic adjustment of gene frequencies in response to recurrent environmental changes. Such a condition will, *e.g.*, apply for species having several generations a year. Dubinin & Tiniakov (1945, 1946) found a seasonal cycle in the frequency of different chromosomal inversions in *Drosophila funebris* as well as a maximum of inversion heterozygotes in summer at the peak of reproduction and population expansion. Manwell & Baker (1970) indicate that a similar mechanism works, when population density creates a selection pressure, a so-called Malthusian selection. Such a process is important in species like locusts, lemmings, voles, quails, etc., which have periodic irruptions.

Shull (1948) defined heterosis "to cover the real, observable phenomena", when unlike gametes are brought together to form a hybrid. In this sense, heterosis is synonymous with hybrid vigour. He also defined heterosis as "stimulus of heterozygosis". Vigour is, however, achievable also through homoalleles, nonallelic and plasmatic effects. He is obviously not being consistent here.

The claim that heterosis only derives from heterozygosity cannot be taken too seriously, since he "definitely intended to include cases in which the effective differences between uniting gametes might be due to anything else than Mendelian genes, as well as the differences caused by such Mendelian genes, if the latter were not individually analyzable as Mendelian genes". He also refers to heterocaryotic vigour (Dodge, 1942) as a case of heterosis.

Obviously, heterozygosity cannot be a necessary condition for phenotypic expressions identified as heterosis. Just as heterozygosity *per se* does not confer heterosis, homozygosity *per se* should not exclude heterosis. This tolerant view on the concept is also consistent with the fact that the hereditary mechanism behind heterosis can often be fixed in a homozygous combination without loss of the phenotypic expression. Further, heterosis is commonly measured in such a general way that heterozygous and homozygous elements are not separated from each other.

Heterosis can thus be taken from at least three different aspects (Table 2). As to its direction, it may be positive or negative. As to its expression, it may be luxuriant (vigour in size, yield, etc.), adaptive, selective and/or reproductive in its advantage or disadvantage. As to its transmissibility to the next sexual generation, it may be unfixable as in connection with freely segregating heterozygosity or fixable either in a balanced heterozygous or a homozygous constitution. All these aspects will be covered by heterosis *sensu lato*. Breeders will tend to follow Shull's original intention and will generally mean a luxuriant, unfixable superiority of the hybrid as compared with the parents. This latter concept could be described as heterosis *sensu stricto*.

From the definitions above, it is clear that heterosis may be expressed in very different, often contradictory ways. This is of course obvious, if it is to reflect relative superiority as well as relative inferiority. But other contradictory situations will also occur. A luxuriant, adaptive or selective heterosis may be combined with sterility and thus confer no reproductive capacity. A luxuriant heterosis may be accompanied with drought susceptibility, too late development with risk of frost induced sterility, etc. and will thus be of inferior adaptive, selective or reproductive value. A selective heterosis may offer advantages in the vegetative but not in the generative phase and so on. It should also be remembered that the different types are overlapping and perhaps more based on the kind of classification needed by the observer.

Table 2. Subdivision of the modern concept of heterosis

A) Direction

1. Positive heterosis
2. Negative heterosis

B) Function

1. Luxuriant heterosis
2. Adaptive heterosis
3. Selective heterosis
4. Reproductive heterosis

C) Transmissibility through sexual phase

1. Labile heterosis
   - i. Labile heterozygous heterosis
   - ii. Heterocaryotic heterosis
2. Fixed heterosis
   - i. Fixed heterozygous heterosis
   - ii. Homoeologous heterosis

True homozygosity for loci with additive allele effects may contribute to phenotypic reactions measured as heterosis but should not be accepted alone as a mechanism behind heterosis.

Of the different types of expression, luxuriant heterosis is often an evaluation restricted to one or a few characteristics. One and the same genotype may even show positive luxuriance for one and negative luxuriance for another character. Adaptive, selective, and reproductive heterosis are concepts always evaluating the phenotypic reaction of the individual, not the character, as the unit. This difference gives luxuriant heterosis restricted value for wild species if not accompanied by one or all of the other types of heterosis. Under domestication, where competition and selection are often relaxed or put under artificial control, pure luxuriance should be of greater value (Dobzhansky, 1952).

From what has been said above, it is clear that heterosis is a measure that may or may not involve competitive ability. Luxuriant, adaptive, and reproductive heterosis may be evaluated on the basis of spaced individuals, in monoculture or in a mixed population. Selective heterosis, however, reflects competitive ability and is thus restricted to studies of populations.

Because of different approaches, there is no generally accepted method of measuring heterosis, *i.e.* of proving its existence. An evolutionist looks upon heterosis as a device for maintaining genetic flexibility and influencing gene frequencies. From this aspect it appears logical to compare hybrid and parental mean, but the breeder is hardly interested in a hybrid which is not superior to both parents. Both measures of heterosis are commonly accepted but under different circumstances. Plant breeders working with autogamous crops ought perhaps to consider even a third criterion, *i.e.* superiority of the hybrid over the best homozygous segregants that can be obtained.

Heterosis in survival is often measured by determining the number of heterozygotes in a population and comparing this with the number predicted under the assumption of Hardy-Weinberg equilibrium at random mating. Such an evaluation is, however, not without errors. Manwell & Baker (1970) point out that an excess of heterozygotes is possible without positive heterosis in connection with differences in gene frequences between the sexes. A deficit of heterozygotes without negative heterosis is possible, if the sample is a mixture of two or more inbreeding populations differing in gene frequency, if assortative mating occurs or if polymorphism is sex-linked.

MECHANISM OF HETEROSIS

Heterosis includes purely recombinative elements (Williams, 1959). This fact may be illustrated by Powers' (1944) classical examples in tomato on heterosis in fruit yield (Table 3). A line with many but small fruits is here shown to combine with a line with few but large fruits in a way that number and size are transgressively improving total yield. This type of heterosis may be based on completely independent units of the genetic system which by themselves may be only intermediate in reaction compared to parents (*cf.* last case in Table 3). As components these units are, however, able to result in heterosis for a complex character of higher hierarchy merely owing to their summation effect.

Table 3. Three crosses in tomato illustrating transgressive heterosis in yield obtained by mere recombination of fruit number and size (Powers, 1944)

	Parental line or hybrid	Mean per ripe plant fruit number	fruit size, grams	fruit yield, grams
P_1:	4109	118.3 ± 12.91	12 ± 0.50	1364 ± 151
F_1:	4109 × 4110	183.2 ± 13.32	16 ± 0.68	2876 ± 177
P_2:	4110	109.1 ± 11.34	17 ± 0.58	1868 ± 149
P_1:	4101	4.3 ± 0.32	119 ± 6.49	513 ± 39
F_1:	4101 × 4103	20.5 ± 2.27	89 ± 3.66	1827 ± 196
P_2:	4103	19.5 ± 2.94	55 ± 2.74	1066 ± 159
P_1:	4102	4.4 ± 0.69	138 ± 12.81	607 ± 86
F_1:	4102 × 4110	44.5 ± 2.52	55 ± 2.93	2428 ± 150
P_2:	4110	109.1 ± 11.34	17 ± 0.58	1868 ± 149

Heterosis with the above-mentioned genetic background is generally referred to as nonallelic heterosis and will include interactions not only of more or less polygenic traits but also of individual loci. Epistatic heterosis is frequently used as a synonym but should have a more restricted meaning, if the term epistasis is taken in its modern, revised connotation of nonallelic phenotypic suppression (*cf.* Hollander, 1955).

Epistatic heterosis must be considered a part only of nonallelic heterosis which also includes transgressive (additive, cumulative) and recombinative (complementary) heterosis (Table 4).

In the general discussion, these various types of nonallelic heterosis are distinctly classified apart from allelic heterosis which has its regulatory system rather within than between individual loci. Reasons for such a separation are clear, but the theoretician ought to be much more annoyed than the practical breeder that such a

Table 4. Subdivision of heterosis as to regulatory systems

a) Genomic heterosis

- 1. Nonallelic heterosis
  - i. Transgressive heterosis
  - ii. Recombinative heterosis
  - iii. Epistatic heterosis
- 2. Allelic heterosis
  - i. Dominant heterosis
  - ii. Overdominant heterosis

B) Plasmatic heterosis

C) Nonheritable maternal heterosis

strict grouping between nonallelic and allelic interaction is impossible in experimental work because of linkage.

Fortunately, however, a distinction set by the degree of linkage is just as rational, since the interactions both at the nonallelic and allelic levels are virtually of the same kind. In both cases they can be described as sheltering, inhibiting, stimulating, complementing, or dosage regulating.

The sheltering and inhibiting effects of allelic interaction lie behind the dominance or dominance of linked genes hypothesis of heterosis (Bruce, 1910; Jones, 1917), according to which deleterious recessives will not be able to exert their negative effects in the presence of their respective dominant alleles. Alleles cooperating in a stimulatory, complementary or dosage-adjusting manner fit into the overdominance hypothesis (Shull, 1908, 1911; East, 1908; Hull, 1945), according to which none of the alternative homozygotes is better than the heterozygote $AA < Aa > aa$).

The suggestion has been made many times that heterosis is not a phenomenon regulated by the genome only. The plasmon should be involved, particularly through an interaction between nuclear genes and cytoplasm. Evidence for this kind of mechanism was extremely sparse (*cf.* Jones, 1952), until Hanson *et al.* (1960) reported that mitochondrial activity in hybrid maize seedlings is correlated with degree of heterosis. This important observation has lately been extensively examined in plants by McDaniel & Sarkissian (1966, 1968), Sarkissian & Srivastava (1969), Hobson (1971), and McDaniel (1972, 1973, 1974), to mention just a few of the contributions.

Polymorphism and complementation of mitochondria have been shown to exist in maize, sorghum, wheat, rye, ryewheat, peas, and cotton. The occurrence of mixed mitochondrial populations in the cells of hybrid organisms and its association with hybrid vigour are also found in microorganisms (Avers *et al.*, 1965; Wagner, 1969) as well as in insects and higher animals (Raicu *et al.*, 1970; Mukherjee *et al.*, 1970; McDaniel & Grimwood, 1971).

Mitochondria are cytoplasmic organelles providing energy, usually in the form of adenosine triphosphate (ATP), for syntheses and other energy-requiring processes. They are heritable entities, reproduce themselves, contain DNA but are partly under the control of the nucleus. Messenger RNA has been found to have a higher level in heterotic hybrids (McDaniel, 1973). Differences in mitochondrial activity of reciprocal hybrids have been observed (McDaniel & Grimwood, 1971), thus indicating possibilities for an unequal parental contribution of mitochondria to the hybrid. New types of mitochondria do apparently arise by some process of hybridization (Sarkissian & McDaniel, 1967).

Mitochondria are important in maintaining intracellular homeostasis (Sarkissian & Srivastava, 1969) and thus provide an explanation of the often greater phenotypic stability of hybrids when exposed to environmental stress and disturbance. Other factors will, however, also intervene.

The fate of a mixed mitochondrial population in a hybrid organism over successive cell divisions as well as over successive generations is insufficiently understood. Mitochondrial segregation and sorting out are observed in early stages of development (*cf.* Wagner, 1969) and may also occur in later stages. Mosaicism with respect to mitochondria is very plausible as well as loss of heterogeneity over generations. It would be interesting to know whether a successive decrease in mitochondrial diversity might partly explain the fact that many new crossbred varieties do not correspond to expectation after some years in use.Such a surmise will of course not exclude other factors already considered such as loss of flexibility owing to homozygotization and purification, changes in climate, disease attacks, etc.

Apart from genomic and cytoplasmic regulatory systems, heterosis might also depend on effects which are generally not considered as true components of heterosis but are often included in measurements of heterosis. Such effects are maternal influences on the progeny acting otherwise than through transmission of hereditary constitutents.

For mammals, conditions *in utero* may be pertinent. Crossing a male of a small breed of livestock with a female from a large breed generally results in larger progeny than would be expected, simply because of the larger size of the uterus (Walton & Hammond, 1938). Thus, when heterosis is measured from the midparent value, maternal effects can easily be critical or, taken from another viewpoint, result in a significant error. Incompatibility between mother and foetus, interaction via milk, etc. are known to have influence. In plants, seed maturation, seed size, endosperm character, seed dormancy, etc. may be critical maternal influences upon the relative superiority of a hybrid offspring.

Artificial manipulations and disturbances may often influence measurements of heterosis but should of course be classified as experimental errors. Such artifacts are especially common at artificial crossing influencing seed set (Hagberg, 1953) and should be taken into account.

It is also important to remember that heterosis is a phenotypic measure and as such not an absolute measure. Direction, extent, degree, and reproducibility of heterosis depend on environment and whenever restricted to a limited number of characters generally on genetic background as well.

Heterosis may show up more universally for a certain genotype, but more often becomes obvious first under certain circumstances. Substantial evidence has accumulated which demonstrates that hybrids often exhibit greater homeostasis by offering alternative genetic pathways (*cf.* Lewis, 1954; Mac Key, 1970). This reaction will generally give them an advantage under conditions of stress such as crowdedness, adverse weather, disease attacks, etc. As a contrast, other expressions of heterosis might not be fully developed, unless the plant or animal is not given the chance to exhibit its genetic capacity.

These conflicting reactions are often met with in connection with plant studies, where preliminary examinations because of restricted seed supply have to be made on a limited number of spaced individuals. Such plants will tiller profusely, extend their vegetation period and give a different response as compared with the reaction in a dense, conventional drill stand.

In fact, modern farm and garden management tends to eliminate much of the need for typical heterosis. Within the limits of economic realism, the grower will minimize weed competition and disease threat and optimize nutrient and water supply. He knows that high yields are never reached under stress. In addition, yield structure analyses have informed him that maximum use of solar energy per unit area of land implies generally a thick and very uniform stand, where each individual plant is limited in its vegetative development and generative productivity.

In modern animal husbandry, similar tendencies can be observed. Resistance against starvation and malnutrition has no longer the same fitness value. The ability to stand social stress like crowdedness, encaging, etc. has become more pertinent. Obviously, conditions in nature and under domestication are diverging more and more.

RELATIVE IMPORTANCE OF MECHANISMS OF HETEROSIS

In early, often very theoretical discussions around heterosis, the possibility of either nonallelic, dominance or overdominance interaction to explain heterosis had their different advocates. Cytoplasmatic contributions were, however, almost ignored. Today enough evidence has been presented (*cf. e.g.* Michaelis, 1951; Dobzhansky, 1952; Gustafsson, 1954; Williams, 1959; Wagner, 1969) to show that all four mechanisms and not just one of them are at work.

The discussion is now centered around their relative importance as regulatory systems for heterosis. How to appreciate plasmatic effects above the search for efficient combinations is not fully understood yet. Neither do we know to what extent observations on *e.g.* mitochondrial activity merely reflect mechanisms of genomic heterosis.

Within this latter type, we know that nonallelic and dominant heterosis rather than overdominance should be taken as end stations in a genetic evolutionary system. Nonallelic interaction as well as dominance can be fixed in homozygous condition, if linkage and repulsion can be overcome. Overdominance is in its strict sense inseparable from heterozygosity and is thus unfixable. It is also possible at least to a large extent to ascribe special combining ability to the direct outcome of nonallelic interaction, while general combining ability is more related to the expression of dominance (Jinks, 1955). The different prevalence of the three categories of genomic heterosis thus has relevance both in evolution and for breeding.

In this evaluation, we ought first to recall that linkage and pleiotropism set overlapping boundaries between the three types of genomic heterosis. A more accurate classification is further possible only by studying oligofactorial heterosis. The biometrical approach to quantitative evaluation is arbitrary in its partitioning and has different possibilities for sorting out different interaction phenomena. Mathematical models have tended to oversimplify and a generalized discussion has often been based on insufficiently representative material.

All this is not to say that the most favoured explanation of heterosis is wrong. The accumulation of completely and partially dominant genes is certainly over-all the most important mechanism behind heterosis (*cf. e.g.* Penny *et al.*, 1962; Gardner, 1963; Paterniani, 1973). Dominance *per se* should in the evolutionary sense of Fisher (1931) indicate advantage in addition to its sheltering effect against deleterious recessives. The recessiveness characteristic of domestication should not influence genetic regulation of luxuriant heterosis.

The strong emphasis on the dominance hypothesis of heterosis is partly explained by the fact that early studies were predominantly based on crosses among open-pollinated varieties of maize of similar origin, and inbred lines selected for average (general) combining ability. We now know that the dominance type of heterosis will prevail in such material (Castro *et al.*, 1968; Paterniani, 1973). The well-known studies in *Drosophila* supported the evaluation. Several deleterious recessive mutant genes per individual are frequently observed in wild populations (Dobzhansky, 1952).

Over the years, a more prudent attitude has, however, developed with respect to overdominance, epistasis, and plasmatic effects which earlier were often considered of limited significance, unimportant or negligible (*cf. e.g.* Gowen *et al.*, 1946; Jinks, 1955; Robinson & Cockerham, 1961).

The appreciation of overdominance has increased as a consequence of more accurate analyses in connection with mutation experiments and biochemical studies of polymorphism. The stabilizing effect of an overdominant system to changing environments has recently been elegantly exemplified by Gustafsson *et al.* (1972, 1973), and distinctive allosteric properties of hybrid enzymes as well as quantitative differences in enzyme activity of heterozygotes are now well documented (*cf. e.g.* Schwartz & Laughner, 1969; Warner *et al.*, 1969; Manwell & Baker, 1970; Fincham, 1972). Marked effects have often been found in connection with monofactorial heterosis, suggesting a considerable contribution already from a fairly limited number of overdominant loci.

With the awareness of risks for genetic erosion in connection with strict inbreeding and the introduction on a large scale of recurrent selection, the contribution of nonallelic interaction to the phenomenon of heterosis has been more appreciated. A clear tendency to widen the genetic pool via exotic combinations has added to this modified standpoint (*cf. e.g.* Gorsline, 1961; Paterniani & Lonnquist, 1963). So also has experience gained more recently in connection with heterosis studies in autogamous crop plants in attempts to produce hybrid varieties. Their reproductive system will hardly allow undesirable recessives to be preserved.

The fairly recent information on mitochondrial heterosis has given new dimensions to the discussion around the relative prevalence of heterotic regulatory systems. Mitochondrial complementation is a real challenge to conventional biometrical models and estimations, since apparently both mitochondrial genes and nuclear genes are involved (Wagner, 1969; McDaniel, 1973).

The attitude to nonallelic and perhaps particularly to the overdominance type of heterosis ought further to change with the advancement of a breeding program for hybrid varieties. In accordance with the principle of stepwise progress, the breeder will tend to improve the vitality of his inbreds without loss of combining ability. Such an aim is favoured by a preference for single crosses in commercial seed production (Weatherspoon, 1970), since this type of cross will allow a maximal exploration of available heterotic potential to be combined with maximal varietal uniformity.

The production of more vital inbreds has mainly to go via homozygotization of favourable dominant factors and elimination of deleterious recessives. Valuable nonallelic gene constellations may be fixed in the same way.

Such a trend will automatically decrease the relative prevalence of the corresponding type of heterosis at hybridization. As in recombination, it will also be less rewarding to try to improve already explored pathways. Since overdominance cannot be homozygotized in the same manner, its relative importance as a heterotic component ought thus to be increased. In addition and according to Mukai's *et al.* (1965) proposed hypothesis of optimal heterozygosity, the development of overdominance should more easily occur, if the number of heterozygous loci is restricted.

This idea of an optimal degree of heterozygosity has also appeal from the viewpoint that a breeding program cannot efficiently handle too wide a variation at a time. Moreover, there are reasons to believe in a gradually diminishing contribution from extra heterozygous loci such that heterosis will asymptotically tend to reach some upper limit (Rasmusson, 1933; Wallace, 1958).

From what has been said above, it is obvious that the dominance, overdominance, nonallelic, and plasmatic regulatory systems behind heterosis may change considerably in their relative importance depending on reproductive system, past selection, the character concerned, etc. The breeder must be aware of this dynamic behind heterosis in order to be able to plan appropriately.

HETEROSIS IN EVOLUTION

The above reasoning has arrived at nothing but the fundamental truth that life has to be based on a balance between creative and preserving forces. Heterosis implies a genetic system for maintaining flexibility (Powell, 1971). As soon, however, as this flexibility becomes unnecessary or can be abandoned for other advantages as often under domestication, heterosis may be described as a primitive genetic system or a short-sighted solution only.

This is the more true, since heterosis creates at each situation and at each generation a genetic élite in the population paid for by the so-called segregational genetic load. Both nature and breeder have interest in diminishing this loss in total productivity, even though heterotic expressions are often related to early ontogenetic stages (Whaley, 1952; McDaniel, 1973) with less risk for the economy of the unavoidable competitive screening process. It is true that the breeder has effective means but at the cost of a more expensive seed production.

This more negative aspect of heterosis is important to visualize. A restriction may be laid upon total number of loci with balanced polymorphism in order to keep the variance of fitness in the population within reasonable limits (*cf.* Sved *et al.*, 1967). Again a support for an optimal degree of heterozygosity. Nature should steadily try to introduce preserving elements in order to avoid an undesirable genetic load. Mostly such corrections have to be made by some compromise with regard to phenotypic flexibility within the population.

A preservation of heterotic expression is possible through either some kind of heterozygous or some type of homozygous balance. The former alternative generally offers a milder compromise, allowing more flexibility to be retained but at the cost of a higher genetic load.

The basic principle is here to maintain the flexibility of heterozygosity but increase the genetic information of the interacting units, *i. e.* to go from the individual locus to larger and larger gene blocks held together by inversion or any other device of non-random or localized chiasma formation. The larger the blocks are, the more will the segregation capacity and flexibility be reduced. If well tuned, the blocks are, however, able to offer different package deals, rapidly exchangable due to the maintained heterozygosity. The earlier mentioned cyclic shift over seasons of different chromosomal inversions in *Drosophila funebris* is an elegant illustration of such a mechanism and its function.

Another example of the same basic principle, but more extreme in its preserving effect is the translocation-induced chromosome ring system built up in *Oenothera* in order to maintain heterozygosity even at self-fertilization (Cleland, 1972). Also this system has in most cases its genetic load, since balanced lethals will prevent the development of homozygotes. In some types, the loss of female gametophytes has, however, been avoided through the so-called Renner effect. The functional genome complex will steadily be present in the egg cell, since competitive ability rather than position in the linear megaspore quartet has become decisive.

An even more perfect system is to be observed in the Australian grasshopper *Moraba virgo,* a pathenogenetic species with a premeiotic doubling of the chromosome number in the oocyte, followed by a meiosis in which synapsis is restricted to sister chromosomes (White, 1966). This system also allows for the complete preservation of a heterozygous constitution. It does in fact function in the same way as apomixis or vegetative propagation.

All these different systems for maintaining heterozygosity allow allelic interaction to operate directly. Although very important as a mechanism behind heterosis and homeostasis, this type of genetic interaction is not the only one, as we have learnt before. Other types of heterosis are directly transferable to a homozygous constitution.

A more polygenic interaction pattern could here further compensate for the loss of heterozygosity by being able also to offer alternative genetic pathways and a better genetic storage capacity. In addition, genetic factors may themselves or by a modified genetic background be improved in their functions if moved from a heterozygous to a homozygous situation. Especially in allogamous species, it appears most likely that the fitness of a new mutant allele is first evaluated in its heterozygous status (Fincham, 1972).

Homozygotes have, however, also developed a system whereby allelic interaction can be fixed without loss of phenotypic expression. The basic principle is here duplication preceded or followed by allelic diversification (Mac Key, 1970). Also this mechanism is able to work from the individual locus by repeating it with one allele at each position *(AA/AA ‹ Aa/Aa › aa/aa)*, via larger and larger isophenic segments up to duplication of whole genomes, *i.e.* up to polyploidy.

Within the level of one or very few loci being duplicated and diversified, the change will probably be registered as a new allele or to be more correct as a pseudoallele. Biochemical analyses particularly of protein polymorphism have furnished numerous examples of this evolutionary solution. Especially well studied is the pseudoallelism behind polypeptide chains of human and animal haemoglobin and myoglobin (*lit. cf.* Spofford, 1969; Manwell & Baker, 1970).

Evidence that the system is frequently utilized for larger segments as well is presented *e.g.* by Rees *et al.* (1966, 1967) and Jones & Rees (1968). They have observed quantitative differences in chromosome length, nuclear dry mass, and nuclear DNA within the genera *Lathyrus, Vicia, Allium* and *Lolium*. They interpreted these differences as being due to duplications scattered around the whole genome. As expected from the need of such arrangements for allelic interaction, autogamous species were found to have longer chromosomes than allogamous types of the same genus. A very elegant demonstration was made in *Lolium,* where hybrids between one outbreeding and one inbreeding species developed unpaired loops and ends at pachytene (Table 5). The evolutionary significance of duplication in autogamous plant breeding has been stressed by Hagberg (1962).

Polyploids will often show heterosis in gigas characters (Stebbins, 1950) as well as in phenotypic stability (Mac Key, 1954). The underlying mechanism will differ in connection with polysomic *versus* disomic patterns of inheritance (Mac Key, 1970), *i.e.* whether corresponding chromosomes from different genomes are able to pair or only strict homologues.

The polysomic system will offer the richer genic interaction pattern potential of polyploidy. Here the whole dosage range or number of alleles set by the number of homologous loci available can be utilized in optimizing heterotic combinations. In situations where the number of alleles is limited, allogamous polyploidy will tend to stabilize an intermediate heterozygous constitution and thus render a characteristic of phenotypic stabilization. Interallelic heterosis is obtained by less genetic load.

Polysomy has, however, these advantages only in connection with allogamy. Inbreeding of a polysomic polyploid will ultimately imply that the choice between either allele has to be made with greater discontinuity. The choice is no longer between two doses of one allele against two doses of another as in the diploid but rather between four or more doses of either kind, depending on the level of polyploidy. The genetic balance will be impaired.

While polysomy will tend to preserve the polyploid status, disomy will break it down. The inhibited pairing between homoeologous sets of genomes will allow reduplicated loci to diversify in a similar way as described above for small and large duplications. The fundamental characteristic of heterozygosity, *viz.* the coexistence

Table 5. Chromosome evolution in *Lolium* (After Rees & Jones, 1967)

Characteristic	*Lolium perenne*	*Lolium temulentum*	F_1
Breeding behaviour	outbreeding	inbreeding	
Growth habit	perennial	annual	
Chromosome number (2n)	14	14	14
Total chromosome length (μ)	51.1	67.5	58.6
Average chromatid width (μ)	0.57	0.59	0.58
Total chromosome volume (μ^3)	25.9	37.1	30.7
Nuclear dry mass ($\times 10^{11}$ g)	3.40	4.67	4.19
DNA content (arbitrary units)	31.3	42.1	36.8

Unpaired loops and ends at pachytene in the F_1:

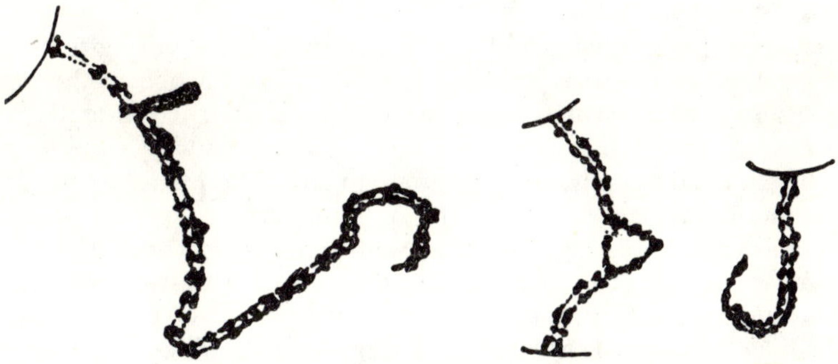

of more than one allele simultaneously, has again been solved for the homozygotes and autogamous species.

The diploid formula for overdominance $AA < Aa > aa$ will correspond to $AA.AA < AA.aa > aa.aa$ for a homozygous allotetraploid (Mac Key, 1968*b*). The point separating each tetraplex set into pairs indicates the boundary fixed by the disomic pairing system. Higher degrees of polyploidy will increase the possible kinds of allelic interaction, including the interference of additional alleles (Table 6).

That overdominance functions in this way is proved for monofactorial heterosis in artificially produced autotetraploids from *Arabidopsis thaliana* (Wricke, 1955), *Silene armeria* (Seyffert, 1959), and from barley (Ellerström & Hagberg, 1967). A diploidization process, *i.e.* a diversification of reduplicated loci at different

Table 6. Overdominance formulae at different levels of ploidy (Mac Key, 1970)

Strict diploidy (infeasible in homozygotes):
$AA < Aa > aa$

Disomic tetraploidy:
$AA.AA < AA.aa > aa.aa$

Disomic hexaploidy (diallelic system):
$AA.AA.AA < AA.AA.aa \lesseqgtr AA.aa.aa. > aa.aa.aa$

Disomic hexaploidy (triallelic system):

$$\begin{matrix} A_1A_1.A_1A_1.A_1A_1 & & A_1A_1.A_1A_1.A_2A_2 & & & & A_1A_1.aa.aa & & \\ \text{or} & < & \text{or} & \lesseqgtr & A_1A_1.aa.A_2A_2 & \lesseqgtr & \text{or} & > & aa.aa.aa \\ A_2A_2.A_2A_2.A_2A_2 & & A_1A_1.A_2A_2.A_2A_2 & & & & aa.aa.A_2A_2 & & \end{matrix}$$

homoeologous genomes, is demonstrated to occur in tetraploid wheat with different trends in different geographic areas (Mac Key, 1967).

It is thus apparent that autogamous disomic polyploidy is able to combine many of the advantages of both the allogamous and the autogamous reproduction system. This may explain the great success of crop plants like wheat, oats, rice, tobacco, cotton, peanut, and coffee (Mac Key, 1970).

It might also be pointed out that an amphiploidization between one autogamous and one allogamous species ought to have certain immediate advantages. In such a combination, lethal recessives of the outbred genome will be buffered by the inbred genome at the same time as a considerable part of the overdominant combination of the former genome can be incorporated into the new pattern. This immediate exploitation will be enhanced, if the allogamous partner contributes an unreduced gamete (Mac Key, 1968*a*).

This process of fixing heterosis via disomic polyploidy is very effective. The so-called homoeologous heterosis is not only directly established in the amphiploid. There is now also a mechanism for successive completion by continued allelic diversification of loci. Loci which at a lower level of ploidy carried functions too vital to be allowed to be changed are now free to mutate in any direction as long as at least one of the homomeric partners maintains the original activity.

This diploidization proceeds of course to different extents and directions according to the genotype and ecological conditions and ought thus to be transgressively enhanced by recombination. Such an idea is supported by the observation that hexaploid wheat appears to show a higher degree of heterosis in crosses than does diploid barley (Wienhues, 1968). If interaction between homoeologous loci is as effective as between true homologous loci and a saturation effect is involved, a considerable part of this heterosis in the wheat crosses ought to be exploited in a conventional breeding program. Thus, the more expensive production of hybrid seed may never be able to compete (Mac Key, 1967, 1973). Or it may be able to do so only temporarily, as long as certain polygenic traits are more quickly combined in a heterozygous than in a homozygous combination.

CONCLUDING REMARKS

I hope to have made it clear that heterosis is a genetic mechanism of the utmost importance for the maximum exploitation of the available genetic variability, particularly in the service of phenotypic flexibility both for the individual and for the population. A toll has, however, to be paid in terms of the genetic load, and whenever possible the species will try to develop mechanisms to escape this sacrifice.

Since breeding is nothing but directed evolution, the same consideration should be valid. The toll will here be a more expensive seed production. Modern management and demand for varietal uniformity will generally make the flexibility components of heterosis less important than in nature. A more elaborate type of heterosis or any other solution diminishing or avoiding the genetic load have even greater justification. The introduction of recurrent selection and rotational crossbreeding are important initiatives along this line by promoting permanent recombination under maintained variation.

The enormous polygenic complexity generally involved behind heterosis as well as obstacles caused by linkage and self-incompatibility will only allow step by step progress. It is a dream to be able to convert an excellent hybrid variety into a self-reproducing product by merely changing it over to an autogamous disomic polyploid. Knowing that the breeder often looks upon his hybrid seed as an efficient grip on his customers, I refrain to say whether such a dream should be called good or bad.

REFERENCES

ANDERS, F., VESTER, F., KLINKE, K. & SCHUMACHER, H. (1962). Genetische und biochemische Untersuchungen über die Bedeutung der freien Aminosäuren für Tumorgenese bei Art- bzw. Gattungsbastarden lebendgebärender Zahnkarpfen *(Poeciliidae)*. *Biol. Zentrabl.* **81,** 45–65.

AVERS, C.J., PFEFFER, C.R. & RANCOURT, M.W. (1965). Acriflavine induction of different kinds of "petite" mitochondrial population in *Saccharomyces cerevisiae*. *J. Bacteriol.* **90,** 481–494.

BRUCE, A.B. (1910). The Mendelian theory of heredity and the augmentation of vigor. *Science* **32,** 627–628.

CASTRO, G.M., GARDNER, C.O. & LONNQUIST, J.H. (1968). Cumulative gene effects and the nature of heterosis in maize crosses involving genetically diverse races. *Crops Sci.* **8,** 97–101.

CLELAND, R. (1972). *Oenothera cytogenetics and evolution.* Academic Press, London, 370 p.

CRAWFORD, R.D. & SMYTH, J.R. (1964). Social and sexual behaviour as related to plumage pattern in the Fayoumi fowl. *Poultry Sci.* **43,** 1193–1199.

DOBZHANSKY, TH. (1950). Genetics of natural populations. XIX. Origin of heterosis through natural selection in populations of *Drosophila pseudoobscura*. *Genetics* **35,** 288–302.

DOBZHANSKY, TH. (1952). Nature and origin of heterosis. In "*Heterosis*" (ed.: J.W. GOWEN), Iowa State Coll. Press, Ames, 218–233.

DODGE, B.O. (1942). Heterocaryotic vigor in *Neurospora*. *Bull. Torrey Bot. Club* **69,** 75–91.

DUBININ, N.P. & TINIAKOV, G.G. (1945). Seasonal cycles and the concentration of inversions in populations of *Drosophila funebris*. *Amer. Nat.* **79,** 570–572.

DUBININ, N.P. & TINIAKOV, G.G. (1946). Natural selection and chromosomal variability in populations of *Drosophila funebris*. *J. Hered.* **37,** 39–44.

EAST, E.M. (1908). Inbreeding in corn. Rpt Connecticut Agric. Exp. Sta. for 1907, 419–428.

ELLERSTRÖM, S. & HAGBERG, A. (1967). Monofactorial heterosis in autotetraploid barley. *Hereditas* **57,** 319–326.

EMSWELLER, S.L., ASEN, S. & UHRING, J. (1962). Tumor formation in interspecific hybrids of *Lilium*. *Science* **136**, 226.

FINCHAM, J.R.S. (1972). Heterozygous advantage as a likely general basis for enzyme polymorphisms. *Heredity* **28,** 387–391.

FISHER, R.A. (1931). The evolution of dominance. *Biol. Rev. Cambridge Phil. Soc.* **6,** 345–368.

GARDNER, C.O. (1963). Estimates of genetic parameters in cross-fertilizing plants and their implications in plant breeding. – In "Statistical Genetics and Plant Breeding" (eds.: W.D. HANSON & H.F. ROBINSON), Nat. Acad. Sci. and Nat. Res. Council, Washington, D.C. Publ. **982,** 225–248.

GORDON, H. & GORDON, M. (1957). Maintenance of polymorphism by potentially injurious genes in eight natural populations of the platyfish, *Xiphophorus maculatus*. *J. Genet.* **55,** 1–44.

GORSLINE, G.W. (1961). Phenotypic epistasis for the quantitative characters in maize. *Crop Sci.* **1**, 55–58.

GOWEN, J.W., STADLER, J. & JOHNSON, L.E. (1946). On the mechanism of heterosis – the chromosomal or cytoplasmic basis for heterosis in *Drosophila melanogaster*. *Amer. Nat.* **80,** 506–531.

GUSTAFSSON, Å. (1954). Mutations, viability, and population structure. *Acta Agric. Scand.* **4,** 601–632.

GUSTAFSSON, Å. & DORMLING, I. (1972). Dominance and overdominance in phytotron analysis of monohybrid barley. *Hereditas* **70,** 185–216.

GUSTAFSSON, Å., EKMAN, G. & DORMLING, I. (1973). Vigour and variability of phytotron-cultivated monohybrid barley. *Hereditas* **73,** 1–10.

HAGBERG, A. (1952). Heterosis in F_1 combinations in *Galeopsis*. I. *Hereditas,* **38,** 33–82.

HAGBERG, A. (1953). Heterosis in barley. *Hereditas* **39,** 325–348.

HAGBERG, A. (1962). Production of duplication in barley breeding. *Hereditas* **48,** 243–246.

HANSON, J.B., HAGEMAN, R.H. & FISHER, M.E. (1960). The association of carbohydrates with the mitochondria of corn scutellum. *Agron. J.* **52,** 49–52.
HOBSON, G.E. (1971). A study of mitochondrial complementation in wheat. *Biochem. J., Proc. Biochem. Soc. 515th Meeting,* Bangor, p. 10.
HOLLANDER, W.F. (1955). Epistasis and hypostasis. *J. Hered.* **46,** 222–225.
HULL, F.H. (1945). Recurrent selection for specific combining ability in corn. *J. Amer. Soc. Agron.* **37,** 134–145.
JINKS, J.L. (1955). A survey of the genetical basis of heterosis in a variety of diallel crosses. *Heredity* **9,** 223–238.
JONES, D.F. (1917). Dominance of linked factors as a means of accounting for heterosis. *Genetics* **2,** 466–479.
JONES, D.F. (1952). Plasmagenes and chromogenes in heterosis. In *"Heterosis"* (ed.: J. W. GOWEN), Iowa State Coll. Press, Ames, 224–235.
JONES, R.N. & REES, H. (1968). Nuclear DNA in *Allium. Heredity* **23,** 591–605.
LEWIS, D. (1954). Gene-environment interaction: a relationship between dominance, heterosis, phenotypic stability and variability. *Heredity* **8,** 333–356.
MAC KEY, J. (1954). Neutron and X-ray experiments in wheat and a revision of the speltoid problem. *Hereditas* **40,** 65–180.
MAC KEY, J. (1967). Physical and chemical mutagenesis in relation to ploidy level. In *"Induzierte Mutationen und ihre Nutzung"*, Erwin-Baur-Gedächtnisvorl. IV, Gatersleben 1966, Abh. Deutsch. Akad. Wiss. Berlin, Kl. Med. 1967(2), 185–197.
MAC KEY, J. (1968*a*). Relationships in the *Triticinae. Proc. 3rd Int. Wheat Genet. Symp., Canberra 1968* Aust. Acad. Sci., Canberra, 39–50.
MAC KEY, (1968*b*). The implication at polyploid level of mating systems for chromosomes and gametes. *Proc. 12th Int. Congr. Genet., Tokyo* **1,** 322.
MAC KEY, J. (1970). Significance of mating systems for chromosomes and gametes in polyploids. *Hereditas* **66,** 165–176.
MAC KEY, J. (1973). Possibility and potential of hybrid wheat. *Cereal Res. Comm.* 1 (**3**), 50–52.
MANWELL, C. & BAKER, C.M.A. (1970). *Molecular biology and the origin of species: heterosis, protein polymorphism and animal breeding.* Biology Series, Sidgwick & Jackson, London, 394 p.
MCDANIEL, R.G. (1972). Mitochondrial heterosis and complementation as biochemical measures of yield. *Nature, New Biol.* **236,** 190–191.
MCDANIEL, R.G. (1973). Genetic factors influencing seed vigor: biochemistry of heterosis. *Seed Sci. Technol.* **1,** 25–50.
MCDANIEL, R.G. (1974). Mitochondrial complementation: a plant-breeding tool for estimating combining ability of wheats and *Triticale. Proc. 4th Int. Wheat Genet. Symp. Columbia 1973* 541–546.
MCDANIEL, R.G. & GRIMWOOD, B.G. (1971). Hybrid vigor in *Drosophila:* respiration and mitochondrial energy conservation. *Comp. Biochem. Physiol.* 38 B:309–314.
MCDANIEL, R.G. & SARKISSIAN, I.V. (1966). Heterosis: complementation by mitochondria. *Science* **152,** 1640–1642.
MCDANIEL, R.G. & SARKISSIAN, I.V. (1968). Mitochondrial heterosis in maize. *Genetics* **59,** 263.
MICHAELIS, P. (1951). Plasmavererbung und Heterosis. *Z. Pfl. zücht.* **30,** 250–275.
MUKAI, T., CHIGUSA, S. & YOSHIKAWA, I. (1965). The genetic structure of natural populations of *Drosophila.* III. Dominance effect of spontaneous mutant polygenes controlling viability in heterozygous genetic backgrounds. *Genetics* **52,** 493–501.
MUKHERJEE, T.K., STEVENS, W.C. & HOOGENDOORN, M.P. (1970). Oxygen uptake of mitochrondrial isolates from two breeds of chickens and their F_1 cross. *Poultry Sci.* **49,** 1130–1131.
PATERNIANI, E. (1973). Recent studies on heterosis. – In *"Agricultural Genetics. Selected Topics"* (ed.: R. MOAV), J. Wiley & Sons, New York, 1–22.
PATERNIANI, E. & LONNQUIST, J.H. (1963). Heterosis in interracial crosses of corn *(Zea mays* L.*). Crop Sci.* **3,** 504–507.
PENNY, L.H., RUSSEL, W.A. & SPRAGUE, G.F. (1962). Types of gene action in yield heterosis in maize. *Crop Sci.* **2,** 341–344.
POWELL, J.R. (1971). Genetic polymorphisms in varied environments. *Science* **174,** 1035–1036.
POWERS, L. (1944). An expansion of JONES' theory for the explanation of heterosis. *Amer. Nat.* **78,** 275–280.
RAICU, P., VLADESCU, B., BORSAN, I. & STAICU, S. (1970). Heterogeneity of mitochondria in the interspecific hybrid *Mesocricetus newtoni* × *M. auratus. Heredity* **25,** 465–469.

RASMUSSON, J. (1933). A contribution to the theory of quantitative character inheritance. *Hereditas* **18,** 245–261.

REES, H., CAMERON, R.M., JONES, G.H. & HAZARIKA, M.H. (1966). Nuclear variation between diploid angiosperms. *Nature* **211,** 828–830.

REES, H. & JONES, G.H. (1967). Chromosome evolution in *Lolium. Heredity* **22,** 1–18.

ROBINSON, H.F. & COCKERHAM, C.C. (1961). Heterosis and inbreeding depression in populations involving two open-pollinated varieties of maize. *Crop Sci* **1,** 68–71.

SARKISSIAN, I.V. & MCDANIEL, R.G. (1967). Mitochrondrial polymorphism in maize. I. Putative evidence for *de novo* origin of hybrid-specific mitochondria. *Proc. Nat. Acad. Sci. USA* **57,** 1262–1266.

SARKISSIAN, I.V. & SRIVASTAVA, H.K. (1969). High efficiency, heterosis, and homeostasis in mitochondria of wheat. *Proc. Nat. Acad. Sci. USA* **63,** 302–309.

SCHWARTZ, D. & LAUGHNER, W.J. (1969). A molecular basis for heterosis. *Science* **166,** 626–627.

SEYFFERT, W. (1959). Untersuchungen über interallele Wechselwirkungen. II. "Superdominanz" bei *Silene armeria* L. *Z. Vererbungsl.* **90,** 231–243.

SHULL, G.H. (1908). The composition of a field of maize. *Rpt Amer. Breeders' Ass.* **5,** 51–59.

SHULL, G.H. (1911). The genotypes of maize. *Amer. Nat.* **45,** 234–252.

SHULL, G.H. (1914). Duplicate genes for capsule form in *Bursa bursa pastoris. Z. ind. Abst. Vererb.* **12,** 97–149.

SHULL, G.H. (1948). What is "heterosis"? *Genetics* **33,** 439–446.

SPOFFORD, J.B. (1969). Heterosis and the evolution of duplications. *Amer. Nat.* **103,** 407–432.

STEBBINS, G.L. (1950). *Variation and evolution in plants.* Columbia Univ. Press, New York, pp. 643.

STERN, C. (1948). Negative heterosis and decreased effectiveness of alleles in heterozygotes. *Genetics* **33,** 215–219.

SVED, J.A., REED, T.E. & BODMER, W.F. (1967). The number of balanced polymorphisms that can be maintained in a natural population. *Genetics* **55,** 469–481.

WAGNER, R.P. (1969). Genetics and phenogenetics of mitochondria. *Science* **163,** 1026–1031.

WALLACE, B. (1958). The role of heterozygosity in *Drosophila* populations. *Proc. 10th Int. Congr. Genet.* **1,** 408–419.

WALTON, A. & HAMMOND, J. (1938). The maternal effects on growth and conformation in Shire horse – Shetland pony crosses. *Proc. Roy. Soc. London, Ser. B* **125,** 311–335.

WARNER, R.L., HAGEMAN, R.H., DUDLEY, J.W. & LAMBERT, R.J. (1969). Inheritance of nitrate reductase activity in *Zea mays* L. *Proc. Nat. Acad. Sci. USA* **62,** 785–792.

WEATHERSPOON, J.H. (1970). Comparative yields of single, three-way and double crosses of maize. *Crop Sci.* **10,** 157–159.

WHALEY, W.G. (1952). Physiology of gene action in hybrids. In *"Heterosis"* (ed.: J.W. GOWEN), Iowa State Coll. Press, Ames, 98–113.

WHITE, M.J.D. (1966). Further studies on the cytology and distribution of the Australian parthenogenetic grasshopper, *Moraba virgo. Rev. Suisse Zool.* **73,** 383–398.

WIENHUES, F. (1968). Long-term yield analyses of heterosis in wheat and barley: variability of heterosis, fixation of heterosis. *Euphytica* **17** (Suppl. 1), 49–62.

WILLIAMS, W. (1959). Heterosis and the genetics of complex characters. *Nature* **184,** 527–530.

WRICKE, G. (1955). Ein Fall von Superdominanz bei einer experimentell hergestellten Autotetraploiden von *Arabidopsis thaliana. Z. ind. Abst. Vererb.* **87,** 47–64.

Population-genetic aspects of heterosis maintenance

J. SVÁB

Hungarian Academy of Sciences, Budapest, Hungary

SUMMARY

A practical and theoretical problem of heterosis breeding, the maintenance of heterosis effect, is discussed in the light of the population-genetic concept. The term "hybrid population", as produced by heterosis breeding, is defined from the population-genetic aspect.

The decline of heterosis effect, often encountered in praxis may be attributed to effects of random drift on labile genetic array of the inbred lines, a phenomenon acting in small-size populations analogously to the mathematical laws operating in small-size sampling. The method of a cyclically renewed inbreeding process, as a possibility to overcome this problem, is discussed. The method is supported by Malécot's theory distinguishing homozygotes with alleles alike in state from alleles identical by descent.

The importance of population-genetics is pointed out as a basis for a better understanding of such inheritance processes as heterosis effect and random drift.

Genetics, the same as many other sciences, started as a unique, integrated domain of science. The immense broadening of knowledge and the improvement of research techniques in late decades resulted in a gradual differentiation into separate fields of genetic research work. One of them deals with the laws of transmission of traits from generation to generation. This field of genetics is referred to nowadays as population-genetics, including quantitative genetics, co-ordinating the laws of inheritance into an over-all mathematical system. The other fields of genetic research are mainly focussed on the realization of heritable traits *within* the individual, including essence, the operating mechanism and interrelationship of genetic factors controlling inheritance.

Heterosis, as a genetic phenomenon of outstanding importance in practical plant and animal breeding, has been given intense consideration in all fields of genetic research. In respect of the realization of the heterosis effect within individuals, genetics has built up a number of hypotheses. These have been and will be discussed intensely at this meeting, too.

On the other hand, in the approach of population-genetics an over-all and clearcut theory, having mathematically expressible principles, has been elaborated for the interpretation and prediction of heterotic conditions and the heterosis effect.

The answer to the question, whether a theory may be considered creative, depends on how far its principles can help further research, in our case plant breeding practice, and whether they offer suggestions and predictions on various "whys and wherefores", as well as the "way ahead".

This is the point of view from which I would like to discuss a particular problem of heterosis breeding in the light of population-genetics, especially from the aspect of one of its fundamental laws: the random drift. This problem is the maintenance of the combining ability of parental inbred lines, the reproduction of hybrid populations, the preservation of what had been already achieved by the breeder.

To make it perfectly clear what I am actually talking about, the term hybrid population, produced by heterosis breeding, should be defined from the population-genetic aspect. A hybrid population of this kind means a generatively reproduced F_1 non-equilibrium population, the genotypic array of which cannot be reproduced by any kind of generative multiplication. That means at the same time that also the parent populations differ from the hybrid population produced. They may be either equilibrium populations or non-equilibrium populations; for instance the parental varieties or inbred lines, or the parental single-crosses etc. Equilibrium parent populations of course must not necessarily be in Hardy–Weinberg equilibrium, as for instance partially allogamous populations, and the term equilibrium is to be understood in the broad and practical sense. That is, some changes may always occur from generation to generation, depending on the genetic stability of the equilibrium population.

From the practical and the theoretical view-point of heterosis breeding of course hybrid populations of genotypic uniformity are the most advantageous, in which all the individuals are analogous heterozygotes for all critical loci. But this criterion can be fulfilled with single-crosses from inbred lines only, the resulting hybrids genetically equalling identical twins. This equality is the precondition for a realization of the heterosis effect with the same direction and on the same level in all individuals of the population, whether the heterosis be based on directional dominance, overdominance, epistasy or other genetic effects.

And now may I come back to the problem pointed to above. This problem arises regarding the reproducibility of heterosis effects in hybrid populations obtained from inbred lines.

In breeding practice and commercial production grave difficulties arise from the frequently experienced fact that with the passing of a certain number of generations of the inbred lines, the heterosis effect realized in the hybrid population decreases or even vanishes, and sometimes new, detrimental characters show up. And that in spite of no phenotypic changes being observed in the parent lines and no crossing-technical faults in the hybrid production. This phenomenon is well known in both plant and animal hybrid production. Unknown are, however, the reasons. There are various empirical ideas and conceptions trying to explain the causes – which rather should not be called theories. Sometimes the blame for decreased heterosis is put upon imperfect hybridization techniques or pollen immigration, in other cases "fatigue" of the lines is talked of. The genetic "degeneration" of the inbred lines is mostly not admitted, the deterioration is attributed to environmental effects. Sometimes the "tired" parent or grand-parent lines get replaced without mention.

It is obvious that no practical solution to this problem can be found until there is an acceptable explanation of the causes of this genetic process. Now, let us look at the explanation population-genetics may offer on this matter.

The parent lines of inbred hybrids are generally maintained with a low number of individuals, in the marginal case with one single individual, only. In lines of that limited size the same mathematical laws operate as in small-size random samples. According to these laws, it depends on mere random chance whether individuals carrying a mutant gene, a chromosomal mutation or having preserved partial heterozygosity in the course of inbreeding, will get included in the line maintained. Such individuals may lead to irreversible changes of the gene frequencies, i.e. the genetic array of the line. In terms of population-genetics this phenomenon is called random drift, or drift in short. In most cases the changes occuring do not appear phenotypically in the line, hence they cannot be controlled by means of selection. The changes in the genetic array of inbred lines induced by the drift process may influence the dominance and epistatic conditions between lines and in consequence their special combining ability, finally the resulting heterosis effect. And even if the heterosis effect does not diminish, new traits may show up in the hybrids in consequence of genetic changes; traits such as fodder-quality, resistance and the like – never observed in the initial population before inbreeding. It may even happen that some detrimental traits will be recognized in commercial production only, because they had not appeared some generations before, at the time of test-crossing and release of the hybrid.

Random drift is by no means a rare exception but a regular phenomenon in small size populations. Nothing can be done to prevent the drift process in such populations.

Looking for a solution of this problem, the idea might be considered to maintain inbred lines by multiplying them up to a great number of individuals. With such a solution the drift effect would in fact be prevented, but in this case natural selection would inevitably lead to changes in the genetic array of the line. That is, for instance mutations may induce within some individuals a heterozygous state in one or more homologous loci and these changes must not necessarily be observable phenotypically. But the fitness of such individuals generally surpasses that of the other individuals within the line. On the other hand, individuals having preserved partial heterozygosity over the inbreeding process also have a greater fitness. In consequence, the number of individuals carrying some heterozygous loci will increase on the account of the homozygotes. The fact that in general the breeder, too, selects quite spontaneously the most fertile and good looking individuals for the maintenance of a line, may furthermore contribute to this process. It follows from all this that the mere enlargement of population size of inbred lines cannot help to prevent changes in their genetic array.

A promising way-out seems to be a system of cyclically renewed inbreeding. The essence of this method is: (1) to increase – at certain generation intervals – the inbred lines to populations of relatively large size, and let them thus recover over some generations by means of back-mutation, crossing-over and like genetic processes; (2) after some generations, but before the consequences of large population size – mentioned above – get through, and while the increased population is still phenotypically inbred-like, the population may be divided again into sub-populations for the development of renewed inbred lines; (3) after having thus obtained renewed inbred lines, the "tired" lines may be substituted with one of these probably regenerated lines. Thereafter this renewing process should be repeated cyclically. Of course, that should not mean that the renewal of lines alone would be a fully safe remedy. Test-crossing with the renewed lines would certainly be indispensable also.

The solution by line-renewal may be supported by the theory of Malécot, which is a fundamental principle of the up-to-date population-genetic concept of inbreeding. According to the terms of Malécot, homodynamic alleles may be either alike in state or identical by descent. The homologous alleles of inbred lines are identical by descent. The coefficient of inbreeding refers to this type of allel-pair within the population. Homologous alleles which are alike in state, though they induce homozygotes in the Mendelian sense, do not influence the degree of inbreeding. Hence, a population may be homozygous for a particular locus, but if the alleles are alike in state only, the population may not be considered an inbred population.

To the best of my knowledge no interpretation has been suggested yet from the aspect of molecular genetics, for the distinction of these two different states of homozygosity. It may be supposed that in this respect there is some basic difference between allogamous and autogamous species, that is between species where inbreeding depression occurs and where it does not occur. Maybe, investigations in other fields of genetics based on Malécot's theory would give a deeper insight into the phenomenon of the heterosis effect within the individual.

In the aspect of Malécot's theory, in case of increasing the size of an inbred population with individuals having alleles identical by descent, the inbreeding coefficient will formally remain unchanged, also for the enlarged population, at least over a certain number of generations. The inbreeding of the population may be recommended with the given coefficient of inbreeding or – as is the usual practice – it may be considered to be an initial population with inbreeding coefficient $F = 0$. The latter case points to the fact that the homologous alleles may have changed in the course of regeneration from alleles identical by descent into alleles alike in state. The alteration of these two possible states of allel-homology gives the theoretical support for the expediency of interrupted, renewing inbreeding.

It follows from the above conclusions that substitution of the so-called "tired" lines by renewed lines showing good special combining ability, may be considered justified or even rather imperative. In the light of population-genetics this practice – seeming unfair from classical breeding aspects – ought to be generally approved as an efficient breeding tool for line-maintenance in heterosis breeding. Furthermore, the conclusion might even be drawn that only those hybrids have a real prospect, where a wide range of substitutionary inbred lines is available. For this purpose, the renewing process should be carried on continuously, in order to have at any time sufficient disposable material.

May I mention by the way that the renewal of inbred lines has much in common with the very usual practice in classical maintenance breeding, to use mother-plants out of stands of commercial production instead from nursery material only. This practice may be considered an empirically founded defense against random drift.

All this suggests a further idea. Namely: with the classical family breeding in allogamous diploid populations all offspring of a given mother-plant of the A-family are half-sibs, all of them originating from the same mother. Hence, their inbreeding coefficient is $F = 1/8$. Multiplied up to B, C and D families the inbreeding coefficient of $F = 1/8$ remains unchanged and the foundation seed produced from the family mixture actually represents an equilibrium synthetic variety, derived from selected families of $F = 1/8$ inbreeding coefficient resulting a slight heterosic effect. In the course of the second and third step of multiplication this slight heterosis effect vanishes, due to regeneration and natural selection, and this is the reason for the necessity of regular seed renewal, quite independent of possible admixture etc.

Developing this thought, recurrent mother-plant selection from successive $A_1, A_2 \dots A_t$ families over t number of generations seems to offer an efficient inbreeding method for self incompatible allogamous species for the induction of higher heterosic effects. In this case, in the sixth generation for instance, the A_6 families reach an $F = 0.5$ inbreeding coefficient, thus the same degree of inbreeding as with one course of selfing – if selfing is feasible at all – but with the great advantage of possible continuous selection over a number of generations. It should be noted that this method is generally applied in animal breeding with the only difference that with the latter not the mother, but the father is the common parent.

Regarding the before-mentioned problems things are somewhat different with variety hybrids. In this case the hybrid population is produced by crossing genetically remote, allogamous populations and the individuals of the parent populations are not inbreds. In such hybrids or variety crosses, the heterosis effect appears in a minor degree as with hybrids of inbred lines. In turn, the maintenance problems mentioned have not to be faced, the parent varieties being large-size equilibrium populations, and in consequence their genetic array is not altered either by drift or by natural selection, at least not within a practically considerable number of generations. It follows from this fact that these variety hybrids may preserve the level of heterosis shown at the time of their release.

Let me finish my discussion with a general notion pointing back to the development of genetic research.

The sensational accomplishments in the field of molecular genetics, the headways marked by Nobel-prizes, make it quite understandable that this field of genetics stands in the limelight of general publicity, of secondary school and university education in biology. On the other hand, the enlargement of knowledge regarding the laws of inheritance, leading to the elaboration of up-to-date principles of population-genetics, might be considered a progress of no less importance, especially for practical breeding and for over-all agricultural production.

From the very fundaments, laid down by Mendel, that the inheritance of traits follows the mathematical rules of the binom $(a+b)^2$, a hard theoretical study was necessary before it has been recognized that the terms of the binom may be considered as relative frequencies and quantitative values and thus the concept finally lead to the principles and methods of modern population-genetics and quantitative genetics. Prominent scientists have contributed to the development of this field of genetics: Hardy, Weinberg, Wright, Fisher, Haldane, Malécot, Lush, Mather, Kempthorne, Falconer – to mention some of the most famous names only. The works of most of these authors have originally been elaborated with a view to animal breeding. They are, however, valid for any allogamously mating population, hence also for plant species. It should be pointed out that classical Mendelian breeding theory and methods served mainly in breeding autogamous species, or at least in the production of pure-bred varieties of allogamous species, while population-genetics serves predominantly in the breeding of species with an allogamous mating system. From this point of view, in many cases, there seems to be much more theoretical analogy between the breeding methods of animals and those of allogamous plant species, than between the breeding methods of allogamous plant species and those of autogamous ones.

The analogy of the laws of inheritance operating in allogamous populations of both the animal and the plant kingdom, seems to support the opinion that closer contacts between plant breeders and animal breeders might be of mutual advantage. Population-genetics offers the common theoretical background for methodological co-operation in the interest of efficient use of theoretical and empirical knowledge accumulated in these two branches of breeding research, especially regarding many aspects of heterosis breeding.

Some theoretical problems of inbreeding

A. BÁLINT

University of Agricultural Sciences, Gödöllő, Hungary

SUMMARY

1. According to the findings of various experiments with different objects, the occurrence of depression during inbreeding is mostly due to the turning of deleterious recessive genes into a homozygous state. Simultaneously, important genes for yielding ability, resistance, etc. also reach the homozygous state, resulting in the uniformity of both the lines and their hybrids.

2. Changes also occur in lines subjected to prolonged inbreeding. Some of these are spontaneous mutations.

3. A part of the changes is considered by our hypothesis as "induced mutation" caused by inbreeding. It manifests itself in different chromosomal aberrations induced by impairments of the nutrient supply to the embryo due to selfing. The appearance and the frequency of appearance of these aberrations show great differences per ear. They occur in larger proportions where the seed setting percentage is very low. Part of the chromosomal mutations manifest themselves also morphologically. There can be among them mutant forms improving viability but the majority seem to show defective characteristics. The establishment of defective mutants does not always lead to a deterioration of combining ability and therefore the "genetic abnormality" of populations can in several cases not be easily determined.

4. In view of our hypothesis and also based on data obtained in practice, the conclusion has to be drawn that hybrid lines are only relatively stable and their maintenance has to be conducted under strict genetic control.

This means that in the course of maintenance, forms which are morphologically unfavorable or show inferior combining ability have to be removed from the lines and the maintenance and improvement of combining-ability must be secured by the constant selection of plus variants.

Since the appearance of the work of East & Jones (1919) and that done by Fisher (1949), no important advances appeared in the field of theory of inbreeding. In the case of existing heterozygous factors, prolonged inbreeding converts them to homozygosity (Fig. 1, Williams, 1964).

The occurrence of defective recessive homozygous genes causes defectiveness, these subsequently cause several damaging effects. Even in the case of viable maize plants this damaging effect remains until the 5–7th generation. Occurrence of harmful recessive genes was studied by Bianchi (1965). By selfing 3750 ears of open

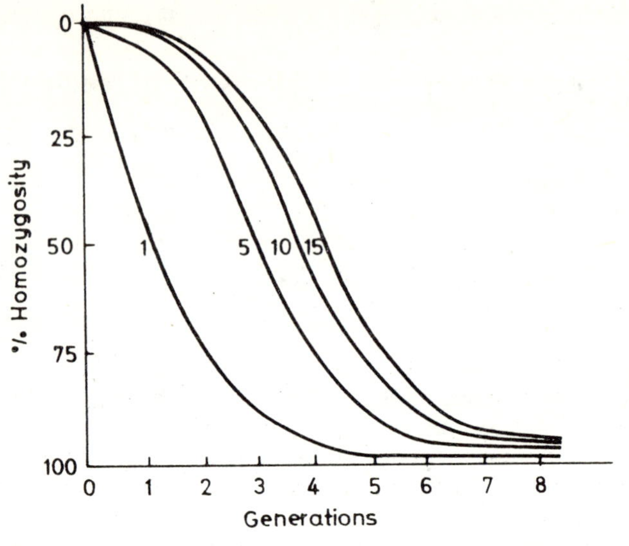

Fig. 1. The comparative formation of homozygosity in the independent gene pairs 1, 5, 10 and 15, in the case of a heterozygote self-fertilized variety (Williams, 1964).

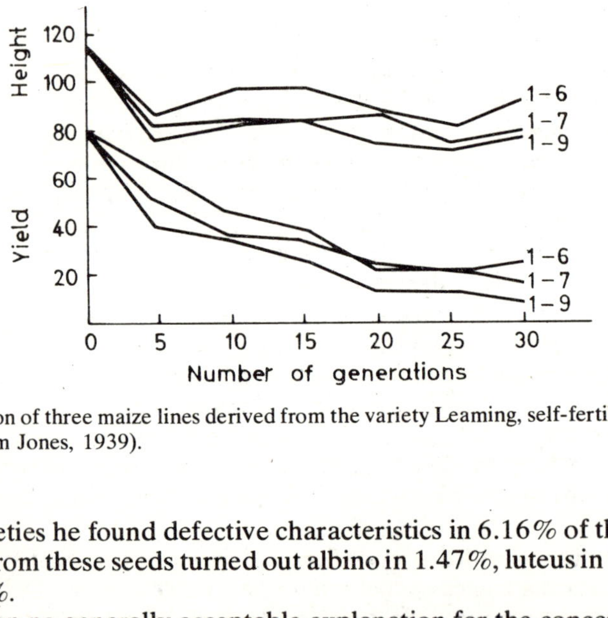

Fig. 2. A comparison of three maize lines derived from the variety Leaming, self-fertilized for 30 generations (adapted from Jones, 1939).

pollinated varieties he found defective characteristics in 6.16% of the seeds. Seedlings obtained from these seeds turned out albino in 1.47%, luteus in 1.84% and pale green in 4.21%.

In our opinion no generally acceptable explanation for the concerned problem is yet available. In 1905 east began experiments of this subject and Jones (1939) completed the work in the same direction. The results obtained were illustrated by Briggs & Knowles (1967).

Figure 2 shows that plant height was not altered after the inbreeding minimum was reached in the 5th year. These figures correspond to the values expected theoretically. As for yield there is quite a different situation. Here the decrease is linear when the line is subjected to inbreeding. Although the genes determining characters became theoretically homozygous, the reduction in yield did not stop.

The mutation rates of the different characters were 10^{-6}. In our opinion, no clear explanation may be given.

WORKING HYPOTHESIS OF OUR EXPERIMENTS

To interprete degenerative changes (positive mutations in a minor percentage only) occurring in the course of inbreeding the theories set up so far are not considered entirely sufficient by us. Recessive lethal factors becoming homozygous during inbreeding are considered by us as a proven fact and as the major cause of the phenomenon. Another reason for the degenerative changes is the formation of spontaneous defective mutants. Especially the linear decrease of yield as an effect of continuous inbreeding, combined with an unchanged level of the assimilative mechanism raises the demand for new theoretical interpretations. The following hypothesis has been elaborated by us.

1. As a response to selfing maize such changes occur in a phase of nutrient uptake, synthesis and translocation which impede the normal development of seeds on the ears. This results in poor fertilization of the ears and in an increase in the number of chromosomal aberrations of the plants emerged from these seeds.
2. Increase in the rate of chromosomal aberrations leads to a similar increase of harmful defective changes.
3. Deleterious mutational changes – though in a slight rate – are accompanied by changes stimulating viability.
4. Recessive lethal genes × normal alleles may increase the superiority of the hybrids.

According to the previous hypothesis some experiments were carried out, the following are some of the most important results (Bálint, 1967). Up-to-date results obtained by other authors will also be considered.

PHYSIOLOGICAL CONSEQUENCES OF SELF-FERTILIZING

It has been found that the deficiency of calcium, nitrogen, phosphorus and potassium (Walther, 1959) out of the plant's fertilizers, cause an increase of chromosomal aberrations. Stubbe & Döring (1938) grew *Antirrhinum* in a media free of P; the plants of the M_2 generation gave a 5 fold higher mutation rate than the control. According to this observation experiments were carried out to find whether the self-fertilized plants showed such an effect, and to study the relation between inbreeding on one hand and the absorption and translocation of the fertilizers on the other.

The use of the ^{32}P isotope gave a good chance for obtaining true results. The experiments were carried out in the radiation experimental field of Gödöllő, the plants were planted in uniform soil, 70 × 70 cm apart. One day before pollination, 100 μCi ^{32}P NaH_2PO_4 were placed 20 cm deep near the roots, and after treatment the plants were irrigated. 10 days after pollination 2 × 10 grains were gathered, and from each sample two subsamples were prepared. These were stuck on aluminium plates, and after drying were measured using a gas feeding GM-tube apparatus. The results are shown in Table 1 (Bálint, 1967). The results indicate that self-pollinated ears showed a phosphorus deficiency compared with those which were open pollinated.

Table 1. The ^{32}P uptake of lines by selfing and cross pollination (Gödöllő, 1963)

Examined material	Number of plants	Cpm/g dry weight in seeds	Values in p %
C5 selfed	30	11.450	100.0
C5 × 014	30	13.315	115.7
M14 selfed	15	9.078	100.0
M14 × WF9	15	14.023	165.1
WF9 selfed	25	28.469	100.0
WF9 × M14	25	48.978	171.8

L. S. D. 5% level: 800

The physiological experiments of Dorohov (1973) indicated that the previously mentioned observations are connected with the whole metabolism of the plant, in the same manner which has been obtained in the case of phosphorus. Dorohov (1973) carried out pot experiments using two levels of fertilizers, one of which was low, I, the other being high, II. The VIR 44 line was used. The yields obtained were compared with those of the line VIR 38. The results are shown in Figure 3.

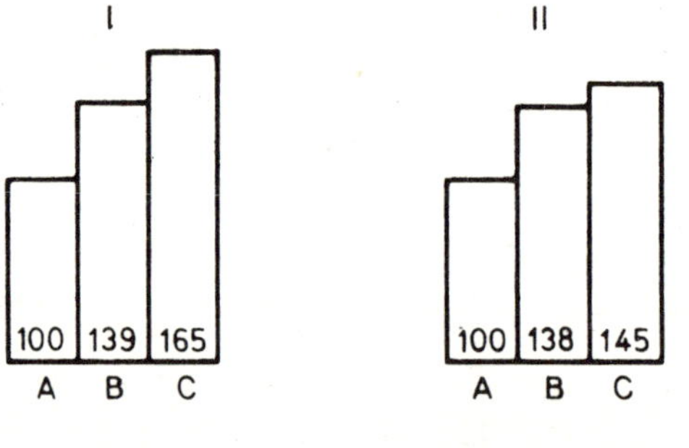

Fig. 3. Effect of different pollination methods on the intensity of photosynthesis of the line VIR 44 (Dorohov, 1973). – I = slight, II = intensive consumption of fertilizers; A = selfing, B = pollination between varieties, C = crossing with the line VIR 44.

The figure clearly shows the useful effect of sib-pollination from the physiological point of view, but the effect of crossing more effective. The trend of the results is about the same, as was shown in Table 1.

Both experiments showed that the disturbance of the metabolic processes sharply changed and may have caused the increase of chromosomal aberrations.

EFFECT OF SELF-FERTILIZATION ON THE CYTOLOGICAL BEHAVIOUR

The increase of chromosomal aberrations of rye due to selfing was concluded by Müntzig & Akdik (1948, *cit.* in Rees, 1955). The cytological changes in maize due to continuous inbreeding were investigated by Morris & Isikan (1963). The lines 289, K 41, Sa 24, M 14, KYS were used. On the cobs of lines selfed over 8–14 generations a high rate of sterility and seed injury was found in 1961. In 1961 seed from the normal cobs and of those exhibiting high sterility were planted separately. Progenies of the normal ears and of partly sterile ones produced sterile pollen in 20% and 30%, respectively. The rate of their occurrence within one line varied strongly.

On 5 plants sampled from each of the different lines the proportion of univalents in metaphase I was determined. It ranged from 1.5 to 22.2% and from 4.7 to 33.7% at diakinesis and metaphase, respectively. This and the aberrations observed provide explanation for the formation of abortive pollen and for sterility in seed setting.

In our experiments with maize the ratio of chromosomal aberrations was studied within the different classes of ears divided according to their percentage of fertile seeds/ear.

Cytological examinations were included in 1963 on C5, in 1965 on T18, and in 1974 on the tetraploid line WF9-t which was self-fertilized over 10 and open pollinated over 5 years. These tests were carried out for the different lines when self-fertilized or when exposed to open pollination. The ears were classified according to their yielding capacity, seeds from 30 ears were germinated for C5, while seeds from only 20 ears were taken in the case of T18. From each treatment 60 root tips were examined.

For these cytological examinations, the seeds were germinated in petri dishes on filter paper made wet by tap water, at 20–23 °C. When the secondary roots began to appear, 1–2 mm of the main root as well as of some secondary root tips were fixed in Carnoy's fixative. After 24 h of fixation the root tips were placed in 30% alcohol for storage. Further cytological analysis of the tips went on continuously. Prior to getting stained the tips were boiled in NHCl at 60 °C for 11–15 minutes so that they might by squashed into a single row of cells. This was followed by staining with fuchsin for 30–40 minutes. The stained tips were mounted on slides in 45% acetic acid, covered with cover slides and squashed. Using an NFDK microscope in 800 × magnification we then studied the cells in mitosis. Table 2 shows the results obtained in maize compared with those of rye. Both plants showed nearly the same trend.

The rate of aberrations was studied on seeds of ears grouped according to the percentage of seed setting. The rate of aberrations (fragments and bridges) of individuals in open pollinated populations proved significantly less than that of selfed C5 cobs with 15% fertilization. In case of 50% fertilization the result was nearly the same. Ears of the line T18 with less than 40% fertilization indicate a similar relationship (due to the good fertilization no group was found with 15% fertilization rate). Besides the differences in generation and in the tested strains, the rate of aberrations was high in both years even in the open pollinated population. Similar experiments were carried out at Gödöllő in 1974, using the tetraploid stock of 1973 to obtain further information concerning their chromosomal aberrations.

The results showed that the cytological instability of tetraploids is larger than that of diploids. This is clearly noticed in the case of cross-pollination (197.3 and 61.0%).

Table 2. Correlation between the number of chromosomal aberrations and seed-setting percentage in seedlings germinated from selfed and open pollinated seeds (Gödöllő 1963, 1965)

Seed-setting	Anaphase tested		Chromosomal aberrations		
	Number	per cent	Number	per cent	$\pm s_x$ per cent
		C5 lines			
15 per cent	992	100	34		3·43 ± 0·57
50 per cent	1343	100	38		2·83 ± 0·44
95 per cent	745	100	14		1·88 ± 0·50
Open pollination	880	100	15		1·70 ± 0·43
	L.S.D. 1 per cent level			1·75	
	L.S.D. 5 per cent level			1·33	
		T18 lines			
Less 40 per cent	1530	100	14		0·89 ± 0·10
More 40 per cent	1995	100	11		1·54 ± 0·10
Open pollination	1758	100	10		0·54 ± 0·17
	L.S.D. 1 per cent level			0·38	
	L.S.D. 5 per cent level			0·32	

Table 3. Cytological analysis of seeds originating from different pollinations (Gödöllő, 1974)

Materials investigated	WF9 (4×) selfed	WF9 (4×) sib	SynOp. selfed	SynOp. sib	WF9 × $SynF_0$ cross	SynOp × $WF9F_0$ cross
Metaphase						
Normal	667	960	994	776	553	790
Abnormal	38	33	53	34	9	31
%	5.38	3.32	5.06	4.20	1.60	3.78
Anaphase						
Normal	202	189	172	331	224	335
Bridge	2	5	–	2	2	2
Fragment	2	7	8	2	2	1
Other Ab.	7	4	3	1	1	4
Total Ab.	11	16	11	5	5	7
% Ab.	5.16	7.80	6.01	1.49	2.18	2.05
Telophase						
Normal	50	630	–	18	146	21
Bridge	1	11	–	1	2	1
Fragment	4	5	–	–	2	–
Other Ab.	1	4	–	–	1	–
Total Ab.	6	20	–	1	5	1
% Ab.	10.71	3.08	–	1.00	3.31	4.76
Total Aberrations						
Number	55	69	64	40	19	39
%	5.65	3.40	5.30	3.50	1.90	3.29
No. of examined cells	974	2,002	1,230	1,165	987	1,184

On account of the rise of aberration in selfed plants, it can be predicted that the proportion of the deleterious, and in a lesser extent, of the vital mutants will increase in the progenies. It must be emphasized, however, that in this relation no uniform response can be expected in either different lines or individual plants of one line.

HEREDITARY CHANGES IN LINES INDUCED BY PROLONGED INBREEDING

It is difficult to show that as an effect of inbreeding new deleterious mutants are formed and not that the existing ones become homozygous. To approach the problem we designed our experiments with the idea of taking a population which included one known deleterious mutant and of showing through the development of the population the establishment of new harmful forms. Two populations containing the albino gene recessively were provided for this purpose. They differed in the quality of the pigment determining the colour of the endosperm but they could be expected to segregate regularly in a 3:1 ratio. Altogether three progenies of the two populations were produced. Segregation ratios of their selfed ears are shown in Table 4.

Table 4. Experiments with albino maize lines (Gödöllő, 1963–65)

Serial number	Year	Population index	Composition of sown population	Number of ears	Number of plants	Number of yellow and white kernels	Ratio of yellow and white kernels
1	1963	A		–	19	4,278 : 1,272	3.36:1
2	1964	A	3.00:1	2	38	10,852:3,337	3.25:1
3	1965	A	3.18:1	10	93	16,665:6,074	2.75:1
4	1963	B		–	4	932:252	3.63:1
5	1964	B	3.20:1	1	17	3,671:868	4.23:1
6	1965	B	3.48:1	4	40	5,918:1,772	3.43:1

Data of the table for both cases suggest that minor deviations arose from the 3:1 segregation (although no reliable conclusion on the 2nd population can be drawn due to the low number of plants).

In the second progeny two plants of population A showed exactly 3:1 segregation. Segregation computed from 17 plants of the progeny differed also in this case from the 3:1 ratio. Even if the first two years are discounted the same tendency of an increased proportion of albino seeds was encountered in the third generation. This increase is especially obvious in population A where a 2.75:1 ratio of yellow and albino seeds was seen, and a similar tendency, but in far lesser rate, for increased albino ratio was also observable in population B.

The cause of the deviations observed in the ratios of the two populations is not known. The change observed in population A is considered significant and characteristic, suggesting a gradual increase in the proportion of albinos compared to the average of the starting population. The rise of the albino proportion appearing in the third generation may be a result of inbreeding over the previous two years.

It was possible to obtain some data from the continuous inbreeding of some lines. From our collection, selfed lines since 1964, a 10 year stable mutation (WF9/9M) and another 10 year old line (2ZE) obtained from the cross of *Zea* × *Euchlaene mexicana* were germinated in petri dishes and from these germinating seedlings the number of chlorophyll defects were calculated. The data obtained are shown in Table 5.

Table 5. Chlorophyll defects of maize seedlings (Gödöllő, 1974)

Material of investigation	Mode of pollination	No. of germinated seeds	% germination after 7 days	No. of albinos	No. of chlorophyll defects	Striata
Diploids						
W7	self-fertilized	98	98	–	14	
N6	self-fertilized	48	96	–	18	
O 118 b	self-fertilized	49	98	–	12	12
BeO 3 b	self-fertilized	10	70	7	–	
WF9 (Mu9)	self-fertilized	48	96	11	–	
22E	self-fertilized	100	100	–	18	
Tetraploids						
WF9 (4×)	self-fertilized	100	41	2	–	
WF9 (4×)	between varieties	100	74	2	–	
WF9 (4×)	open pollinated	100	94	–	–	
Synthetic open pollinated	self-fertilized	200	93	3	1	
Synthetic open pollinated	between varieties	200	96	–	1	
Synthetic open pollinated	open pollinated	100	98	–	1	
WF9 × SynOp.F_1	crossing	100	97	–	1	
SynOp. × WF9F_1	crossing	100	93	–	2	1

The cytological examinations of tetraploid lines and hybrids are shown in the previous table. In the diploid lines the defects obtained were of a wide scope, albino plants were especially obvious in two lines. In the case of the F_1 tetraploid generation defect forms were obtained, while albinos were only found in the lines. Strelchuk (1968) also concluded that inbred tetraploid hybrids show the same direction of changes as did our material.

DISCUSSION

Employing statistical genetics for obtaining explanations for heterosis and its relation with inbreeding is considered insufficient in the opinion of many authors. Biochemistry and physiology draw nearer some explanations, but these explanations only serve for a certain direction. To get clearer theoretical conclusions, these separated explanations have to be drawn together. Physico-chemical and biophysical tests gave new explanations (Shakhbazov, 1974).

From the results of our experiments concerning the inbreeding of a single maize plant we worked out a hypothesis, the most important part of which is that the degenerative effect of prolonged inbreeding after having reached the inbreeding minimum is partly due to the formation of induced deleterious mutant genes. The individual ears obtained from self-fertilization are exposed to a disturbed metabolism resulting in the appearance of chromosomal aberrations and some morphological mutants.

REFERENCES

BÁLINT, A. & KOVÁCS, G. (1960). Study in plants with ^{32}P on the genetics and physiology of heterosis. *Növénytermelés* **9,** 1, 15–26.

BÁLINT, A. (1967). *Heterosis and mutation in maize.* Akadémiai Kiadó, Budapest. (In Hungarian).

BIANCHI, A. (1965). *Some aspects of mutagenesis in maize. Induction of mutations and the mutation process.* Publ. House of the Czechoslovak Academy of Sciences, Prague 30–37.

BRIGGS, F. N. & KNOWLES, P. F. (1967). *Introduction to plant breeding.* Reinhold Publ. Corp. New York, Amsterdam, London.

DOROHOV, B. L. (1973). Change in photosyntetic activity in maternal plants by different pollination. "Stiintsa" Kisinow. (In Russian).

EAST, E. & JONES, D. F. (1919). *Inbreeding and outbreeding, their genetic and sociological significance.* Washington Square Press, Philadelphia.

FISHER, R. A. (1949). *The theory of inbreeding.* Oliver and Boyd. Edinburgh and London.

JONES, D. F. (1939). Continued inbreeding in maize. *Genetics* **24,** 462–473.

MORRIS, R. & ISIKAN, M. H. (1963). Cytological study on maize inbred lines. ***Maize Gen. Coop. News*** *Letter* **37,** 107–108.

REES, H. (1955). Heterosis in chromosome behaviour. *Proc. Roy. Soc. (London)* 150–159.

SHAKHBAZOV, V. G. (1974). On physico-chemical mechanisms of inbred depression and heterosis. *Genetics (USSR)* **10,** 4, 153–164. (In Russian.)

STUBBE, H. & DÖRING, H. (1938). Untersuchungen über experimentelle Auslösung von Mutationen bei Antirrhinum majus. VII. (Über den Einfluss des Nährstoffmangels auf die Mutabilität.) *Z. f. indukt. Abst. u. Vererbungst.* **75,** 341–351.

STRELCHUK, S. J. (1968). Combined use of induced mutations and polyploidy in breeding of plants. *Russian Cytology and Genetics* **2,** 6, 501–506. (In Ukrainian.)

WALTHER, F. (1959). Fertilitätsuntersuchungen beim Roggen. *Z. f. Pflanzen* **41,** 1–32.

WILLIAMS, W. (1964). *Genetical principles and plant breeding.* Blackwell Scientific Publications, Oxford.

EUCARPIA VII (1976)

Expression of homoeologous molecular systems in wheat alloploids

F. GARCÍA-OLMEDO, P. CARBONERO, C. ARAGONCILLO, R. FERNÁNDEZ DE CALEYA AND J.V. TORRES

Department of Biochemistry, Polytechnical University of Madrid
E. T. S. Ingenieros Agrónomos
Madrid-3, Spain

SUMMARY

Allopolyploidy is widespread in the plant kingdom, where it has been of considerable evolutionary significance. Although the existence of heterotic interactions between the genomes that make up an alloploid have been generally assumed, the precise nature of these interactions has not been extensively investigated.

Presently available evidence about metabolic integration of the wheat genomes is examined in search of new insights about the different modes of genome interaction. Although additive expression seems to be the case for many homoeologous systems, more complex patterns of integration have become evident. Examples of enzyme subunit complementation, gene compensation and other dosage effects, holoprotein completion, and complementation of metabolic pathways are discussed.

INTRODUCTION

One of the most important processes which have affected the evolution of higher plants is allopolyploidy. An alloploid is in fact a "permanent heterozygote" in which positive and negative heterotic interactions between homoeoalleles are effectively fixed.

It is accepted that, in terms of measurement of specific characters, heterosis implies values outside the parental range. In the case of alloploids, inter-genome heterosis is difficult to assess because it is superposed with positive and negative effects associated with the change in ploidy level. It could be argued that in an allotetraploid the terms of reference should be the genome donors at the tetraploid level. This criterion would be hard to apply to the case of an allohexaploid. Characters controlled by only two homoeogenes could be investigated at the homozygous tetraploid level in the appropriate nulli-tetrasomics. In most cases, however, ploidy and heterosis effects have to be considered jointly.

A variety of theories have been put forward to explain heterosis. These can be generally grouped into dominance and overdominance theories or, in other words, those based on intergenic complementation and those based on interallelic complementation (see, for example Finchham, 1966). It has been acknowledged for a long time that the two types of theories are not mutually exclusive and that both types of phenomena are probably involved in the observed association of heterosis and heterozygosity.

The biochemical bases of heterosis are becoming better known. The studies with hybrid antigens and, more conclusively, those with hybrid proteins have clarified the nature of inter-allelic complementation; while numerous examples of complementation at the level of metabolic pathways have accumulated (for a review see Manwell & Baker, 1970).

The study of inter-genomic interactions should be easier to undertake at the molecular rather than at the morphological or physiological levels of gene action. The purpose of this paper is to examine available data about homoeologous biochemical systems in wheat alloploids in the above-mentioned context.

METABOLIC INTEGRATION OF HOMOEOLOGOUS MOLECULAR SYSTEMS IN WHEAT

A number of homoeologous molecular systems have been studied in wheat. Those to be discussed here are listed in Table 1. The list includes homoeosystems which have been studied at the protein level, others described at the level of enzyme activities and finally a few characterized by their metabolic product profiles. Available

Table 1. Homoeologous molecular systems in wheat

SYSTEMS	CHROMOSOMAL LOCATIONS
Proteins	
Glutenins	group 1
Gliadins	groups 1 and 6
CM-Proteins	groups 7 and 4
Purothionins	group 1
Enzymes	
Acid phosphatase	4Aβ, 4BS, 4DL
Alcohol dehydrogenase	4Aα, 4BL, 4DS
Alpha-amylase	6Aβ, 6BL, 6Dβ
	7AL, 7BL, 7DL
Aminopeptidase	6Aα, 6BS, 6Dα
Esterase	3Aβ, 3BS, 3Dα and β
Peroxidase	group 1
Metabolic products	
Anthocyanins	group 7
Phenolic glycosides	5A, 6A, 2B, 3B
and glycoflavones	5B, 1D, 7D
Sterol esters	group 7

quantitative data about their expression will give us some clues of the different ways in which they are metabolically integrated.

Early biochemical studies with alloploids showed that, in general, biosynthetic activity of the component genomes was maintained in them, although some proteins present in one or both parents were not detected (Hall, 1959). "Hybrid substances", present in the alloploid but absent in the genome donors, were looked for in these studies but not unequivocally demonstrated until later, when heterooligomeric enzymes were shown to exist. The impression of strictly additive metabolic activity was reinforced by observations of reserve seed protein by electrophoresis in ditelosomics and nulli-tetrasomics, where chromosome dosage effects were evident (Shepherd, 1968). This seems also to be the case for aminopeptidase and acid phosphatase (Hart, 1973).

However, when other systems have been studied, more complex patterns of genome interactions, such as enzyme subunit complementation, gene dosage compensation, holoprotein completion and metabolic complementation have been shown to exist.

ENZYME SUBUNIT COMPLEMENTATION

The presence of hybrid enzymes in heterozygotes was first demonstrated by Schwartz (1960) for the E_1 esterase of maize. Some years later he presented evidence of single gene heterosis for maize alcohol dehydrogenase (1969). The Adh_1F allele specifies an active but somewhat labile dimeric enzyme (FF) and the Adh_1 Cm allele produces a less active but more stable enzyme (Cm Cm). In the heterozygote a heterodimer (F Cm) is formed which is both active and stable.

Barber *et al.* (1968) obtained evidence of wheat esterase subunit structure and, more recently, Hart (1970) has demonstrated intergenomic heterodimers in wheat alcohol dehydrogenase. His results are consistent with the existence of one locus in each of the group 4 chromosomes that produce about equal amounts of their specific subunits (α, β, δ) which by random association generate the six possible dimers in the expected proportions ($^2/_9$ each heterodimer; $^1/_9$ each homodimer). This illustrates how additive genetic activity of the genomes can result in non additive functional effects. Indeed, additive electrophoretic patterns obtained under dissociating conditions do not rule out non-additive interactions in vivo.

GENE DOSAGE COMPENSATION AND OTHER DOSAGE EFFECTS

Although the relative amounts of homoeoallelic products are very often proportional to the number of doses of the chromosome(s) carrying each homoeoallele, this is not always the case. At least two types of non additive dosage effects have been demonstrated: gene dosage compensation and inhibition of protein synthesis by extra doses of chromosomes non-homoeologous with that carrying the structural genes.

We have shown that two electrophoretic components of endosperm protein soluble in chloroform-methanol, CM1 and CM2, are controlled by chromosomes 7D and 7B respectively (García-Olmedo & Carbonero, 1970; Aragoncillo, 1973). Densitometric measurements of these components in different ditelosomic lines are shown in Fig.1.Suppression of CM1 synthesis in the appropriate ditelosomic is compensated by increased synthesis of CM2 and vice versa. However, suppression of CM1 synthesis by extraction of the whole D genome does not yield the same result. It seems that the level of this type of proteolipid determined by some regulatory factor associated with the D genome or resulting from genome interaction.

A similar situation has been reported by May *et al.* (1973) for the esterases controlled by group 6 chromosomes. Suppression of chromosome 6A esterase (Est-6AL-1) is compensated by increased synthesis of Est-6DL-1 and vice versa. However, Est-6BL-1 was not under the same type of regulation. This has been explained by postulating that the regulatory genes of Est-6AL-1 and Est-6DL-1 are unchanged while those of Est-6BL-1 have diverged. The alpha-amylases of groups 6 and 7 may be regulated in the same way (Nishikawa & Nobuhara, 1971).

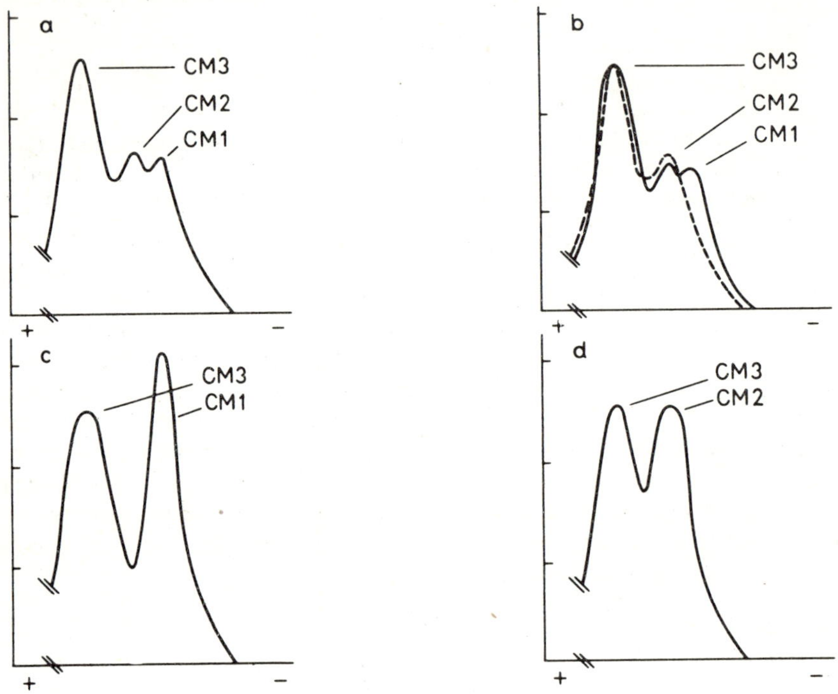

Fig. 1. CM-pattern densitograms of a) euploid Chinese Spring, b) Thatcher and tetra-Thatcher (– – –), c) ditelo 7BL Chinese Spring, d) ditelo 7DS Chinese Spring.

That extra doses of one chromosome can inhibit synthesis of proteins coded by genes located in a chromosome belonging to a different homoeologous group has been shown by Shepherd (1968) for the gliadins and by Orth & Bushuk (1973) for the glutenins. Four doses of chromosome 2A inhibited synthesis of chromosome 6D gliadins. In a similar way, four doses of either 2B, 3B, or 6B inhibited synthesis of certain glutenins coded by 1D.

HOLOPROTEIN COMPLETION COMPLEMENTATION

An interesting case of genome interaction is that of lipopurothionin synthesis, presently under study in our laboratory (Carbonero & García-Olmedo, 1969; F. de Caleya *et al.*, in prep.). Purothionins are basic polypeptides that can be extracted with light petroleum from wheat endosperm while non covalently associated with polar lipids, but are insoluble in this non polar solvent as the free apoproteins (Carbonero *et al.*, 1969; Fernández de Caleya *et al.*, in prep.; Balls *et al.*, 1942; Nimmo *et al.*, 1968; Fisher *et al.*, 1968; García-Olmedo *et al.*, 1968). Two electrophoretic variants of the apoprotein are known, named α and β.

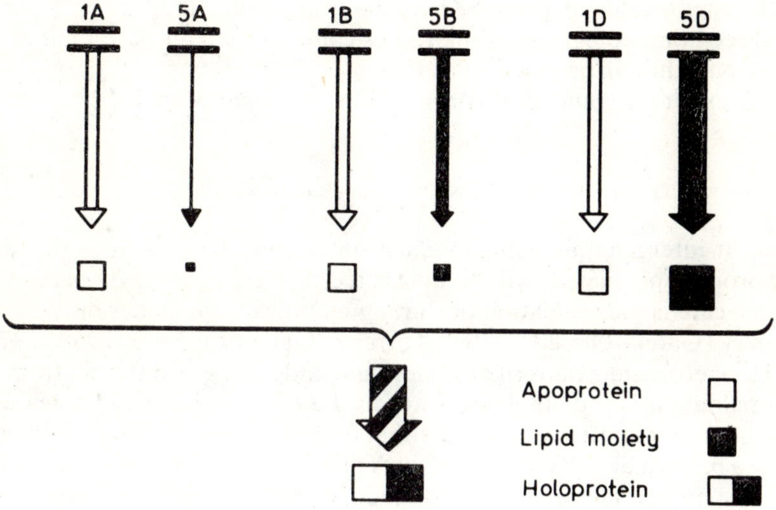

Fig. 2. Tentative model of genetic control of lipopurothionins synthesis.

Preliminary results of an electrophoretic analysis of the purothionin system are summarized in Tables 2 and 3. A tentative model based on these data is outlined in Fig. 2. The results are consistent with location of the apoprotein structural genes

Table 2. Amount of lipopurothionins in different genotypes

Euploid ~ n 5A t 5B ~ n 5A t 5D ≥ n 5B t 5D ≫ n 5B t 5A > n 5D t 5B > n 5D t 5A
hexaploid wheats ≫ tetraploids
synthetic *T. spelta* ≥ *Ae. squarrosa* ≫ *T. carthlicum*
Prelude ≫ tetra-Prelude
Rescue ≫ tetra-Rescue
Thatcher ≫ tetra-Thatcher

Table 3. Ratio of α and β purothionins in different genotypes

	α/β
Tetraploid wheats	1
Hexaploid wheats	2
Euploid C.S.	2
nulli 1A – tetra 1B	no β, α enhanced
nulli 1A – tetra 1D	no β, α enhanced
nulli 1B – tetra 1A	½
nulli 1B – tetra 1D	2
nulli 1D – tetra 1A	½
nulli 1D – tetra 1B	2
Other nulli-tetrasomics	2

in homoeologous chromosome group 1 and of some regulatory factors, possibly affecting lipid moiety synthesis, in group 5. The gene for β purothionin would be located in chromosome 1A and two genes for α purothionin would be located in 1B and 1D respectively. While the three apoprotein genes seem to be about equally productive, and thus fully compensate for each other in the nulli-tetrasomics, this is not the case for the genes located in group 5. The results in Table 2 indicate non-compensation. The three homoeologous chromosomes could be ranked as to their affectiveness in lipopurothionin production as follows: 5D > 5B ≫ 5A.

A more complete account of this research is in preparation.

COMPLEMENTATION OF METABOLIC PATHWAYS

Evidence of intergenomic complementation can be derived from the study of metabolic product profiles as will be illustrated in the following discussion.

One such a case is the regulation of sterol metabolism, which has been studied in our laboratory (García-Olmedo, 1968; Torres & García-Olmedo, 1968; 1974, and in press). The sterol esters pattern of *T. aestivum* endosperm is different from that of *T. durum:* palmitate is the dominant ester in the first species and linoleate in the second. *Ae. squarrosa* pattern resembles that of *T. aestivum.* A few *T. aestivum* varieties show a pattern like that of *T. durum.* This phenotypic difference is controlled by a single Mendelian factor (García-Olmedo, 1968), which is located in chromosome 7DS (Torres & García-Olmedo, 1974). Palmitate synthesis takes place in the late stages of endosperm development and drains the high level of free sterol that exists in the early stages of development. The recessive allele determines blockage of palmitate synthesis, high level of free sterol in the mature kernel, and some accumulation of steryl glycosides (Torres & García-Olmedo, in preparation). Data in Table 4 show that the activity of the dominant allele is high enough to cause the esterification of sterol resulting not only from D genome activity but also from that of genomes A and B. No dosage effect is observed.

Gale & Flavell (1971) have studied the genetic control of anthocyanin biosynthesis using Hope-Chinese Spring substitution lines and their hybrids. They found homoeologous loci in chromosomes 7A and 7B that controlled flavonoid class profiles. In this case a linear chromosome dosage effect was demonstrated.

May *et al.* (1973) have tentatively concluded that the two-step transformation of the o-glycoflavone saponarin into wyomin is controlled by two genes of chromosome 5A. This system would probably behave in the same way as that of the sterol esters.

Table 4. Regulation of sterol palmitate and free sterol in wheat endosperm

Material		Free sterol*	Sterol palmitate*
T. aestivum	var. Aragon 03	3.3	11.8
T. aestivum	var. Mara	21.3	1.7
T. aestivum	var. Chinese Spring	1.4	6.3
T. aestivum	var. C.S. nulli 7D tetra 7A	4.0	0.1
T. aestivum	var. C.S. nulli 7D tetra 7B	4.0	0.3
T. durum	var. Senatore Capelli	5.6	0.6

* mg of sterol per 100 g dry matter.

REFERENCES

Aragoncillo, C. (1973). *Dr. Ing. Agr. Thesis.* Polytechnical University of Madrid.

Balls, A.K., Hale, W.S. & Harris, T.H. (1942). *Cereal Chem.* **19,** 279.

Barber, H.H., Driscoll, C.J., Long, P.M. & Vickery, R.S. (1968). *Nature* **218,** 450.

Carbonero, P. & García-Olmedo, F. (1969). *Experientia* **25,** 1110.

Fernández de Caleya R., Hernandez, C., Carbonero, P. & García-Olmedo, F. (in preparation)

Finchham, J.R.S. (1966). *Genetic Complementation.* W.A. Benjamin Inc.

Fisher, N., Redman, D.G. & Elton, G.A.R. (1968). *Cereal Chem.* **45,** 48.

Gale, M.D. & Flavell, R.B. (1971). *Genet. Res. Camb.* **18,** 237.

García-Olmedo, F., & Carbonero, P. (1970). *Phytochemistry* **9,** 1495.

García-Olmedo, F. (1968). *Nature* **220,** 1144.

García-Olmedo, F., Sotelo, I. & García Faure, R. (1968). *An. Inst. nac. Invest. agron.* **17,** 433.

Hall, O. (1959). *Hereditas* **45,** 495.

Hart, G.E. (1970). *Proc. Natl. Acad. Sci.* **66,** 1136.

Hart, G.E. (1973). *4th Wheat Genet. Symp.* (preprint).

Manwell, C. & Baker, C.M.A. (1970). *Molecular Biology and the Origin of Species.* Sidgwick and Jackson.

May, C.E., Vickery, R.S. & Driscoll, C.J. (1973). *4th Wheat Genet. Symp.* (preprint).

Nimmo, C.C., O'Sullivan, M.T. & Bernardin, J.E. (1968). *Cereal Chem.* **45,** 28.

Nishikawa, K. & Nobuhara, M. (1971). *Jap. J. Genet.* **46,** 345.

Orth, R.A. & Bushuk, W. (1973). *Cereal Chemistry* **51,** 118.

Schwartz, D. (1960). *Proc. Natl. Acad. Sci.* **46,** 1210.

Schwartz, D. & Laughner, W.J. (1969). *Science* **166,** 626.

Shepherd, K.W. (1968). *Proc. 3rd Int. Wheat Genet. Symp.* 86.

Solari, R.M. & Favret, E.A. (1970). *Bol. Genet. Inst. Fitotec. Castelar* **7,** 23.

Torres, J.V. & García-Olmedo, F. (1974). *Plant Science Letters* (in press).

Torres, J.V. & García-Olmedo, F. (in preparation).

Wrigley, C.W. & Shepherd, K.W. (1973). *Ann. N.Y. Acad. Sci.* **209,** 154.

New aspects of heterosis in tomato selection

D. D. BREZHNEV
N. I. Vavilov All-Union Institute of Plant Industry, Leningrad, USSR

SUMMARY

Studies on different types of sterility of tomato have shown plants with functional male sterility of the John Bear form and stamenless forms to be most valuable.

The methods have been elaborated in order to obtain functionally sterile analogues of cultivars possessing good combining ability for the principal biological and economical characters. Number of hybrids have been developed, some of which are already in commercial production, many others under state testing.

In order to eliminate manual labour in the course of the process of pollination, experiments have been initiated by employing bumble bees (*Bombus agrorum* L.) that showed promising results.

For many decades the heterosis problem of tomatoes has been studied on all continents of the world. All the investigators obtained approximately identical results, proving the importance of this breeding method and the advisability of wide utilization of hybrid tomato seeds in outdoor and indoor vegetable production.

In many countries the pairs for crossing that produce seeds ensuring the highest effectiveness in F_1 have been selected. To a considerable extent hybrid seeds are produced only in some countries.

However, as a whole, the problem of using heterotic tomato seed has not achieved such a wide range as that of maize and even onion. The main reason for this situation is that up to now factors necessitating excessive manual labour for heterotic seed production have not been eliminated. Therefore, it is natural, that many scientists continue to work out new methods that would minimize the expense of manual labour.

Our research work is conducted in two directions: to locate the sources and on these bases to develop sterile maternal plants not requiring expenses for manual labour for flower emasculation.

It is known, that up to now tomato plants have no forms and/or varieties possessing cytoplasmic male sterility. This is why we, as well as other investigators, use cultivars possessing other types of sterility.

The most valuable of these for solving the first task, proved to be plants with functional male sterility of John Bear form and stamenless forms. But by themselves they are not fit for hybridization, as they lack a number of economically valuable characters and properties necessary for breeding new commercial tomato varieties. The sterile strains also have to be of high combining ability. But neither the cultivar John Bear, with FMS, nor stamenless forms possess these properties. Using them as a basis it became possible however to develop new forms and cultivars suitable for the production of hybrids, possessing a great number of commercially valuable characters, with the least expense of manual labour for hybrid seed production.

First for easing the hybridization process some characters had to be transferred to the sterile plants of a John Bear form, such as a flower with a long style along with a marker character, and to stamenless plants: a high percentage of fruit set with smooth and even fruit surface.

Many years of research mostly based on hybridization and selection enabled us to work out methods for obtaining plants combining the John Bear sterility form with a marker character. There are two ways for obtaining these types of plants: either by developing sterile long style plants incorporating a marker character, or by distinguishing fertile long style plants with a marker character incorporating the John Bear sterility form. Using this method we developed a number of new sterile strains (Sterilnaya 1, Sterilnaya 3, Sterilnaya 4 and others). All these strains have long style flowers and are mainly determinant bush type. By using them, it is possible to develop dwarf hybrids with valuable commercial characters. But utilization of these strains in hybridization for heterosic seed production does not always guarantee the necessary effectiveness from the point of view of the vegetable grower. Thus, it was a basic requirement to work out methods in order to obtain functionally sterile analogue cultivars with high combining ability. These methods have been developed and are quite simple. An analogue of any tomato cultivar having John Bear type sterility, long style flower, and potato-type leaf (recessive character) is developed in seven generations, using the backcross method of paternal fertile cultivars with high combining ability. Sterile plants are used as maternal forms. The first hybrid generation is completely fertile, the second segregates into fertile and sterile plants at an approximate ratio of 3:1. Starting from the second generation the sterile plants are twice backcrossed with the parental cultivar, and the sterile plants obtained are propagated by artificial self-pollination. Methods concerning this problem were published and given to breeders and on these basis hybrids were developed; some of them are in commercial production and many others are under state testing. The commercially cultivated hybrids include Hybrid 12, Salatny, Khersonsky, Virovsky skorospely and others.

Stamenless plants are one of the most suitable forms for hybridization. They are easily pollinated, but the percentage of fruit set under artificial pollination is very low (up to 6%) and as a rule the fruits are extremely misshapen. Elimination of these faults and development of stamenless forms suitable for hybridization with the aim of producing high yielding heterotic seeds is the second part of our investigations. For its achievement we conducted a number of experiments to work out methods of producing stamenless forms satisfying these demands.

From all tested methods for eliminating the main faults of initial stamenless plants (low fruit set as a result of artificial pollination and their misshapen form, for example cultivar Stamenless) the best proved to be a double-triple backcross of stamenless plants, distinguished in F_2, to a maternal fertile cultivar (Table 1).

Table 1. Scheme of the development of stamenless tomato strains with high percentage of smooth fruit setting

Years	Generations	Parental forms and their type ♀	♂
1	—	Stamenless (stam.)*	"A" cultivar (fert.)**
2	F_1	Stamenless × "A" cultivar (fert.)	Self-pollination
3	F_2	Selection of stamenless plants with a high percentage of fruit setting and smooth surface fruits	Stamenless (fert.)
4	BC_1	Selection of stamenless plants with a high percentage of fruit setting and smooth surface fruits	Stamenless (fert.)
5	BC_2	Selection of stamenless plants with a high percentage of smooth fruit setting	Stamenless (fert.)
6	BC_3	Selection of stamenless plants with a high percentage of smooth fruit setting	Propagation (stam.)

(stam.)* = stamenless plants
(fert.)** = fertile plants

For that purpose the stamenless plants (cultivar Stamenless) were crossed to fertile plants of the cultivar from which it is derived to obtain a new stamenless plant form. The first generation is completely fertile and in the second generation we select the stamenless plants which are backcrossed two or three times to the maternal fertile cultivar. In this case we always use stamenless plants with a high percentage of smooth fruit setting for backcrossing, as it is shown on the scheme of Table 1.

The elimination of the principal faults in stamenless plants can be combined with simultaneously obtained analogues, which is clearly seen in scheme 2 (Table 2).

To this end triple backcrossing of stamenless plants of F_2 hybrids to the paternal variety is carried out. When growing BC_1, BC_2, and BC_3 plants under cover, stamenless analogues can be obtained within 6 years.

The method of backcrossing to paternal and maternal fertile varieties enables us to obtain stamenless strains possessing high fruit setting (70–100%) and smooth surface fruits.

According to economic characteristics the stamenless plants of Fireball variety are preferable to the stamenless plants of the stamenless variety for hybridization.

To speed up the breeding of new industrial hybrids it is necessary to know the combining ability of tomato cultivars for the principal biological and economic characteristics. The study undertaken enabled us to obtain groups of cultivars with high combining ability on yielding, early ripening, chemical composition of fruits, fruit size and other characteristics. Thus, Heinz 1350, Fireball, Deteniske salatove, Jubileyniy 261 and Orlovsky 5 have been found to possess high combining ability for total and for marketable yield, Ventge, Meteor, Beliy naliv, Brianskiy 2 and Deteniske salatove early ripening, Ottawa 39, Jubileyniy 261, Line XXIVa, Hemus 13a, Bonita, Maikopskiy urozhainiy 2090, Matrosov 48 and others have favourable chemical composition of the fruits.

Table 2. Scheme to develop stamenless tomato strains suitable for hybridization with simultaneous obtaining of stamenless analogues

Years	Generations	Parental forms and their type	
		♀	♂
1	—	Stamenless (stam.)*	"A" cultivar (fert.)**
2	F_1	Stamenless × "A" cultivar (fert.)	Self-pollination
3	F_2	Selection of stamenless plants with a high percentage of fruit setting and smooth surface fruits	"A" cultivar (fert.)
4	BC_1	F_2 (stamenless × "A" cultivar) × "A" cultivar (fert.)	Self-pollination
5	Self-pollination BC_1	Selection of stamenless plants with a high percentage of fruit setting and smooth surface fruits	"A" cultivar (fert.)
6	BC_2	F_2 (stamenless × "A" cultivar) × "A" cultivar (fert.)	Self-pollination
7	Self-pollination BC_2	Selection of stamenless plants with a high percentage of smooth surface fruit setting	"A" cultivar (fert.)
8	BC_3	F_2 (stamenless × "A" cultivar) × "A" cultivar (fert.)	Self-pollination
9	Self-pollination BC_3	Selection of stamenless plants with a high percentage of smooth surface fruit setting	Propagation (stam.)

(stam.)* = stamenless plants
(fert.)** = fertile plants

According to our data, it is necessary to breed maternal sterile lines and paternal fertile strains possessing a complex of valuable economic characteristics.

Our work resulted in the elimination of manual labour in flower castration when producing hybrid heterotic tomato seeds. Though the process of flower pollination is considerably improved manual labour is not eliminated with this operation. In this connection we have undertaken some experiments to study the pollination of tomato plants by honey bees and bumble bees. In all crossing combinations we used the varieties with potato-type leaf as the maternal component. This recessive characteristic was a genetic marker. When crossing such maternal cultivars to others with the common-type leaf it is possible in the first generation to reject uncrossed plants at the seedling stage.

The study showed that under controlled conditions bumble bees were more active tomato pollinators than honey bees. The most promising results were obtained when using *Bombus terrestris* L. and especially *Bombus agrorum* L. When pollinating tomato plants by these species the output of hybrid seeds, according to preliminary data, was 56.6–80.1%.

This study has been undertaken quite recently but we hope that in the near future we shall succeed in producing heterotic tomato seed material practically without input of manual labour by using bumble bees, new cultivars with functional male sterility in combination with long styled as well as stamenless forms.

EUCARPIA VII (1976)

New patterns of heterosis in *Fragaria* hybrids

R. BAUER

Max-Planck-Institut für Züchtungsforschung,
Köln-Vogelsang, FRG

SUMMARY

The common varieties of *Fragaria ananassa* ($2n = 56$) are an example of heterosis caused by hybridization of two wild octoploid species, *F. chiloensis* and *F. virginiana,* adapted to extremely opposite climatic conditions. The yearly growing cycle of strawberry varieties is controlled by daylength and temperature. Depending on the fact whether the "selected genotype" of a variety is more related to *F. chiloensis* or to *F. virginiana,* the degree of heterosis changes as well as the degree of adaptation to a local climate. This high sensitivity to climatic factors, daylength and temperature, is a strong limit for breeding widely adaptable strawberry varieties.

In crossings including a third species of *Fragaria,* the runnerless *F. vesca semperflorens* ($2n = 28$), decaploids could be selected after backcrossing the F_1 to the *F. ananassa* parent.These decaploids combine 8 genomes of *F.ananassa* and 2 genomes of *F.vesca* in an amphidiploid state. They are named "Vescana"-hybrids.

In these hybrids the heterosis is increased in all vegetative characteristics, but contemporaneously the sensitivity to short day conditions of the *F. ananassa* parent is decreased. The combination of the short day sensitivity of *F. ananassa* with the day-neutral reaction of *F. vesca* leads to changes in the growing pattern of the Vescana-hybrids compared with their parents.

The genomes of the parents present inthe amphidiploid hybrids can act independently of each other due to distinct changes caused by the factors, daylength and temperature.

Varieties of the octoploid garden strawberry *Fragaria ananassa* which are cultivated today demonstrate heterosis. This heterosis comes from the combination of the genomes of two octoploid American species, *F. chiloensis* and *F. virginiana,* which show extremely different ecotypical patterns and from the chance distribution of their chromosomes into the different existing strawberry varieties.

Developmental growth patterns of *F. ananassa* varieties are genetically determined and are controlled by temperature and daylength. Thus, for best heterosis effects one needs to select a variety for each location according to the best adaptation to the local temperature and daylength conditions. Breeding of such locally adapted strawberry varieties can only be successful when one uses as parental material vari-

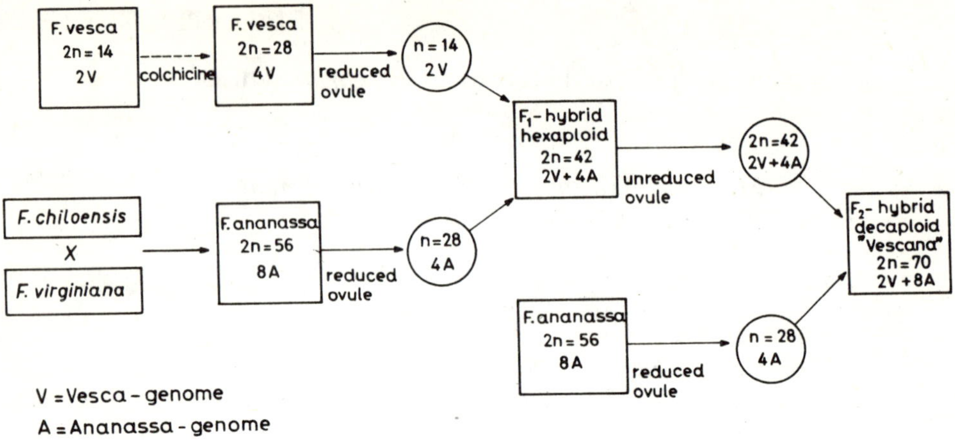

Fig. 1. Scheme for crossing Vescana-hybrids

eties belonging to two basic groups, one with most of the genes of *F. chiloensis* and the other with most of the genes of *F. virginiana.* Many successful strawberry varieties are examples of genetically fixed heterosis after new combinations of the respective gene complexes.

Crossing *F. ananassa* with a third species of *Fragaria* the runnerless *F. vesca semperflorens* which was made tetraploid by colchicine treatment ($2n = 28$) gave a sterile F_1, but after backcrossing the F_1 to the *F. ananassa* parent, fertile decaploid progenies could be selected. These amphidiploid decaploids ($2n = 70$) combine 8 genomes of *F. ananassa* and 2 genomes of *F. vesca semperflorens.* They are called "Vescana"-hybrids (Fig. 1).

Introducing the third genome of *F. vesca semperflorens* results in enhanced heterosis effects. Among other things, it changes the response of the hybrid to daylength and to temperature. While the octoploid *F. ananassa* varieties are adapted only to local climates (adaptive heterosis), heterosis effects in the decaploid Vescana-hybrids are not so strongly limited to certain climatic conditions.

Apparently, the Vescana-hybrids combine the genetically determined requirement for flower development of *F. ananassa* for short daylength with the insensitivity to daylength of *F. vesca semperflorens,* and the responsible genes function differentially during the different seasons of the year.

Under long day condition, the vegetative growth of the amphidiploid hybrid is greatly enhanced as compared with pure *F. ananassa* varieties. This is indicated by the strong production of runners instead of induction of flowers. This suggests the expression of the *ananassa* genome. The *vesca* genome further enhances vegetative growth under these conditions.

In the *F. ananassa* parent short days in the fall and temperatures below 15 °C induce the generative phase (inhibition of stolons, rosette formation of the leaves and induction of flowers). Short days also inhibit metabolic processes which finally reach a very low level during the rest period of the winter months – "hibernation".

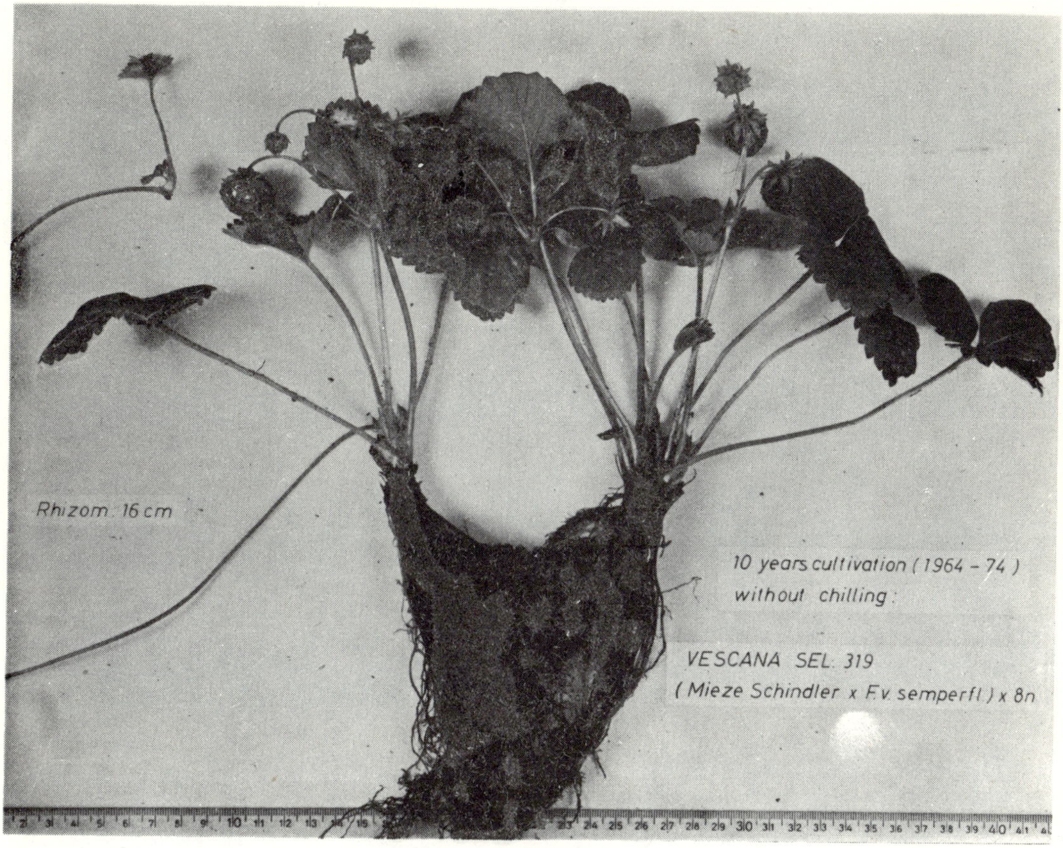

Fig. 2. 10 years cultivation without chilling of Vescana Selection 319

In the Vescana-hybrids this induction occurs later and apparently does not depend on the daylength. Runners are produced until late October, i.e. when days are already short. Similarly, rosettes are formed later or not at all.

Greenhouse experiments (without additional light) showed that the Vescana-hybrids do not "hibernate" when temperatures are kept between 5 °C and 10 °C. Under these conditions they form new leaves, flowers and fruits already in February. With increasing temperatures and under long day conditions they regain full capacity for vegetative growth. They continue to flower throughout the following season (Fig. 2).

Outside of the greenhouse in the field, however, where temperatures regularly fall below 5 °C the vegetative phase is induced and with increasing temperatures in spring all parts of the plant start growing luxuriously. Note, that the daylength was the same as in the greenhouse (Figs 3–5). The continuous activity during the short winter days suggests that the corresponding *vesca* genes in the amphidiploid hybrid

Fig. 3. Different Vescana-hybrids in the field (with chilling). Each plot is the vegetative progeny of one mother-plant (the first plan in front of each plot)

are active. "Hibernation" of the Vescana-hybrids in the field depends mainly on the low temperature not on the daylength. This reaction is similar to the reaction of *F. vesca semperflorens.*

Presumably, the chromosomes of the two parental species show little homology; the F_1 hybrids are sterile. Thus, we assume that the complete genomes of both parental species function independently. Certain conditions inactivate gene function of only one parental genome, while they have no effect on expression of the genome from the other parent. This can be illustrated by two manufacturing systems which occupy the same building and can work jointly or independently.

Fig. 4. "Meadow-culture" of Vescana Selection 505 (a first step to mechanical harvesting).

Fig. 5. 3 fruit clusters from the "meadow-culture" of Vescana Selection 505 showing the uniformity in length of stems and in fruit development.

REFERENCES

BAUER, R. (1960). Aspekte der Polyploidiezüchtung in der Gattung *Fragaria. Extr. Bull. Inst. Agronom. Stat. Rech. Gembloux.* Hors série **II,** 994–1006.

BAUER, R. (1960). Grundlagen und Methoden der Züchtung bei der Gartenerdbeere (*Fragaria ananassa* Duch.). *Z. Pfl.zücht.* **44,** 73–100.

BAUER, R. &. BAUER, A. (1967). Neue Wege in der Erdbeerzüchtung. *Der Erwerbsobstbau.* **9,** 83–85.

SCOTT, D.H. (1951). Cytological studies on polyploids derived from tetraploid *Fragaria vesca* and cultivated strawberries. *Genetics,* **36,** 311–331.

Session II

PHYSIOLOGY OF HETEROSIS

Chairmen:

DR. A. LEIN AND DR. Á. KISS

The physiological basis of heterosis in wheat

F. G. H. LUPTON

Plant Breeding Institute, Trumpington
Cambridge, England

SUMMARY

The exploitation of heterosis in wheat involves the identification of parental combinations showing highly specific combining ability. Measurement of heterosis at the sub-cellular level, by determining the oxidative efficiency of mixtures of the mitochondria of potential parents, gives a measure of the yield potential of the crosses concerned. The results obtained may, however, be invalidated by the incidence of disease, or other agricultural limitations to crop performance.

Differences in the photosynthetic activity of seedling plants, obtained by measuring the increase in dry weight of detached leaf segments, were not significantly correlated with yield heterosis.

Comparative growth analyses of F_1 combinations showed marked heterosis for total dry weight production and leaf area index. This was associated with heterosis for photosynthetic rate shortly before anthesis. The dry weight of the F_1 hybrids fell during the last weeks of grain filling, possibly because of inadequate capacity of their ears to utilize the carbohydrate available.

INTRODUCTION

Despite much early enthusiasm, progress in the development of "hybrid" crops in the small grained cereals has been slow. Many wheat hybrids show heterosis for vegetative vigour but the exploitation of this vigour in the form of economic yield has proved very difficult and requires the identification of parental combinations showing highly specific combining ability.

Heterosis may be expressed at any stage in crop development, and the purpose of the breeder is to identify those stages at which heterosis may most conveniently be assessed, and when correlation with yielding capacity is greatest. Our work at Cambridge has been concerned with studies of heterosis at the sub-cellular level, measuring the oxidative efficiency of the mitochondria; at the single leaf level, measuring the photosynthetic efficiency of leaf segments detached from seedling plants; at the whole plant level, by growth analysis of plants growing under field conditions; and with estimates of rates of photosynthesis and carbohydrate translocation under field conditions. In all cases, the estimates have been combined with measurement of the yielding capacity of the crosses considered.

The mitochondria are cytoplasmic organelles concerned with the conversion of energy, derived from photosynthetic activity or from reserves of carbohydrate in the endosperm, to a form available for plant growth, that is to ATP. McDaniel (1972) showed that there was a significant correlation between plant growth and mitochondrial energy conservation expressed as number of ADP molecules oxidised for each oxygen atom consumed. Previously, McDaniel & Sarkissian (1966) had demonstrated that the oxidative activity of mixtures of mitochondria from different inbred lines of maize sometimes exceeded that of the lines considered separately. They also suggested that this heterotic effect, which they called mitochondrial complementation, might be used to select useful parental combinations without the need to make crosses which might subsequently prove to be useless. This paper was followed by demonstrations by McDaniel and his colleagues of mitochondrial complementation in wheat and barley (Sarkissian & Srivastava, 1967, 1969; McDaniel, 1970, 1972), but other workers have had difficulty in repeating their results. My colleagues, Sage & Hobson (1973), were however able to demonstrate mitochondrial complementation in wheat, though the level of heterosis they reported was very much lower than that found by McDaniel and his colleagues.

Sage and Hobson assessed the mitochondrial complementation in an incomplete diallel set of crosses. The female parents were three male sterile lines obtained by back-crossing the varieties Kloka and Maris Ranger and a selection TB 306/45 (Jufy × Svenno) to male sterile Bison, a source of *T. timopheevi* cytoplasm. They were each crossed with three restorer lines, Maris Beacon, Primepi and a selection TW 74/34/3 (Minister × Els). The mitochondria of each parent and of 1:1 mixtures of the parents were extracted and their ADP:O ratios assessed using a 2 oxoglutarate substrate as described by Sarkissian & Srivastava (1969). The data obtained (Table 1) suggests a moderate level of mitochondrial complementation in some of the parental combinations, though in no case was the ADP:O ratio of a mixture significantly higher than that of either of the parents. Sage and Hobson compared their ADP:O data with those from a yield trial in which the F_1's and their parents were grown at a spacing of 5 × 20 cm. The comparison was suggestive of a relation between yield heterosis and mitochondrial complementation, but the correlation coefficient between the two sets of data was not significant. It was, however, observed in the yield trial that the cross Kloka × Primepi was very severely attacked by yellow rust *(Puccinia striiformis)* and eyespot *(Cercosporella herpotrichoides)* whereas the parents and the other crosses were much less severely attacked. If this cross is omitted in the calculation, a significant correlation of 0.845*** was obtained between yield heterosis and mitochondrial complementation. This was further improved to 0.921*** if the cross TB 306/45 × Maris Beacon, which showed incomplete restoration of fertility, was also omitted.

This result suggests that mitochondrial complementation provides a useful indication of the yield potential of a cross. It cannot, however, be expected to discriminate between crosses which fail to achieve this potential for agronomic reasons such as disease susceptibility or susceptibility to lodging, or for genetic abnormalities such as hybrid necrosis or incomplete restoration of fertility.

Sage and Hobson's results have now been repeated by my colleagues D.H.P. Barrett and R.B. Flavell, who also found that the levels of heterosis, though significantly correlated with the yield data, were very much lower than those reported by McDaniel. They also investigated the mitochondrial activity of a range of wheat varieties. These include varieties which have in turn been widely grown in Britain during the past forty years and have thus proved themselves to be well adapted agronomically, and newly introduced varieties which have yielded well and not yet succumbed to any major disease. When listed in order of ADP:O ratio, these varieties fall into an order agreeing remarkably well with a yield ranking based on official trials (Flavell & Barrett unpublished). This finding demonstrates the possibility of using estimates of ADP:O ratios as an initial screen, independent of such complications as regional adaptation or disease susceptibility, in assessing the potential of varieties as parents for wheat breeding.

Table 1. Mitochondrial heterosis in wheat

(a) ADP:O ratio of parental mixture
(b) Grain yield, as percentage of the higher yielding parent

	TW 74		Primepi		Maris Beacon	
	a	*b*	*a*	*b*	*a*	*b*
Kloka	99	104	99	81	97	98
TB 306/45	94	100	101	107	98	97
Maris Ranger	107	110	94	98	103	106

PHOTOSYNTHESIS BY LEAF SEGMENTS

As a second approach to the selection of parental combinations likely to show heterosis for yield, we have used a leaf segment technique similar to that developed by Natr & Spidla (1961).

Seedlings of F_1 plants and their parents were vernalised in a low temperature cabinet and then grown under controlled conditions until four leaves had developed on the main stem. They were then kept in the dark for 48 hours to allow reserves of carbohydrate to be translocated away from the sites of photosynthesis. At the end of this period, two 3 cm lenghts were cut under water from the central part of the lamina of the third leaf of each plant. One segment was immediately dried in an oven at 95°C and weighed; the other was floated on a water trough, with the ends held under water to prevent wilting, and exposed to a uniform light of 32000 lux. After 16 hours exposure to light, these leaf segments were also dried and weighed. The position on the mother plant of the segments exposed to light or weighed immediately were alternated at random, and their weights compared in order to determine the photosynthetic activity of the leaves concerned.The experiment was laid out with five replications sown successionally.

The results obtained (Table 2) show significant differences in photosynthetic activity between the parent lines and between the F_1 progenies. They also suggest that some parental combinations, notably Maris Ranger × Primepi and TB306/45 × Maris Beacon, are heterotic for photosynthetic efficiency. The correlation between the heterotic expression of photosynthetic rate and yield heterosis was, however, not significant, even when the two crosses which were unlikely to give a full expression of their potential yield were omitted. It thus appears that the leaf segment technique, though useful in distinguishing between the photosynthetic capacity of varieties, is less valuable in assessing parental combinations likely to show heterosis for yielding capacity.

GROWTH ANALYSIS OF F_1 HYBRIDS

The growth and development of four of the F_1 hybrids which had previously been considered was followed in yield trials grown in 1971 and 1972. Both trials were sown in October at a spacing of 5 × 10 cm, using a randomised block lay out with five replications. Sub-samples were taken from each plot for measurement of leaf area index and dry weight increment, from mid April until maturity in August. Counts of shoot number/m^2 were also made at intervals during crop development.

Table 2. Percentage increase in dry weight of leaf segments after 16 hours illumination at 32,000 lux

		TW 74	Primepi	Maris Beacon
		31.8	25.4	29.5
Kloka	30.3	31.9	28.7	23.8
TB 306/45	27.8	30.8	25.8	31.3
Maris Ranger	26.8	29.6	28.9	29.9

L.S.D. (P=0.05) 4.06

All the crosses showed a marked heterosis for total dry weight production and for leaf area index in both years, but it was found that while the dry weight of the parental varieties continued to increase until maturity that of the F_1 hybrids reached a maximum about six weeks before the final harvest, and subsequently declined. This effect was most clearly seen in the second year of trial and is shown in Fig. 1. The tiller number of the hybrids was in all cases intermediate between the parents, and there were no clear differences in the proportion of tillers surviving to bear ears at harvest.

All four crosses exhibited heterosis for grain yield in the first year of trial, but only the cross TB 306/45 × TW 74/34/3 in the second year (Table 3).

The fall in total dry weight of the F_1 hybrids was very clearly seen in each of the hybrids examined, and suggests that the capacity of the hybrid plants to utilise the carbohydrates being formed during the later parts of the grain filling period may have been limiting. This may have arisen because of inefficient fertility restoration by the male parents, though counts of grain number per ear gave no indication of poor grain set (Table 4), nor was there any heterosis for 1000 grain weight, except in the cross TB 306/45 × TW 74/34/3. There was no evidence of tiller death between anthesis and grain maturity, so that the loss in dry weight of the F_1 hybrids must have been due

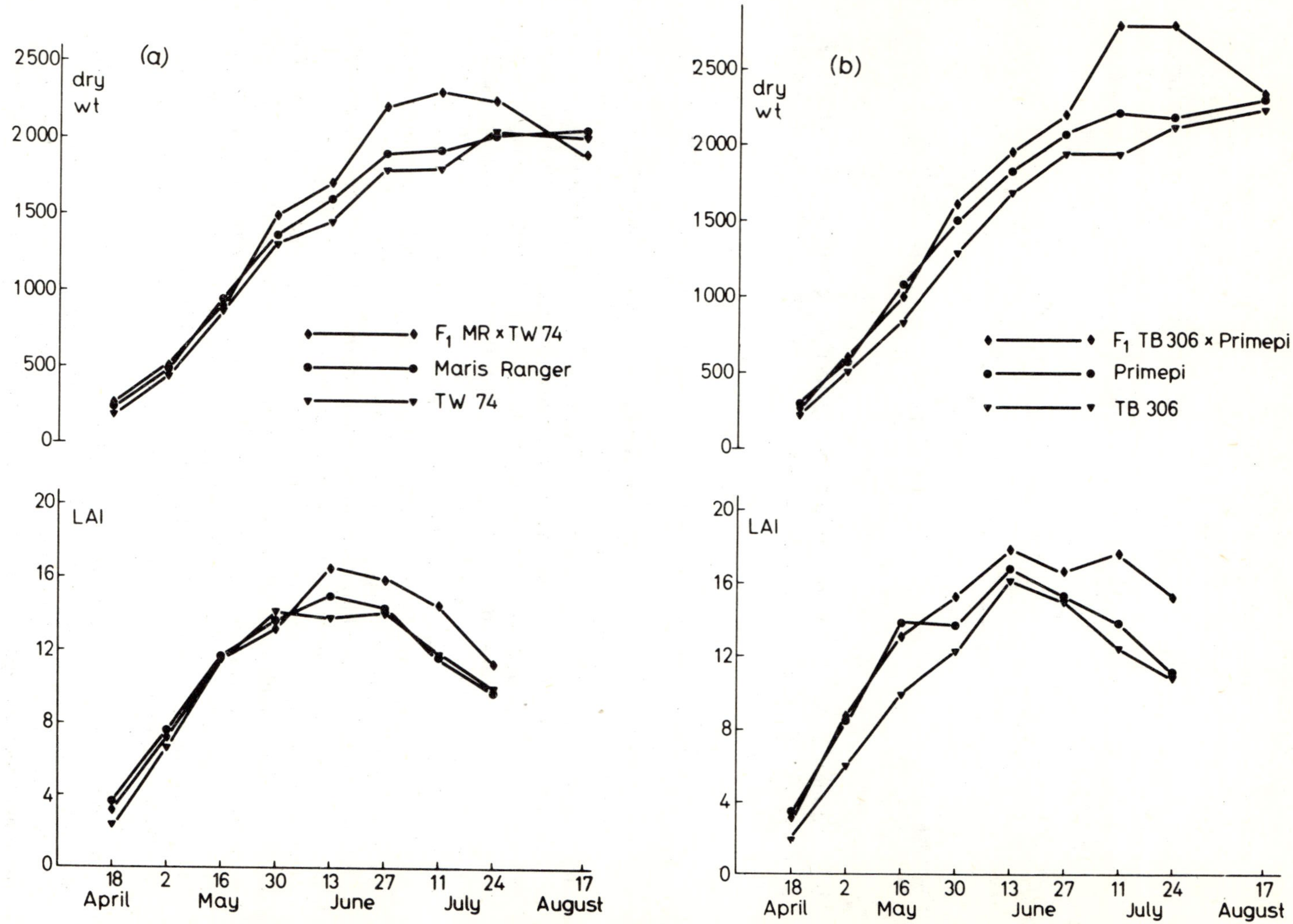

Fig. 1. Growth analysis 1972: Leaf area index and dry weight development of *(a)* F_1 Maris Ranger × TW 74 and *(b)* F_1 TB 306 × Primepi in comparison with their parents.

Table 3. Grain yields for 1971(a) and 1972(b) from F_1 growth analysis experiments (g/m^2)

			TW 74		Primepi	
	a	*b*	*a*	*b*	*a*	*b*
			795	739	577	1075
TB 306/45	600	789	947	934	867	1001
Maris Ranger	782	856	936	850	785	797

Table 4. Grain number per ear(a) and 1000 grain wt(b) in F_1 growth analysis experiment 1972

			TW 74		Primepi	
	a	*b*	*a*	*b*	*a*	*b*
			25.2	45.6	25.6	59.9
TB 306/45	28.4	47.3	25.3	53.2	23.0	55.4
Maris Ranger	25.6	57.9	24.1	53.3	25.5	57.0

to respiration or to translocation to the roots, which were not sampled. The latter seems very unlikely, however, as root growth normally ceases at about the time of anthesis (Troughton, 1962).

MEASUREMENT OF PHOTOSYNTHESIS AND TRANSLOCATION

The rate of photosynthesis and pattern of translocation of carbohydrate of one of the F_1 hybrids, Maris Ranger × TW 74, and its parents were measured in plots of a yield trial growing under field conditions. Flag leaves of groups of three shoots, still attached to their parent plants, were temporarily sealed in a Perspex container through which air was passed at a rate equivalent to approximately 10 changes per minute. Rates of photosynthesis at a range of light intensities, obtained by placing filters of different density over the assimilation chamber, were measured using an infra red gas analyser. The assimilation chamber was then switched to form part of a closed circuit containing also an air pump, a gas scintillation cell connected to a recording rate meter, a condenser to remove transpired water vapour and a small flask containing ^{14}C labelled sodium carbonate. When conditions were ready, excess lactic acid was added to the sodium carbonate, releasing sufficient $^{14}CO_2$ to increase the total carbon dioxide concentration in the circuit to approximately 600 ppm. The leaves were then left to assimilate until the concentration fell to 150 ppm when the circuit was flushed with air. The dosed stems were left in the growing crop until the grain was ripe, the ^{14}C distribution within them was then determined using a dry planchet counting system at infinite thickness. Rates of photosynthesis and translocation were measured 12 days before anthesis, at anthesis, and 14 and 28 days after anthesis.

An approximately linear response of photosynthesis to light intensity was observed over the range of light intensities used in these experiments (15 to 40 cal cm^{-2} h^{-1}). Regression lines were fitted for each set of readings and the rate of photosynthesis at an incident light intensity of 30 cal cm^{-2} h^{-1} calculated (Table 5). The results

Table 5. Rate of photosynthesis (mg CO_2 dm^{-2} h^{-1}) measured under field conditions at 30 cal cm^{-2} h^{-1}; linear regression coefficient (b) of rate of photosynthesis on light intensity; and percentage of assimilated carbon recovered in the grain at maturity

	Anthesis −12 days	Anthesis	Anthesis +14 days	Anthesis +28 days
Maris Ranger				
Rate of photosynthesis	17.80	17.20	19.60	8.70
Regression coefficient	0.59	0.41	0.44	0.18
Percentage carbon recovered in the grain	1.50	8.80	67.30	75.70
TW 74				
Rate of photosynthesis	17.40	18.30	13.40	11.10
Regression coefficient	0.55	0.58	0.34	0.26
Percentage carbon recovered in the grain	2.80	8.00	36.50	61.30
F_1 Maris Ranger × TW 74				
Rate of photosynthesis	19.40	18.10	16.50	9.30
Regression coefficient	0.64	0.57	0.34	0.13
Percentage carbon recovered in the grain	6.30	11.00	48.10	66.20

obtained followed the pattern indicated by the growth analysis curves (Fig. 1). There was a marked heterosis for photosynthetic rate 12 days before anthesis, at the time when the growth rate of the F_1 hybrid was markedly greater than that of its parents. The response of the F_1 to increasing light intensity was also greater at this stage. Subsequently, however, the photosynthetic rate of the F_1 fell to a level intermediate between those of its parents.

Only a small proportion of the ^{14}C synthesised before anthesis was translocated to the grain, but the F_1 hybrid translocated rather more than either of its parents. There was little varietal difference in translocation pattern at anthesis, but after anthesis, Maris Ranger translocated more ^{14}C than TW 74 and the F_1 was intermediate in this respect. As grain carbohydrate is almost entirely derived from photosynthates assimilated after anthesis (Thorne, 1965), the grain yield of the F_1 was also intermediate between the parents (102% of the mid parent and only 92% of the higher-yielding parent, Maris Ranger).

DISCUSSION

Engledow & Wadham (1923) suggested the need for plant breeders to define a "touch stone" by which the yielding capacity of a crop could be determined; they were, however, many years ahead of their time, and it was not until the development of the modern concepts of growth analysis by Watson and his colleagues (Watson, 1952; Watson *et al.* 1963) that breeders were able to define the characters upon which the growth of their crops depended. Even these developments were of limited application, as they were difficult to use in selection though they enabled the breeder to define his objectives more precisely, and hence to choose the parents for his crossing programmes on a more logical basis. Perhaps the best example of the application of the results of growth analysis in practical breeding has been Bingham's demonstration of the value of selection for tiller economy in breeding of the varieties Maris Huntsman and Maris Nimrod (Bingham, 1971).

The need for a touch stone is even greater in choosing parents for a hybrid wheat breeding programme than in a conventional breeding programme. The possible exploitation of mitochondrial complementation as a guide in the choice of parents is therefore of the greatest interest to the hybrid wheat breeder, and it is much to be hoped that the improved techniques now being developed will make its application both easier and more reliable. As I have attempted to show, however, techniques of this sort may be of very great value in eliminating undesirable parental combination, but cannot be expected to account for their agronomic defects. These may sometimes be anticipated by a knowledge of the pathological characters or geographical distribution of the parents concerned, but a considerable wastage of parental combinations must be anticipated.

Much has been written in recent years of the interaction of "source" and "sink" in determining the yielding capacity of cereal crops. In a recent publication (Lupton *et al.*, 1974), we showed that over 90 per cent of yield variance of a group of F_2 plant progenies could be explained by the simultaneous consideration of photosynthetic capacity and ear primordium development though either component alone could only account for about half this variance. Even in the absence of agronomic defects, estimates of mitochondrial activity or of seedling photosynthesis can only be concerned with total crop growth, that is with the "source" component of yield. The "sink" component must be improved by other considerations, of which one of the most important is adequate fertility restoration of the F_1 plant. Sink size may, however, be determined much earlier in crop development, and there is, I consider, considerable scope for research on the relative lengths of successive phases in ear and crop growth, particularly in relation to crops grown at the low seed rates which may be used when sowing expensive "hybrid" wheats.

Considerable progress has been made in studies of the physiology of hybrid wheats, but, as in so many lines of research, the results gained have exposed many new problems to investigate. Let us hope that this Congress may stimulate increased research on these most important subjects.

ACKNOWLEDGEMENT

I would like to thank my colleagues W.G. Hughes, R.H. Oliver and G.C.M. Sage for permission to use their unpublished data on growth analysis and on the photosynthesis of detached leaf segments.

REFERENCES

Bingham, J. (1971). Plant Breeding: Arable Crops. In *Potential Crop Production* (Ed. P.F. Wareing & J.P. Cooper), 273–294. London, Heinemann

Engledow, F.L. & Wadham, S.M. (1923). Investigations on yield in the cereals, I. *Journal of agricultural Science, Cambridge* **13,** 390–439.

Lupton, F.G.H., Oliver, R.H. & Ruckenbauer, P. (1974). An analysis of the factors determining yield in crosses between semi-dwarf and taller wheat varieties. *Journal of agricultural Science Cambridge,* **82,** 483–496.

McDaniel, R.G. (1970). Mitochondrial heterosis in barley. *Proceedings of the 2nd International Barley Genetics Symposium* 323–337.

McDaniel, R.G. (1972). Mitochondrial heterosis and complementation as biochemical measures of yield. *Nature, New Biology* **236,** 190–191.

McDaniel, R.G. & Sarkissian I.V. (1966). Heterosis: complementation by mitochondria. *Science* **152,** 1640–42.

Natr, L. & Spidla, J. (1961). Application of the leaf disc method to the determination of photosynthesis in cereals. *Biologia Plantarum (Praha)* **3,** 245–57.

Sage, G.C.M. & Hobson, G.E. (1973). The possible use of mitochondrial complementation as an indicator of yield heterosis in breeding hybrid wheat. *Euphytica* **22,** 61–69.

Sarkissian, I.V. & Srivastava, H.K. (1967). Mitochondrial polymorphism in maize, II. Further evidence of correlation of mitochondrial complementation and heterosis. *Genetics* **57,** 843–850.

Sarkissian, I.V. & Srivastava, H.K. (1969). High efficiency, heterosis and homeostasis in mitochondria of wheat. *Proceedings of the National Academy of Science of the USA* **63,** 302–309.

Thorne, G.N. (1965). Photosynthesis of ears and flag leaves of wheat and barley. *Annals of Botany* **14,** 55–68.

Troughton, A. (1962). The roots of temperate cereals (wheat, barley, oats and rye). *Mimeographed Publications of the Commonwealth Bureau of Pastures and Field Crops* **2,** 91.

Watson, D.J. (1952). The physiological basis of variation in yield. *Advances in Agronomy* **4,** 101–145.

Watson, D.J., Thorne, G.N. & French, S.A.W. (1963). Analysis of growth and yield of winter and spring wheats. *Annals of Botany* **27,** 1–22.

Physiological and biomechanical aspects of heterosis

V. G. KONAREV

N. I. Vavilov All-Union Institute of Plant Industry, Leningrad, USSR

SUMMARY

The essence and genetic causes of heterosis are briefly reviewed according to literature data and recent hypotheses.

The importance of investigations into the physiology and biochemistry of heterosis, especially of the analysis of the results obtained in the light of contemporary notions of molecular genetics and phenogenetics is emphasized.

Recent findings and ways ahead in the physiology and biochemistry of heterosis hybrids and selfed lines are briefly discussed. Importance of different physiological (photosynthesis, respiration, ferment activity, mitochondrion activity, isoenzyme spectrum, composition of a number of endogenic regulators) and molecular-genetic (synthesis intensity of DNA, RNA, and genome activity)processes is treated, with regard to the use of such information in the choice of couples, evaluation of heterosis effect and prognostication.

Finally, some methods of line evaluation for combining ability and prediction of heterosis levels are discussed.

HETEROSIS AND YIELD

Heterosis is a complex biological phenomenon manifested in the superiority of hybrids over parent forms according to the rate of development of one or complex characters. In contrast to the common recombination effect, heterosis is displayed mainly in the F_1 progeny and falls off sharply in the following generations. It is expressed very clearly in many interspecies and in a number of intervarietal hybrids. The best results, however, are achieved due to definite combination of inbred lines, obtained as a result of directed selection on combining ability according to commercially valuable characters. Thanks to this possibility heterosis may be genetically controlled in a number of crops.

As a result of numerous investigations it was revealed (Fedorov, 1968; Brezhnev, 1969; Brezhnev *et al.,* 1969) that heterotic hybrids are notable for their intensive accumulation of nutritional substances up to the beginning of fruit setting, and for their very energetic utilization of the substances, in the process of fruit and seed development. They have a well-developed leaf area, and a high rate of accumulation of substances and of transportation of metabolites, as compared with the self-pollinated lines. The intensified growth of the heterotic hybrid vegetative organs (e.g. leaves) takes place mainly due to the increase of cell number (Uchimiga & Takahachi, 1973).

Heterosis is mainly manifested in quantitative characters and is expressed in vigour of vegetative organs, in increased grain yield, and in resistance to diseases, climatic factor fluctuations, etc. Corresponding heterosis is mainly manifested as increase in total plant productivity, but has nearly no influence on plant quality, including the correlation of the chemical components in the yield. Occasionally observed superiority in ascorbic acid and sugar content (Brezhnev, 1969; Lukovnikova & Gomolyakov, 1969) in vegetable crops over their original lines may be explained by the physiological condition of the fruits of heterotic plants. Heterotic cereal hybrids are often inferior to the original lines in protein content of the grain. This is due to the common phenomenon of the negative correlation between yield and protein content. Heterosis is expressed in increased content of these components that are mainly responsible for yield increase. In the caryopsis of cereals the carbohydrates have this function.

Thus, correlation of the nutritional substances in the yield is only indirectly connected with heterosis. The content of separate chemical components is inherited according to normal genetic principles as a polygenic character. Therefore, to obtain heterotic hybrids with the desired content of the nutritional components it is necessary to have selfed lines not only of good combining ability but also as good sources of yield quality.

THE RELATIONSHIP OF HETEROSIS MANIFESTATION DEPENDS ON PLANT GROWTH CONDITIONS

According to numerous data this relationship is close. Heterosis in yielding capacity is realized to the full under favourable conditions. In comparison with inbred lines heterotic hybrids are more responsive to mineral fertilization, to irrigation, etc. At the same time they are more resistant to unfavourable conditions. But some cases are known when the latter cause negative heterosis (Nettovich, 1968; Burton, 1968).

Despite some gaps in the knowledge of the nature of heterosis, the genetic methods of the practical utilization of this phenomenon in a number of crops are rather well elaborated. All of them are based on the principles of heterozygosis and obligatory cross pollination, and in the majority of cases – on the utilization of self-sterility and cytoplasmic male sterility *(CMS)*.

The primary task for the realization of the heterotic effect in plant industry is the selection of lines and clones of high combining ability, i.e. capable of producing high yielding progeny as a result of hybrid combinations. The greatest success in this work was achieved with maize and with some vegetable crops (Brezhnev, 1969; Turbin, 1971).

The development of heterotic hybrids of wheat, sorghum, barley, rye and other crops is carried out on a large scale. At the same time it should be noted that the choice of parents and the evaluation of the heterosis effect are predominantly empirical. In the main, the success in production of high yielding heterotic hybrids in plant industry was achieved by the efforts and skill of a breeder. It is necessary to look for more efficient ways for the realization of heterosis effect in plant industry, and to use more reliable methods for evaluating lines and clones for their combining abilities, and for heterosis prognostication. This can be achieved through knowledge of the genetic nature of heterosis and of the biochemical and physiological mechanisms of its manifestation.

THE NATURE OF HETEROSIS

The nature of heterosis has not yet been revealed, though there is no doubt that its causes are enclosed in the sphere of genetic processes. According to the existing hypotheses, its effect in a hybrid organism may appear as a result of (Turbin, 1966, 1971; Dubinin, 1967):

- suppression of negative action of some recessive genes by dominant ones,
- combined action of favourable dominant factors,
- compensation of a defective recessive (e.g. mutant) gene of one parent form by a normal gene of the other,
- superiority of heterozygotes *(Aa)* over homozygotes *(AA,aa)*,
- forming of complete genetic complexes by including additional genes and arising surplus information in the form of the duplicates of structural genes (James, 1961; Severin, 1967),
- changing of the genetic balance of a genome in the process of interaction of different genotypes (Turbin, 1966, 1971; Mather, 1955).

All the enumerated genetic causes of heterosis are worth mentioning. It is most likely that they are all involved in heterosis, but their appearance and the scale of manifestation may be different. As a genetic idea, capable of binding all available data into a general theory of heterosis the hypothesis of genetic balance and genetic complex completeness may be suggested. The first proceeds from the assumption that the performance of any character is determined by the relative influence of many factors. Some of them stimulate, i.e. increase the character, and others act in the reverse way. The degree of manifestation of any character is a result of genetic balance in the action of differently directed factors. According to this point of view inbreeding results in the development of forms that are not balanced genetically. The crossing of these forms favours hybrids with greatly improved genetic balance, promoting the development of simple or complex characters. According to the second hypothesis the causes of heterosis are additive and modified gene actions, resulting in an appearance of surplus information, etc. This is in compliance with the contemporary notion about the realization of a genotype into a phenotype – the development of any character being determined not only by primary activity of its gene, but also by the condition of the whole genetic system.

According to existing data (Koelle, 1973) heterosis is mainly expressed in increased development of separate characters such as height of plant, weight of roots, leaf area etc. Sometimes heterosis is displayed in a complex of characters, but it almost never concerns directly the whole plant organization. It may be concluded that sources of heterosis are formed and located in the spheres of separate genetic systems of the hybrid organism.

These may be represented in general outline in the following way.

Photosynthesis is a main factor of plant productivity. It is the most frequently used for heterosis evaluation. In the works of a number of authors it is shown, that continuous inbreeding induces depression of photosynthetic productivity. This phenomenon is mostly typical of lines with low combining ability (Baranovski & Semenova, 1968). A number of investigators found that the heterotic hybrids of vegetable crops and maize exceed the selfed lines in intensity and productivity of photosynthesis (Smirnova, 1966; Rubtsova, 1964; Litvinenko, 1968; Serbanescu, 1965, 1968). But there is no strict regularity in this phenomenon (Serbanescu, 1968). The comparative analysis of the content of chlorophyll, carotene, xanthophyll and lycopin in the leaves has not revealed any clear regularities (Lukovnikova & Gomolyakov, 1968; Serbanescu, 1968; Orlyuk, 1968).

In the context of the heterosis problem *respiration* is very often treated as a factor causing loss of the substances necessary for crop formation. In this aspect a certain regularity has been revealed – in the majority of cases the intensity of respiration of heterotic hybrids is lower than that of selfed lines. Accordingly the photosynthesis/respiration ratio of the hybrids is higher, and may be one of the essential physiological indices of heterosis (Baranovski & Semenova, 1968; Serbanescu, 1965, 1968; Ovechkin, 1968; Golinskaya *et al.*, 1966; Marsalek & Vorisek, 1968).

The importance of *enzyme activity* as an intensity index of the metabolic processes was studied mainly on oxidoreductase and transferases. Numerous investigations have shown that in most cases heterotic hybrid plants are either intermediate in enzyme activity between the parental forms or are inferior to them (Serbanescu, 1965, 1968; Marsalek, 1969, 1972; Hageman *et al.*, 1962; Boos & Sarkissian, 1968; Berzinya, Berzite & Maurinya, 1973). It is interesting that in the monoecious plants (maize, cucumber and others) heterotic intensification is noted in cases when a difference in the oxidation-reduction potential of generative organs of parental forms is clearly defined (Berzinya *et al.*, 1973). As is generally known, the oxidation-reduction potential of generative organs is one of the indices of the degree of sexualization in plants. This suggested the use of plants with intensified male and female sexualization as parental components (Maurinya, 1973). Positive results have been obtained even through experimental intensification of sexualization – female sexualization was intensified by treating plants with 0.01% $KMnO_4$ solution, male – with 0.1% hydroquinone solution (Maurinya, 1973).

In comparative studies on different chemical components in hybrids and inbred lines, some phosphorus compounds have turned out to be the most interesting, especially substances such as ATP, with energy-rich phosphate bonds as well as enzymes involved in energy transformation. There are many facts indicating the high activity of these enzymes and the synthetic intensity of high energy bonds in heterotic hybrids (Yakovleva & Ovchinikova, 1968; Ching, 1973).

The data obtained by Sarkissian & Srivastava (1969) on *mitochondria* are of a special interest. They found that hybrid plants possessed a higher ATP-ase activity. In experiments with isolated mitochondria of wheat and maize, it was established that in comparison with inbred lines and non heterotic hybrids mitochondria of heterotic hybrids absorbed more oxygen and had a higher P/O index. This indicates that a high level of oxidizing phosphorylation and synthesis of high energy bonds of the ATP type, creates favourable conditions for biosynthetic processes and is an important requisite for heterotic development (Sarkissian & Srivastava, 1969, 1971; McDaniel & Sarkissian, 1968).

Mitochondrion activity of heterotic hybrids is believed to increase as a consequence of mitochondrial complementation of the parental forms, partly at the expense of new growth of organelles specific to the hybrid. In the experiments of Sarkissian and co-workers, the activity of mitochondrial mixtures of inbred lines producing a heterotic hybrid was higher than the activity of the initial lines, but somewhat lower than the activity of the mitochondria of the hybrids themselves. It is to be supposed, that the increase in mitochondrial activity in heterotic hybrids occurs not only as the result of an additive effect but also as the result of complementation in a hybrid genome. Cytoplasmic structures such as mitochondria and chloroplasts are known to have their own genetic systems, but at the same time are controlled by the cell nucleus (Sager, 1972; Henningsen *et al.*, 1972; Wild & Schwahn, 1973).

The reported phenomena of mitochondrial heterosis and mitochondrial complementation, as is seen, comprise a rich source of information on the mechanisms of the heterotic effect, but require further studying (Ellis *et al.*, 1973).

For understanding the nature of heterosis and the mechanism of its development the results of a comparative study of the *isoenzyme* spectrum and composition of a number of *endogenic regulators* of metabolism and formative processes are of great interest. The extensive studies on the esterase isoenzymes have been carried out by Schwartz (1959, 1960, 1967). He has found three isoenzymes of this enzyme in maize, synthesis of which is controlled by the allelic genes E^F, E^N and E^S. When crossing the plants homozygous for these genes, heterotic forms arise, in which are found the esterases of the parents and a new hybrid type of the enzyme more active than a common esterase.

In the maize hybrids have been found 7 allelic esterase genes, 4 forms of leucine-aminopeptidase, 3 isoenzymes of alcohol-dehydrogenase, 3 hybrid forms of catalase, some isoenzymes of peroxidase, dehydrogenase of malic acid and so on (Holson, 1967). Many enzymes have a quaternary structure and when crossed are capable of producing heterozygous lines with hybrid molecules with a higher enzyme activity (Longo & Scandalios, 1969; Kraevoy & Litvinenko, 1969).

As can be seen isoenzymes are of interest by helping us to comprehend to some extent the nature of interallelic and intergenic complementation. Moreover, they give grounds to believe that hybrid vigour is conditioned not only by increase in activity of separate enzymes but by the more effective organization of the enzyme system as a result of the enrichment of the set of isoenzymes in the hybrid plant.

In heterotic development the endogenic regulators of metabolism and morphogenesis should be prominent. Among the facts available the following deserve a special attention. The increased response of self-pollinated lines to gibberellin treatment (Fedorov, 1968) and the different reaction of inbred line and heterotic hybrid seedlings to polyphenol solutions (Paskar, 1968) are well known. Special attention was given to vitamins of the group B. It was demonstrated that inbreeding leads to reduction in their general activity (Baranovski & Semenova, 1968). As it turned out (Manzyuk & Zakrevskaya, 1968; Zakrevskaya, 1969) inbred lines differ in their ratio of the single components of this group. Differences in the content of pantothenic and nicotinic acids, thiamine, pyridoxine, biotin and inositol between the lines producing heterotic hybrids are especially significant. The latter are distinguished by a high overall activity of the whole vitamin group, but do not exceed their parents in any of these factors.

On the basis of studying the reaction of heterotic hybrids and parental forms to physiologically active substances an idea has been made up, that under heterotic condition nothing is developed but a more balanced system of stimulators and inhibitors of metabolism due to the formation of a more balanced genotype in hybrids (Konarev *et al.,* 1971).

BIOCHEMICAL AND MOLECULAR-GENETIC ASPECTS OF THE HETEROSIS PROBLEM

There is much information on heterosis biochemistry and physiology. Most of it, however, concern only a general characteristic of metabolism and formation of heterotic hybrid yields. Those components and elements of metabolism that are directly connected with a genetic apparatus of the cell are only slightly affected, namely nucleic acids, proteins, biosynthesis of nucleic acids and proteins, and finally, the processes of protein realization as primary products of the gene in metabolism and morphogenesis, i.e. the processes of character development, yield formation etc.

At present, it is essential to pick out those elements of the molecular-genetic cell apparatus with which heterosis phenomena are particularly closely connected.

The analysis of available information gives ground to believe that heterosis may be connected with structural and regulatory genes as well. In the first instance the evidence is obtained from the work with isoenzymes described above.

The arguments in favour of the participation of regulator genes and mechanisms of the genetic apparatus of the cell are as follows

– Heterotic development depends greatly on hybrid growth conditions.
– Heterosis is expressed mainly as a change in the rate development of the characters already available in parents.

Therefore, if the sources of heterosis are involved in genome structure, then its manifestation should be connected with gene activation and refer to a category of phenomena that are directly or indirectly related to mechanisms of the genetic character regulation at ontogenesis (Konarev *et al.,* 1971; Demarly, 1972). In this connection the data on genome structure and the activity of the genome or its separate genetic systems should be of definite value to clear up the nature of heterosis and to work out the effective methods of initial material analysis for combining ability and heterosis prediction.

There are no reliable methods available for evaluation of molecular structure and functional activity of the genome. At the same time a certain comparative analysis of hybrids and their original lines in this direction may be possible. Such analysis of genome structure, more exactly its separated functioning districts or sites, was carried out in two directions: due to the contents of isoenzymes and by means of the nucleic acid molecular hybridization method.

As was shown above, in crossing inbred lines it is possible, first to enrich enzyme systems as a consequence of adding the structural genes of the parents coding for different isoenzymes, and secondly – and this is especially important – to form hybrid molecules of protein with a higher activity as the result of favourable combinations of structural genes in the heterozygote (Spofford, 1969).

Accordingly the isoenzyme spectrum and presence of protein hybrid molecules give an opportunity to speak about the structure and character of interallelic and intergenic complementation in relation to the genetic system of the corresponding enzyme. In the simplest case when the enzyme molecule is a dimer it consists of two identical or different subunits (polypeptides) and the original lines are homozygous for genes coding these subunits. The genetic structure of such an enzyme in a hybrid organism may be represented thus:

Allelic genes	$A_1A_1 \times A_2A_2$	→	$A_1A_1 + A_2A_2 + A_1A_2$
Isoenzymes	$P_1P_1 \quad P_2P_2$		$P_1P_1 \quad P_2P_2 \quad P_1P_2$
	Isoenzymes Parent forms		Possible isoenzymes of heterotic hybrid
			P_1P_2 = hybrid isoenzyme's molecule

The structure of such genetic systems are more complicated when enzyme molecules consist of many subunits (Spofford, 1969).

There is no doubt that this approach will be extended also to genetic systems coding for the synthesis of the other proteins including the constitutive proteins of the cytoplasm and nucleus. There are some prerequisites to such an opinion (Jamaguzina *et al.*, 1970; Dimitrov *et al.*, 1972).

For the comparative genetic structure evaluation of the "working" genome areas of heterosis hybrids and their original lines the method of competitive hybridization of new-formed RNA with DNA immobilized by filters was used (Konarev *et al.*, 1971; Gilyazetdinov *et al.*, 1971). One of the variants consists in the following (Gillespie & Spigelman, 1965): DNA was hybridized with heterologic RNA. Afterwards it was treated with RNA-ase and the activity counted (^{32}P–RNA); the repeated hybridization was carried out but with homologic RNA. The increase of ^{32}P activity displayed the presence of DNA loci in the genome and showed their nonfunctioning or absence in plants from which RNA was taken for the first hybridization.

In another variant of this method separate fractions of RNA were used for hybridization. In this case there was a possibility to reveal messenger-like RNA populations. In combination with other characteristics, i.e. nucleotide composition and chromatographic profile of components, this method makes it possible to point out the composition of the population of newly-formed RNA molecules.

The results obtained by these methods on seedling and mature maize plants, showed that the newly-formed RNA composition of maize hybrids differed from that of their parents to a less degree than the parents differed between themselves.

The analysis of numerous combinations showed that heterotic hybrids produced from the crossing of inbred lines differed clearly from each other in content of m-RNA molecule population and consequently possessed different structures of the genome's "working" sites. It may be supposed that interaction between such genomes creates favourable conditions for additivity, modified gene action and mutual compensation of functionally active loci in heterozygotes.

It is quite possible that heterosis results from the great ability of hybrid plants to form surplus information at the expense of somatic polyploidization of nuclei, formation of replications and extra copies of DNA within the scale of genes and elementary genetic systems (James, 1961; Severin, 1967; Pavan & Da Cunha, 1969; Ali-Zade & Aliev, 1973). This can be confirmed by the fact that the DNA of heterotic hybrids possesses high replicative activity, contains more repetitions of r-RNA loci per metaphase nucleus and more repeated sequences in the functioning genome loci (Konarev *et al.,* 1971; Ali-Zade & Aliev, 1973).

It has been shown that genome activity may be considered as a display of heterosis. For its estimation two principles may be used. The first is the measurement of DNA and RNA synthesis rate incorporating labelled precursors, the second is the evaluation of chromosome functional activity or gene activity of DNA according to their structural state (Konarev, 1970, 1973; Konarev & Tuterev, 1971). In the second case the following index may be used: the correlation of active (labile) and blocaded by proteins (stable) DNA (Konarev & Tuterev, 1971), the correlation of fractions of diffusional and compact chromatin (Frenster *et al.,* 1963), fluorescence spectra of acridine orange and DNA complexes taken directly from chromosomes or chromatin on microscopic slides (Konarev, 1970; Serdyuk & Konarev, 1964; Rigler, 1966).

The non-histone contents and modifications of histones may be used as indices of the chromosome structural state. For example histone phosphorylation and acetylation are tightly connected with chromosome structure transition and with genic activation and inactivation of DNA.

The research work that was carried out by these methods on seedlings and mature plants showed the following:

The parental forms of heterotic hybrids differ in degree and character of genome activation. This can be seen in the rate of DNA and RNA synthesis, in the correlation of diffusive and compact chromatin and in the contents of metastable, i.e. functionally active fraction of DNA. Parental pairs that produce heterotic hybrids also differ in speed of r-RNA and m-RNA synthesis as well as in population composition of newly formed m-RNA molecules.

According to the DNA synthesis intensity, heterotic hybrids excel, as a rule, initial forms and in RNA synthesis they stand at the level of the better parent or to some extent excel it. According to genome activation they take very often an intermediate position between the parental forms.

As one can see the problem of heterosis has a number of biochemical and molecular genetic aspects. Among them the evaluation of genome activity should be especially stressed, since the display of heterosis is directly connected with the intensity and character of heterotic hybrid genotype realization within the morphogenetic systems and the analysis of those genome areas. They are closely connected with the hormone regulation of metabolism and morphogenesis of plants.

As was remarked heterosis resembles the effect of stimulation arising under the influence of biologically active substances or some other factors. This makes us think that hormone factors of gibberellin, kinin and phytochrome types are the binding links between genetic source of heterosis and its manifestation in metabolism and morphogenesis. At present it is known that a number of substances of this kind perform a regulating function through the genetic apparatus of the cell (Konarev *et al.,* 1971; Bonner, 1965; Konarev & Elsakova, 1965).

Existing data on the hormones and biologically active compounds of the vitamin type give an opportunity to suppose that complementation of such systems on the metabolic level may be one of the essential factors of heterosis.

At the same time the hormone regulation of plants in this direction is not properly studied. We know little about the mechanical action of phytohormones and even less about the genetic control of their biosynthesis. The study of the nature of heterosis depends to a great extent both on the working out of this problem and on the problem of nucleic acids and molecular genetic processes.

MODEL OF HETEROSIS

To some extent the nature of heterosis may be understood on the ground of contemporary conceptions concerning genetic control of the metabolism and morphogenesis on their different levels.

First of all two aspects may be distinguished in a complete study of the heterosis phenomenon – first, sources or causes of heterosis and secondly its manifestation, that is its realization on all levels of organization of the hybrid organism.

Undoubtedly the sources of heterosis are enclosed in the genetic apparatus of the cell. They represent changes in the structure of separate genetic systems as a result of the interaction of heterogenous genotypes. These changes are directed first to the completion of the genetic system and are mainly concerned with the complementation and mutual intensification of structural genes, by different ways including the changes of gene dose, gene position effect, production of extra DNA copies, etc. Secondly the changes in the structure of the elementary genetic systems are concerned with improvement of the balance of genetic factors controlling the regulating mechanism of the cell.

The heterosis manifestation may be traced at all levels of the hybrid plant organization.

On the molecular level heterosis is manifested by increased DNA reduplication, transcription and translation, in the formation of surplus genetic information in the form of DNA extracopies of separate genetic systems, in increased efficiency of the enzymes and other regulatory factors due to completeness of the isoenzyme set and the formation of hybrid protein molecules with higher enzymatic or other functional activity.

The particular characters of the heterotic organism are formed at the molecular level. For example the completeness of the enzyme systems and the presence of isoenzymes with hybrid molecules determine not only the high enzyme activity but also the adaptation of the systems to the environmental conditions. The rich set of isoenzymes itself creates favourable conditions for enzyme adaptation. It is also known that hybrid isoenzymes can possess a high adaptation value due to the stability of oligomeric molecules and their ability to preserve functional activity over a wide range of such factors as temperature, pH, ionic strength, etc.

In the realization of heterosis at the metabolic path level and in the coordination of the organoid's functions the enzymes and hormones play an exclusively essential role. Heterosis at this level is expressed in the forming of more effective regulatory systems of metabolism and morphogenesis. It takes place on the basis earlier described of heterosis on the molecular level, by means of complementation of separate genetic systems in the hybrid organism, including mitochondrial, plastid, genome-mitochondrial and other complementations. For example as a consequence of mitochondrial and genome mitochondrial heterosis arises mitochondrial complementation (Sarkissian & Srivastava, 1969; McDaniel & Sarkissian, 1968). It is manifested in increased coupling of oxidation to phosphorylation (P/O) and in enhanced ATP synthesis.

On the cell level heterosis is expressed in the increased mitosis number, that is registered on the molecular level as the intensity of DNA reduplication (Konarev *et al.*, 1971).

On the organism level, as was shown, heterosis is manifested in increased growth of vegetative organs, in intensification of synthesis and accumulation of nutritional substances, transportation of metabolites in the plant and their utilization in the processes of yield formation.

Since heterosis is mainly manifested in the developmental scale of separate characters or their complexes it may be supposed that the prerequisites of heterosis arise in the sphere of separate genetic systems. Correspondingly the manifestation of heterosis on the organism level is in every concrete case dependant on the position and significance of the "heterotic" genetic systems of the cell in relation to its general genetical and morphogenetical systems of organism. In this respect heterosis must be more effective in those genetic systems which code key positions of metabolism and morphogenesis. Particularly one may include with them genetic systems of enzymes, hormones, plastids and mitochondrial apparatus.

Accordingly the heterosis types named by Gustafsson (1946) are connected with the structure of genetic systems involved in heterosis. In particular "Luxuriance" – one of the heterosis types – may be an example of a disbalanced metabolic and morphogenetic system on the level of tissues and organs.

LINE EVALUATION AND HETEROSIS PREDICTION

The prospect of discovering the nature of heterosis is connected with hope of working out of quick and precise methods of line evaluation for combining ability and heterosis prediction. Lately biochemists and physiologists have proposed a number of methods including: the method of biotest, based on the selection of line pairs whose mixed seed extracts produce the best effect of yeast growth (Matskov & Manzyuk, 1961), a serologic method based on comparative electrophoretic analysis of protein extracts from vegetative organs or seeds of heterotic hybrids and their inbred lines (Jamaguzina *et al.*, 1970; Dimitrov *et al.*, 1972), the evaluation of heterosis according to electrokinetic properties of cell nuclei (Shabkhazov *et al.*, 1973), selection of parental components according to the degree of shift of male and female plant sexualization with employment of those showing oxidizing and reoxidizing potential of the generative organs, etc. (Berzinya-Berzite & Maurinya, 1973; Maurinya, 1973).

Since the increased synthesis of DNA, RNA and genome activity associated with heterotic effect is manifested within the first 24 hours, hybrids showing these effects can be used for early prediction of heterosis (Konarev *et al.*, 1971; Gilyazetdinov *et al.*, 1971; Ivleva & Gilyazetdinov, 1971). Seedling differences in composition of newly formed RNA and other nucleotide characters can be used as criteria for parental pair selection for combining ability (Gilyazetdinov *et al.*, 1971).

In this respect the considered principle of molecular hybridization and hybridologic analysis of m-RNAwith DNA of active areas of inbred lines and hybrid genomes is very valuable.

No doubt, the molecular-genetic methods for initial line evaluation and early heterosis prediction will be the best. Especially the molecular-genetic analysis of the systems coding the key positions of metabolism and morphogenesis will be promising.

CONCLUSION

Heterosis arises as a result of changes in the genome structure due to interaction of different genotypes on the level of separate genetic systems. It is expressed at all levels of metabolism and morphogenesis.

Genetic sources of heterosis: completeness of structural and regulatory genes, interallelic and intergenic complementations, improved genetic balance, especially of functioning genome loci, effect of gene position, gene dose etc.

Sources and manifestations of heterosis on the molecular level: The ratio of reduplication, transcription and translation, structural state of DNA and chromosomes, repetition r-RNA loci, enrichment of newly formed m-RNA populations.

Efficiency of the enzyme or other systems due to:

- wide range of isoenzymes,
- formation of hybrid molecules with high functional (enzyme) activity and adaptation value.

On the metabolic and functional levels: The formation of more complete enzyme and hormonal regulation of the metabolism and morphogenesis:

- on the basis of "molecular" heterosis,
- due to complementation of separate genetic systems coding metabolic pathways and functions such as mitochondria, plastids, mitochondrial-genome, plasmone-genome and other complementations.

On the cell and organism levels: cell division, growth and differentiation on the basis "metabolic and functional heterosis", interactions and complementation of the differentiated cells, tissues and organs. Enrichment and balance of metabolic factors according to the accumulation and utilization of nutritional substances in yield formation. Resistance to climatic factor fluctuations.

Heterosis-types are connected with the structure of genetic systems involved in heterosis. It may be more effective in those genetic systems which code key positions of metabolism and morphogenesis. Close study of such systems is the main task of contemporary biochemistry, physiology and molecular genetics of heterosis.

REFERENCES

ALI-ZADE, M.A. & ALIEV, R.T. (1973). *Dokl. A.N. SSSR* **29,** 1. 72–74.

BARANOVSKI, P.M. & SEMENOVA, D.I. (1968). Heterosis in Field Crops. (In Russian) *Kolos* 205–209.

BERZINYA-BERZITE, R.V. & MAURINYA, H.A. (1973). *Uchen. zapiski* **183,** 5–11.

BONNER, J. (1965). *The Molecular Biology of Development.* California.

BOOS, E.E. & SARKISSIAN, I.V. (1968). *Crop. Sci.* **8,** 683–686.

BREZHNEV, D.D. (1969). *Vestnik s/h nauki* **7,** 29–36.

BREZHNEV, D.D., FEOFANOVA, N.D. & LUKOVNIKOVA, G.A. (1969). *Selskohos. biologia* **4,** 4, 510–520.

BURTON, G.W. (1968). *Crop. Sci.* **8,** 3, 365–368.

CHING, TE-MAY (1973). *Seed Sci. Technol.* (formerly "Proc. Int. Seed Test. Assoc.") **1,** 1, 73–88.

DEMARLY, G. (1972). *Ann. Amelior. Plant,* **22,** 2, 143–166.

DIMITROV, P., PETKOVA, S., NACHKOV, D., NACHKOVA, C. & MARINKOV, E. (1972). *Theor. and Appl. Genet.* **42,** 7, 306–309.

DUBININ, N.P. & GLEMBOCKI, J.L. (1967). Population-genetics and Breeding. (In Russian) Nauka.

ELLIS, J.R.S., BRUNTON, C.J. & PALMER, J.M. (1973). *Nature* **241,** 5384, 45–47.

FEDOROV, P.S. (1968) Biochemical and Physiological Principles of Heterosis (In Russian) Frunze, Kirgistan.

FRENSTER, J., ALLFREY, V.G. & MIRSKY, A.E. (1963). *Proc. Nat. Acad. Sci.* **50,** 1026–1036.

GILYAZETINOV, S., IVLEVA, L.A., KAMALETDINOVA, M.A. & KONAREV, V.G. (1971). *Voprosi biokhimii heterozisa u rasteni,* Ufa.

GILLESPIE, D. & SPIGELMAN, S. (1965). *Journ. Mol. Biol.* **12,** 4, 829–842.

GOLINSKAYA, E.L., GRITSENKO, E.M., MIKHALKO, S.N. & STETSENKO, N.M. (1966). *Fiziologia rasteni,* **14,** 2.

GUSTAFSSON, A. (1946). *Hereditas* **32,** 2.

HAGEMAN, R.H., ZIESERL, J.F. & LENG, E.R. (1962). *Nature* **197,** 213.

HATHER, R. (1955). *Proc. Roy. Soc.* **144,** 915, 143–150.

HENNINGSEN, K.W., BOYNTON, J.E., WETTSTEIN, D. & BOARDMAN, N.K. (1972). Biochem. Gene Express. High Organisms

HOLSON, O. (1967). *Ann. Rev. Genet.* **1,** 245–268.

IVLEVA, L.A. & GILYAZETDINOV, S.J. (1971). *Voprosi biokhimii heterozisa u rasteni* Ufa, **6,** 41.

JAMES, K.C. (1961). *Amer. Naturalist* **95,** 885, 345–364.

JUMAGUZINA, H.A., ISTIRYAKOVA, F.K. & KONAREV, V.G. (1970). *Voprosi biokhimii heterozisa u rasteni,* Ufa, FAN, USSR, 81–87.

KOELLE, G. (1973). *Z. Pflanzenzücht.* **70,** 2, 166–170.

KONAREV, V.G. & ELSAKOVA, T.N. (1965). Regulators of Plant Growth and Nuclein Metabolism (In Russian). *Nauka* **5,** 26.

KONAREV, V.G. (1970). The Nucleus and its Ultrastructure (In Russian). *Nauka* **28,** 38.

KONAREV, V.G. (1973). *Trudi prikladnoy botaniki, genetiki i selekcii* (Leningrad) **52,** 1, 5–28.

KONAREV, V.G., AKHMETOV, R.R. & GILYAZETDINOV, S.YA. (1971). *Selskohoz biologia* **6,** 5, 653–662.

KONAREV, V.G., AKHMETOV, R.R. & GILYAZETDINOV, S.YA. (1971). *Voprosi biokhimii heterozisa u rasteni* 5–35.

KONAREV, V.G. & TYUTEREV, S.L. (1971). Biochemical and Cytochemical Methods of Nucleinacid Determination in Plants (In Russian). *Kolos.*

KRAEVOY, S.A. & LITVINENKO, A.I. (1969). *Dokl. A. N. SSSR* **189,** 2, 403–406.

LITVINENKO, A.I. (1968). Selekcia i semenevodstvo *(Sbornik)* **10,** 155–163.

LONGO, P. & SCANDALIOS, J.G. (1969). *Proc. Nat. Acad. Sci., USA* **62,** 1, 104–111.

LUKOVNIKOVA, G.A. & GOMOLYAKOV, L.G. (1969). Heterosis in Vegetable Crops. (In Russian). *Kolos* 86–92.

MATSKOV, F.F. & MANZYUK, S.G. (1961). *Fiziologia rasteni* **8,** 1, 92–100.

MANZYUK, S.G. & ZAKREVSKAYA, L.E. (1968). Heterosis in Field Crops. (In Russian) *Kolos* 210–214.

MARSALEK, L. (1969). *Savremena poljopr* **17,** 9, 887–889.

MARSALEK, L. (1972). *Acta Univ. Agr. A.* **20,** 2, 211–217.

MARSALEK, L. & VORISEK, V. (1968). *Z. Pflanzenzücht.* **59,** 3, 259–272.

MAURINYA, H.L. (1973). *Uchen. zapiski, Lato. Y-ta* **183,** 92–101.

MCDANIEL, R.G. & SARKISSIAN, I.V. (1968). *Genetics* **59,** 4, 465–475.

NETTOVICH, E.D. (1968). Heterosis in Field Crops. (In Russian). *Kolos* 3–15.
ORYLUK, A.P. (1968). *Selskohoz biologia* **3,** 6, 842–846.
OVECHKIN, S.K. (1968). *Zhurn. NJJ. rastennievodstva i selekcii, Kiev* **10,** 128–139.
PASKAR, S.D. (1968). Phenol Compounds and their Biological Function. (In Russian). *Nauka* 296–301.
PAVAN, C. & DA CUNHA, A.B. (1969). *Genetics* **51,** 1, Part 2. 289–304.
RIGLER, R.J. (1966). *Acta Physiologica Scandinavica* **67,** Suppl. 267, Stockholm.
RUBTSOVA, M.S. (1964). *Fiziologia rastenij* **11,** 3, 473–479.
SHABKHAZOV, V.G., LOBINTSEVA, G.S. & ATRAMENTOVA, L.A. (1973) *Dokl. A.N., SSSR* **210,** 5, 1203–1205.
SAGER, R. (1972). *Cytoplasmic genes and organelles.* New York, Acad. Press.
SARKISSIAN, I.V. & SRIVASTAVA, H.K. (1969). *Proc. Nat. Acad. Sci. USA* **63,** 3, 302–309.
SARKISSIAN, I.V. & SRIVASTAVA, H.K. (1971). *Biochem. Genet.,* **5,** 1, 57–63.
SCHWARTZ, D. (1959). *Planta* **54,** 151–161.
SCHWARTZ, D. (1960). *Genetics* **45,** 10, 1419–1427.
SCHWARTZ, D. (1967). *Proc. Nat. Acad. Sci, USA* **58,** 2, 568–575.
SERBANESCU, E. (1965). *Studii si cercetari biol. ser. bot.* **17,** 1, 53–62.
SERBANESCU, E. (1968). *Studii si cercetari biol. ser. bot.* **20,** 1, 61–67.
SERDYUK, L.S. & KONAREV, V.G. (1964). Biology of Nuclein Metabolism of Plants. (In Russian) *Nauka* 207–209.
SEVERIN, V.I. (1967). *Citologia i genetika* **1,** 5, 17–26.
SMIRNOVA, N.P. (1966). Heterosis in Field Crops. (In Russian). *Stavropol* 82–90.
SPOFFORD, J.B. (1969). *The American Naturalist* **103,** 932, 407–432.
SHUPAK, K.D. & SEREDINSKAYA, A.F. (1968). Heterosis in Vegetable Crops. (In Russian). *Kolos* 127–138.
TURBIN, N.V. (1966). Sbornik "Aktualnie voprosi sovremennoy genetiki". *Mosk. Univ.* 434–466.
TURBIN, N.V. (1971). Genetic Principles of Plant Breeding. (In Russian). *Nauka* 112–155.
USHIMIGA, H. & TAKAHACHI, H. (1973). *Ann. Bot.* **37,** 150, 247–252.
WILD, A. & SCHWAHN, M. (1973). *Biol. Zbl.* **92,** 3, 237–305.
YAKOVLEVA, A. P. & OVCHINIKOVA, M. F. (1968). *Selskohoz. biologia* **3,** 5, 676–680.
ZAKREVSKAYA, L.E. (1969). *Tr. Harkovskogo s/hos. Inst.* **78,** 115, 72–76.

Heterosis, combining ability and photosynthesis in bread wheat (*Triticum aestivum* L.)

G. PLANCHON

Laboratoire d'Amélioration des Plantes
École Nationale Supérieure Agronomique,
Toulouse Cedex, France

SUMMARY

The two factors:

– large chlorophyll *a* content as measured in the flag leaf,

– moderate net CO_2 exchange decrease at high temperatures, a trait strictly associated with photorespiration, are responsible for the good combining ability of some progenitors.

New advances made in order to obtain high heterosis for yielding performance would be possible by the association of two partners, both having a low photorespiration rate, and one a high chlorophyll *a* content.

It has now been demonstrated that heterosis does exist in bread wheat. Experiments based on crosses between old local varieties, Rouge de Bordeaux, Touzelle blanche de Provence, Bladette de Puylaurens, and modern French cultivars, Etoile de choisy, Capitole, Moisson, showed that yield can exceed the better parent by up to 50% in low density conditions (100 plants per square meter).

In this investigation, repeated over several years, the progenitor Rouge de Bordeaux was found to exhibit constantly the best performance and Bladette de Puylaurens the worst, in crosses with improved varieties. The 1969 and 1970 results (Table 1) substantiate this observation.

In other words, the combining ability of the latter two varieties is respectively high and poor in such conditions.

On the other hand, it is known that, in wheat, the photosynthetic activity of the flag leaf blade during the blossoming-maturity period determines the yielding capacity to a large extent (Thorne, 1965; Lupton, 1968; Planchon, 1968). In order to obtain a clearer understanding of the combining ability of our various cultivars, we analyzed the photosynthetic activity of the flag leaf blade in the parents as well as in F_1 hybrids. photosynthetic activity being defined as the product of three factors: – net CO_2 exchange rate per unit surface area, flag leaf surface and activity duration from blossoming time on. A previous study (Planchon, 1973) has shown that amongst the three above mentioned factors, the former accounts to the largest extent for the observed yield differences between parents and F_1's.

Table 1. Yield of F_1 winter wheat hybrids
(values are expressed as percent of the better parent)

Hybrids F_1	1969	1970
Capitole × Rouge de Bordeaux	123.7*	149.2*
Capitole × Touzelle blanche de Provence	102.2*	121.6*
Capitole × Bladette de Puylaurens	91.3	117.3*
Moisson × Rouge de Bordeaux	116.8*	134.2*
Moisson × Touzelle blanche de Provence	105.9	–
Moisson × Bladette de Puylaurens	–	102.3
Etoile de choisy × Rouge de Bordeaux	130.5*	–
Etoile de choisy × Touzelle blanche de Provence	116.8*	–
Etoile de choisy × Bladette de Puylaurens	111.1*	115.7*

*Significantly different from its better parent at P = 5% level.

POTENTIAL PHOTOSYNTHETIC RATE; ITS DEPENDENCE ON CHLOROPHYLL *a* CONTENT

The net CO_2 exchange rate of the flag leaf blade in optimum temperature and light intensity conditions (22°C, 50,000 lux, i.e. saturation value) is called the potential photosynthetic rate. Preliminary investigations showed that this differs largely amongst varieties. To interpret these differences we undertook an analysis on the relationship of the above-mentioned index, which we found to be very closely related to the chlorophyll *a* content of the flag leaf blade prior to blossoming time (Fig. 1). The varieties which have the highest chlorophyll *a* content, and therefore the highest net photosynthetic rate, are all red-headed varieties.

PHOTORESPIRATION AND PHOTOSYNTHESIS

We also found that the net change in the CO_2 exchange rate is a function of temperature. Its decrease with increasing temperature is steeper in some varieties, Bladette, Fec, Goya, but slower in others, Capitole, Rouge de Bordeaux, Touzelle blanche de Provence, Moisson (Figs 2 and 3).

As the net CO_2 exchange rate depends upon the combination of gross photosynthesis and photorespiration, we analyzed the latter in varieties representing each group namely Capitole and Bladette. Photorespiration largely accounts for the net decrease in the CO_2 exchange rate at high temperatures for the Bladette group (Figs 4 and 5). Thus, photorespiration is an important factor capable of limiting the intensity of photosynthesis and consequently the yielding capacity in a number of varieties.

In order to improve the net photosynthetic rate per unit surface area progenitors should be sought for which have:

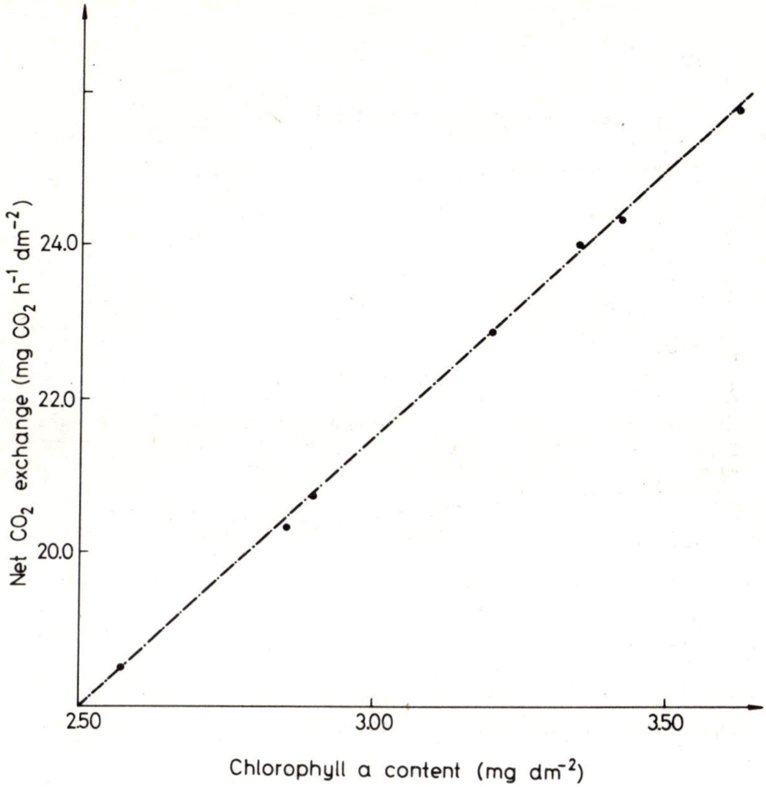

Fig. 1. Relationship of net CO_2 exchange and chlorophyll *a* content.

1 – a high chlorophyll *a* content in the flag leaf,
2 – a low CO_2 exchange rate decrease at high temperature, a trait associated with a low photorespiration index.

Using such progenitors for wheat breeding implies a knowledge of the inheritance of these two characters.

GENETIC INVESTIGATIONS

A) Chlorophyll a content

The chlorophyll *a* content per surface unit in the F_1's studied is equivalent to, or even better than, its value in the better parent (Table 2).

An analysis of the F_1, F_2 and backcrosses of the Capitole × Rouge de Bordeaux hybrid shows that a high chlorophyll *a* content is dominant over a low content. The character, segregating 3:1 in F_2 and 1:1 in blackcrosses, is monogenic (Table 3).

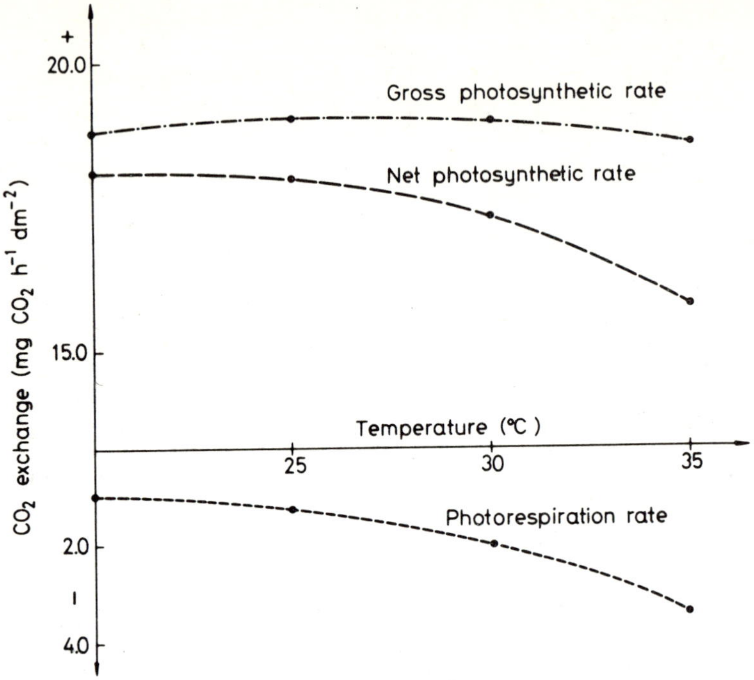

Fig. 2. Changes of photosynthesis and photorespiration with temperature for the variety Capitole.

Table 2. Chlorophyll *a* content of F_1 hybrids and their parents

	Chl. *a* mg/dm^2
Capitole	2.65 ± 0.04
Moisson	2.85 ± 0.04
Rouge de Bordeaux	3.37 ± 0.06
Touzelle blanche	2.92 ± 0.04
Bladette	3.20 ± 0.05
Capitole × Rouge de Bordeaux	3.34 ± 0.05
Capitole × Touzelle blanche	2.94 ± 0.05
Capitole × Bladette	3.44 ± 0.04
Moisson × Rouge de Bordeaux	3.33 ± 0.05
Moisson × Touzelle blanche	2.90 ± 0.04
Moisson × Bladette	3.35 ± 0.05

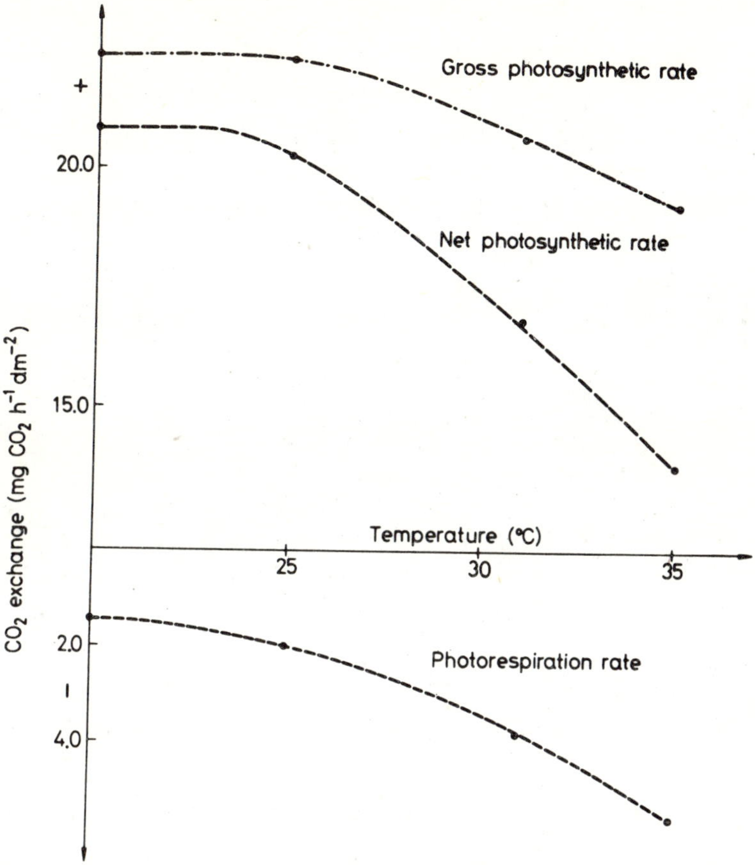

Fig. 3. Changes of photosynthesis and photorespiration with temperature for the variety Bladette.

We have already mentioned that varieties with a high chlorophyll *a* content are all red-chaffed. In any cross between a red-chaffed and a white-chaffed variety the F_1 is red-chaffed and its chlorophyll *a* content in the flag leaf is high. In the F_2 progeny of the Capitole × Rouge de Bordeaux hybrid both traits segregate simultaneously (Table 4).

B) Net CO_2 exchange rate as a function of temperature

a) In F_1 hybrids, the parents of which have a similar temperature response, the F_1 behaves like both its parents (Fig. 6).

b) In F_1 hybrids, each parent of which has a different temperature response, the F_1 behaves like the parent displaying the steeper net decrease in the CO_2 exchange rate at high temperature (Fig. 7).

Thus, F_1 hybrids with a high photosynthetic potential can be obtained by the association of the following two traits: high chlorophyll *a* content, and low CO_2 ex-

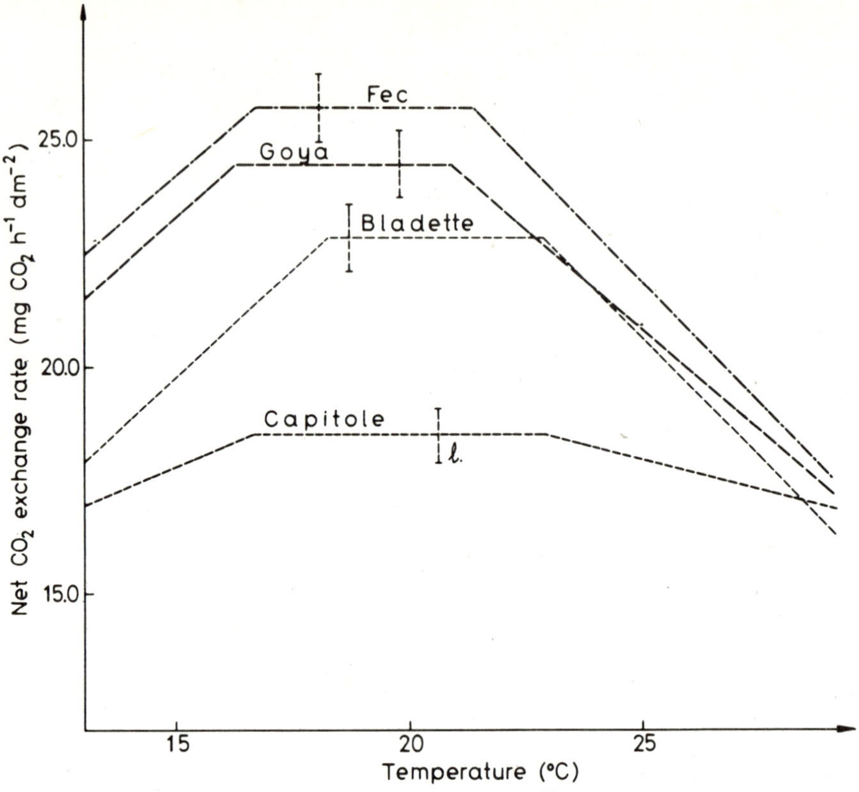

Fig. 4. Net CO_2 exchange rate as a function of temperature. I = confidence interval at P = 0.05.

Table 3. Inheritance of high chlorophyll *a* content in the cross Capitole × Rouge de Bordeaux

Number of plants \ Chl. *a* content	2.30	2.40	2.50	2.60	2.70	2.80	2.90	3.00	3.10	3.20	3.30	3.40	3.50
Capitole		2	8	7	2								
Rouge de Bordeaux										1	10	8	1
F_1									1	11	7	1	
			13							37			
F_2	1	2	4	4	2		3	6	6	8	8	4	2
		24								26			
Back-cross	1	8	8	7				2	4	8	5	3	4

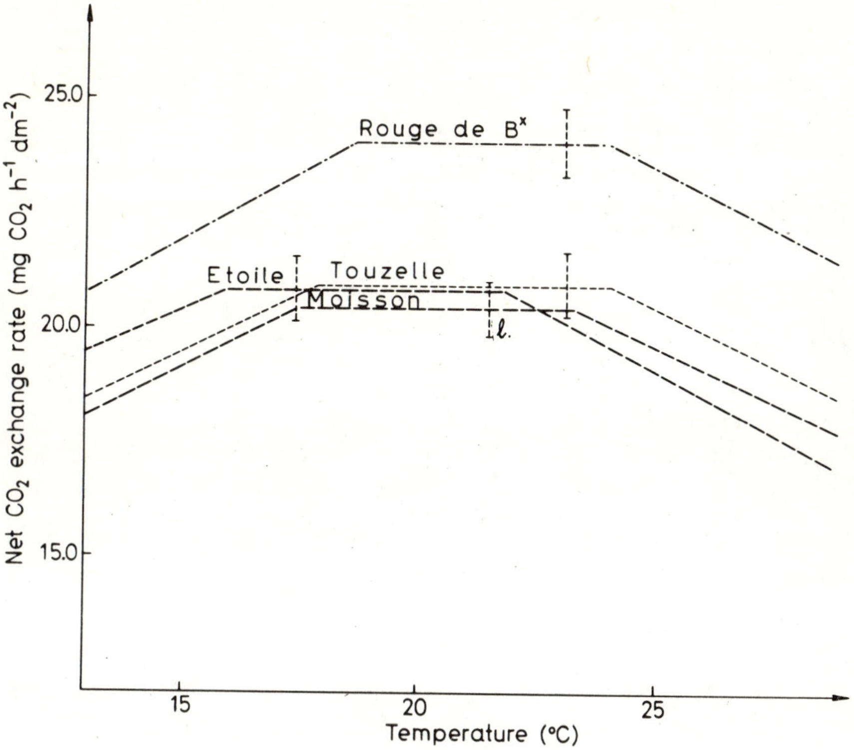

Fig. 5. Net CO_2 exchange rate as a function of temperature. I = confidence interval at $P = 0.05$.

Table 4. Segregation for red colour of the chaff and chlorophyll *a* content

Chl. *a* content (mg/dm²)		2.30	2.40	2.50	2.60	2.70	2.80	2.90	3.00	3.10	3.20	3.30	3.40	3.50
F_2	Total number	1	2	4	4	2		3	6	6	8	8	4	2
	Red-headed plants								1	1	2	5	3	1
	Half-red-headed plants					1		3	5	5	6	3	1	1
	White-headed plants	1	2	4	4	1								

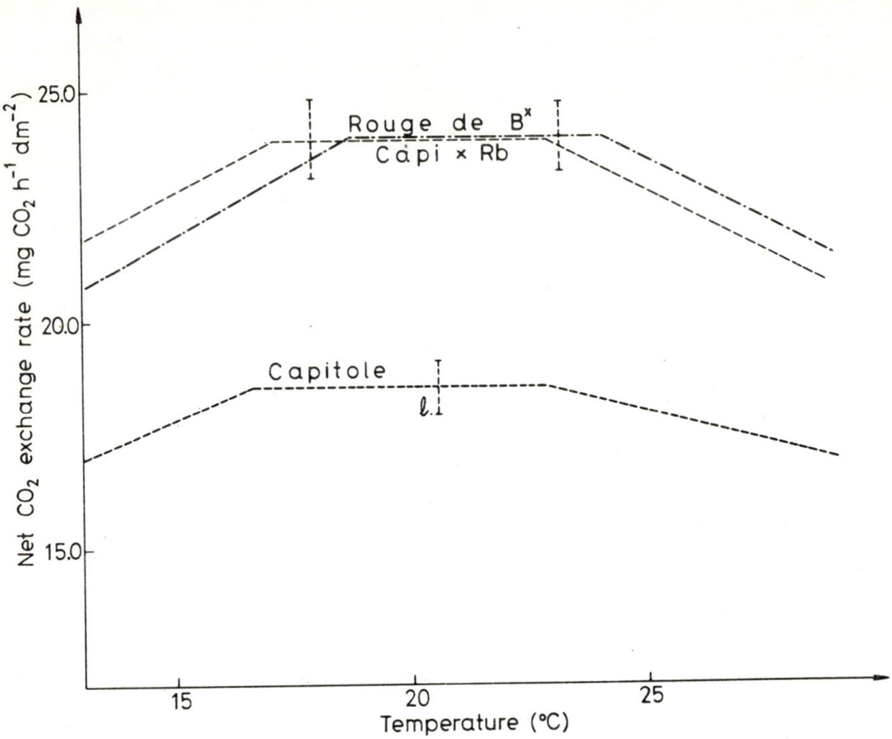

Fig. 6. Net CO_2 exchange rate as a function of temperature for the Capitole × Rouge de Bordeaux hybrid. I = confidence interval at $P = 0.05$.

change rate decrease at high temperature, a character correlated with a low photorespiration index.

Such a combination accounts for the good combining ability of Rouge de Bordeaux, whereas Bladette de Puylaurens, the CO_2 exchange rate of which decreases rapidly with increasing temperature, manifests a poor combining ability.

We conclude that, by crossing two varieties with low photorespiration, one of which at least having a high chlorophyll *a* content, new progress can be expected as far as heterosis for yield is concerned.

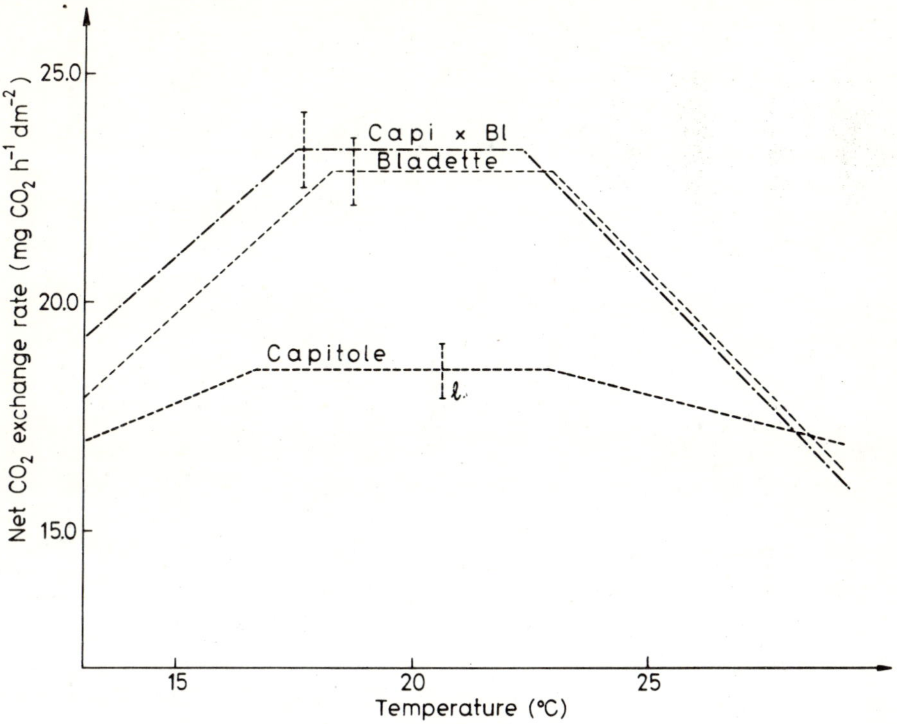

Fig. 7. Net CO_2 exchange rate as a function of temperature for the Capitole × Bladette hybrid. I = confidence interval at $P = 0.05$.

REFERENCES

LUPTON, F. G. H. (1968). Physiological parameters of yield in wheat. *Genetica agraria.* Eucarpia, Ed. G. C. CHISCI, G. HAUSSMANN, Milan, 457–464.

PLANCHON, C. (1968). Activité photosynthétique et rendement en grain chez le Blé tendre (*Triticum aestivum* L.). *C. R. Acad. Sci.* Paris, 2011–2014.

PLANCHON, C. (1973). Productivité, hétérosis et photosynthèse chez le Blé tendre (*Triticum aestivum* L.). *Thèse Doct. Sc. Nat.* Université Paul Sabatier, Toulouse, 130 p.

THORNE, G. N. (1965). Photosynthesis of ears and flag leaves of wheat and barley. *Ann. Bot.* **29,** 317–329.

Fertility restoration capacity of different Rf-sources in *Triticum timopheevi* cytoplasm

J. MILOHNIĆ AND M. JOŠT

Institute for Breeding and Production of Field Crops
Faculty of Agriculture, Zagreb University,
Yugoslavia

SUMMARY

The fertility restoration capacity of 21 sources of Rf-genes in male-sterile lines with *T. timopheevi* cytoplasm was studied in 3 cms-tester lines during 3 years. The mean capacity of fertility restoration ranged from 0.1 to 94.1%. Primepi (94.1) surpassed all others, followed by: CH1 × 9E1[3] (90.9) and PI 277013 (84.1). Differences in the mean capacity of fertility restoration of all the Rf-sources studied at various cms-tester lines – were not significant (27.5; 29.1; and 27.5%), whilst differences between years were significant (23.7; 28.8 and 31.6%).

INTRODUCTION

Complete restoration of fertility in F_1 generations is now essential in wheat hybrids based on cytoplasmic male sterility. Fertility-restoring genes in male-sterile lines with *T. timopheevi* cytoplasm are found in various tetraploid and hexaploid wheats and in their derivatives (Wilson & Ross, 1962; Schmidt & Johnson, 1963; Livers, 1964; Oehler & Ingold, 1966; Zeven, 1968; Joppa & McNeal, 1969; Miri *et al.*, 1970; Lukyanenko, 1971; Mihalev, 1973; Rajki & Rajki, 1968 and many others). Fertility restoration capacity depends primarily on the effects of the Rf-genes present and their interaction with the cytoplasm of the male-sterile line (Kihara, 1968). However, it may be considerably influenced by the environment and by the nucleus of the male-sterile line (Wilson, 1968). Therefore, detailed investigations on fertility restoration capacity of restorer sources should be conducted using various sources of cytoplasmic sterility and at different environments.

A good knowledge of restorer sources is indispensable for improving R-lines. Thus, three years ago we initiated our studies on fertility restoration capacity using all the Rf-sources available at that time.

MATERIALS AND METHODS

The Rf-sources used in these investigations were tested on three genotypically different A-lines with *timopheevi* cytoplasm, during three years.

Fertility restoration capacity is expressed by the percentage of seed set on bagged and unbagged heads of F_1 hybrids in relation to the seed set by the corresponding R-line. 20 heads of F_1 hybrids and 10 heads of each R-line were analyzed.

Table 1. Average fertility restoration ability in F_1 test-hybrids (Means of three years and three cms-testers)

Rf-sources		Seed set (%) Bagged heads	Unbagged heads
Primepi	R 6	94·1 *a*	95·4 *a*
CH 1 × 9E1³	R 10	90·9 *a*	95·3 *a*
PI 277013	R 20	85·1 *	87·4 *
PI 277016	R 21	71·5 *	83·9 *
Rw Mir³ × Prof. March.	R 9	62·2 *b*	82·0 *abc*
Bison	R 3	61·3 *b*	83·4 *abc*
Sel. from Lot-2	R 14	51·4 *b*	75·1 *cd*
Palmaress	R 15	46·5 *b*	56·1 *ef*
Kansas	R 4	24·8 *c*	56·5 *ef*
Sel. from Lot-2	R 14/1	24·3 *c*	55·1 *ef*
Lot-1	R 1	21·9 *c*	58·2 *de*
Nebraska	R 5/1	16·5 *c*	52·2 *ef*
Lot-2	R 2	13·9 *cd*	51·9 *ef*
CH 4-II-2	R 12	9·9 *cde*	36·6 *fg*
CH 3-I-6	R 11	9·0 *cde*	49·4 *efg*
Nebraska	R 5	2·1 *def*	49·4 *efg*
IBo — 460/616	R 19	1·3 *ef*	37·3 *fg*
Rw Bezostaja 1⁴S⁴	R 8	1·1 *ef*	27·2 *g*
IBo — 544/616	R 18	0·3 *f*	34·5 *fg*
IBo — 1029/604	R 16	0·3 *f*	41·5 *fg*
IBo — 519/616	R 17	0·1 *f*	37·5 *fg*

Values followed by the same letter are not significantly different at P = 0·01.
* = Means of only two years.

RESULTS AND DISCUSSION

Table 1 shows fertility restoration capacity of the Rf-sources studies. The mean percentage of seed set on bagged heads of F_1 hybrids varied from 0.1(IBo-519/616) to 94·1 (Primepi) among three tester-A-lines and Rf-sources. The difference in fertility restoration capacity between Primepi the best Rf-source and CH 1 × 9E1³ was not significant.

The mean fertility restoration capacity for all the Rf-sources tested is presented in Table 2. Differences between cms-tester lines were not significant (percentages of seed set on bagged heads: 27·5; 29·1 and 27·5), whilst significant differences between years were found (23·7; 28·8 and 31·6%). The same refers to the results of seed set on unbagged heads. Thus, it may be concluded that in our studies environmental conditions had a greater effect on fertility restoration than the genotype of the cms-tester lines.

Figure 1 shows Rf-sources classified according to fertility restoration capacity of the cms-tester lines, averaged for 3 years. By the same arrangement of Rf-sources, the mean values for all the three cms-tester lines are presented in Figure 2 for each year of testing. Comparing these results, a conclusion can be drawn that the arrangement of restorer sources according to an average fertility restoration capacity

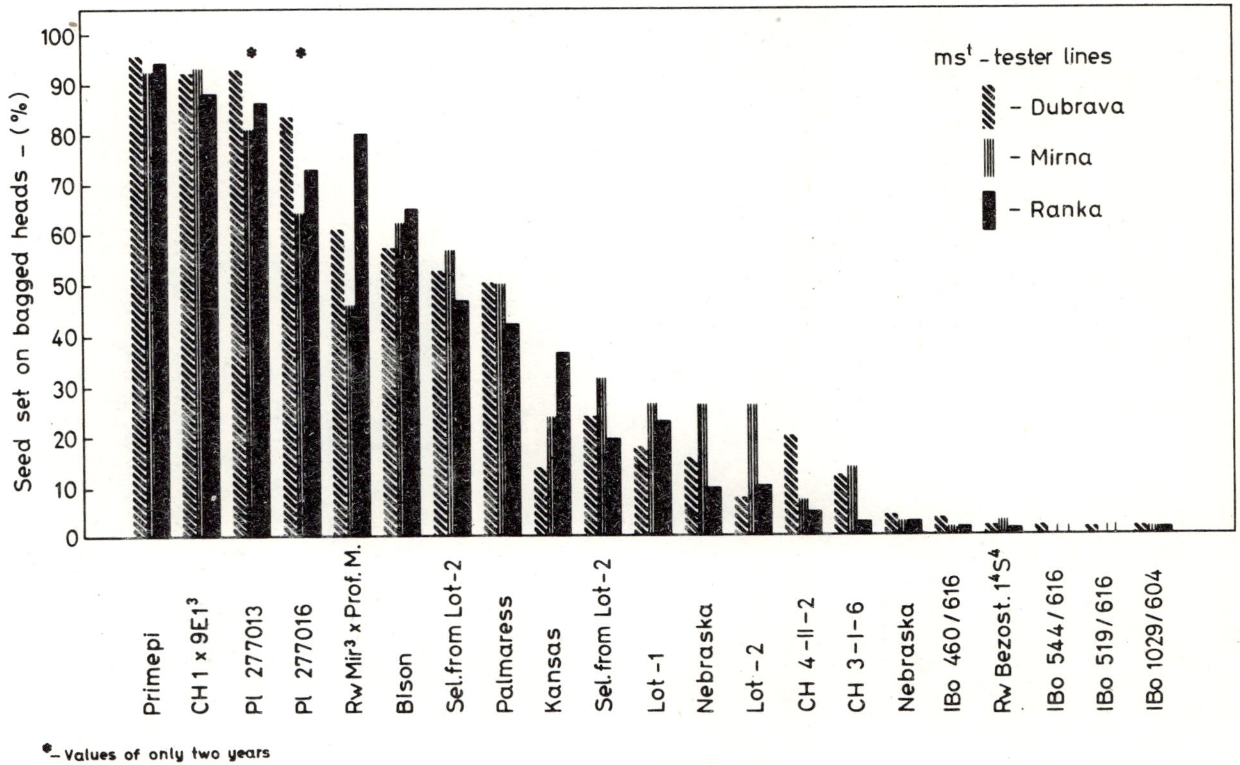

Fig. 1. Fertility restoration ability of 21 Rf-sources (means of three years).

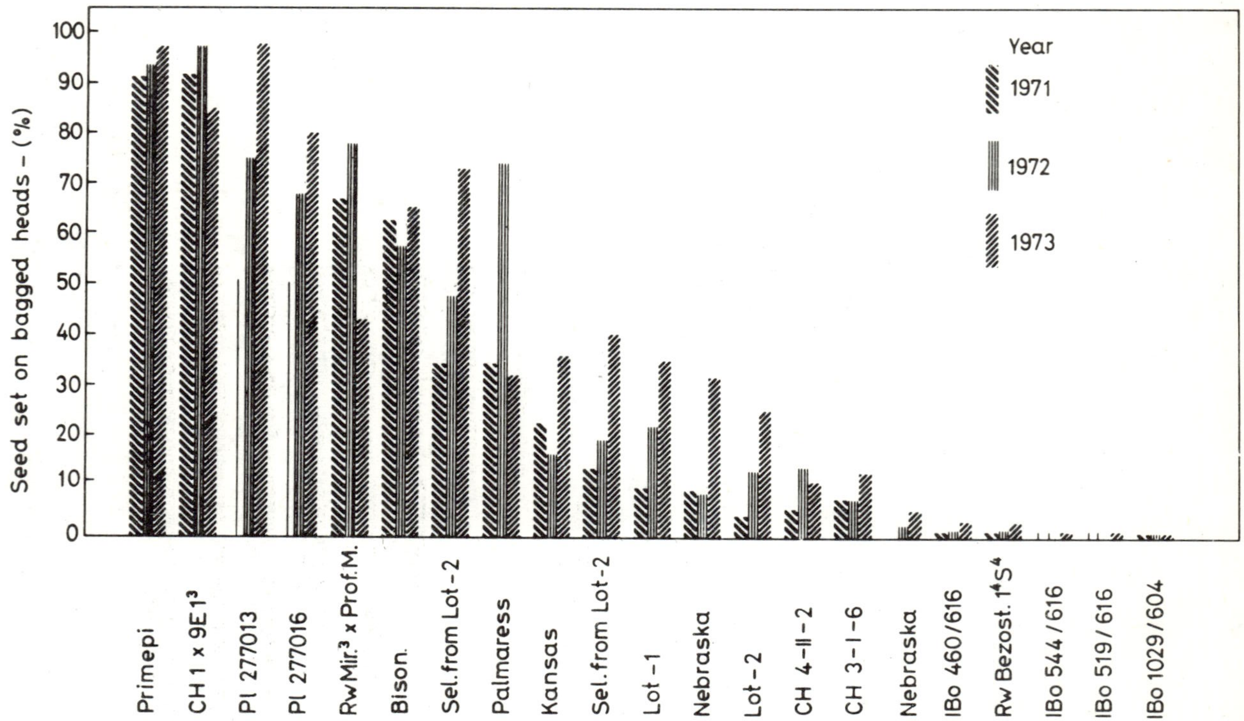

Fig. 2. Fertility restoration ability of 21 Rf-sources (means for three ms-testers).

Table 2. Average fertility restoration ability for 21 Rf-sources in three years and three cms-tester lines

Years \ Tester-lines	Seed set (%)						Average for years	
	cms^{t}-Dubrava		cms^{t}-Mirna		cms^{t}-Ranka			
	$\bar{x}_{(1)}$	$\bar{x}_{(2)}$	$\bar{x}_{(1)}$	$\bar{x}_{(2)}$	$\bar{x}_{(1)}$	$\bar{x}_{(2)}$	$\bar{x}_{(1)}$	$\bar{x}_{(2)}$
1971	24·2	73·3	21·8	55·1	25·2	44·9	23·7 *a*	57·8 *a*
1972	29·9	54·5	29·3	52·1	27·3	54·5	28·8 *ab*	53·7 *a*
1973	28·5	52·1	36·1	60·5	30·1	72·0	31·6 *b*	61·5 *b*
Average for cms-tester lines	27·5 *a*	60·0 *a*	29·1 *a*	55·9 *a*	27·5 *a*	57·1 *a*		

$\bar{x}_{(1)}$ — Mean for bagged heads
$\bar{x}_{(2)}$ — Mean for unbagged heads
Values followed by the same letter are not significantly different at $P = 0{\cdot}01$

remains nearly the same without regard to nucleus of tester line and environment for growing the F_1 hybrids. This is in congruity with Kihara's opinion (1968) "that it is the cytoplasm that reacts to the Rf-gene".

According to the results obtained in these investigations one might conclude that it would suffice to determine the fertility restoration capacity of Rf-sources in the cytoplasm of one tester line, but under different environmental conditions.

REFERENCES

JOPPA, L. R. & MCNEAL, F. H. (1969). Pollen fertility restoration by three spring wheats (*Tr. aestivum* L.) of the male sterility induced by *T. timopheevi* cytoplasm. *Crop Sci.* **9,** 839–840.

KIHARA, H. (1968). Cytoplasmic Relationships in the Triticinae. *Proc. 3rd Int. Wheat Genet. Symp. Canberra, 1968* 125–134.

LIVERS, R. W. (1964). Fertility restoration and its inheritance in cytoplasmic male-sterile wheat. *Science* **144,** 420–421.

LUKYANENKO, P. P., TIMOFEEV, V. B. & KOLESNIKOV, F. A. (1971). A new method of breeding fertility restorers for winter bread wheats with cytoplasmic male sterility. *Dokl. Vses. Akad. Selskohas, Nauk* **9,** 2–4.

MIHALEV, I. M. (1973). A new fertility-restorer source in common wheat. *4th Int. Wheat Genet. Symp. Columbia, 1973.*

MIRI, R. K., AMAWATE, J. S. & JAIN, H. K. (1970). NP 839, NP 883 and NP 880: new sources of fertility restoration in male sterile wheat. *Wheat Infor. Serv.* **31,** 9–11.

OEHLER, E. & INGOLD, M. (1966). New cases of male sterility and new restorer source in *T. aestivum. Wheat Infor. Serv.* **22,** 1–3.

RAJKI, E. & RAJKI, S. (1968). Research work on hybrid wheat at Martonvásár, *III. Euphytica Supplement* No. **1,** 43–48.

SCHMIDT, J. W. & JOHNSON, V. A. (1963). Sterility-fertility interaction in wheat. Proc. 2nd Int. Wheat Genet. Symp., *Hereditas* **2,** 119–120.

WILSON, J. A. & ROSS, W. M. (1962). Male sterility interaction of the *T. aestivum* nucleus and *T. timopheevi* cytoplasm. *Wheat Infor. Serv.* **14,** 29.

WILSON, J. A. (1968). Problems in hybrid wheat breeding. *Euphytica Suppl.* No. **1,** 13–33.

ZEVEN, A. C. (1968). Cross pollination and sources of restorer genes in wheat and a semi-hybrid variety. *Euphytica Suppl.* No. **1,** 75–81.

Ethrel induced male sterility in wheat (*Triticum aestivum* L.)

W.G. HUGHES

Plant Breeding Institute, Cambridge, England

SUMMARY

Work on Ethrel as a chemical means of inducing male sterility in wheat seems justified despite two major disadvantages – 1) the critical stage of ear development at which the chemical has to be applied to induce full male sterility, and 2) the associated failure of full ear emergence at effective sterilising concentrations. The problems are particularly evident where there is a range of tiller age within and between plants.

In a glasshouse experiment using the spring wheat cultivar Sirius, age differences between tillers was reduced by sowing at high density, so that each plant formed only one ear. The experiment showed that full emergence of sterilised ears was produced by spray concentrations of between 1,000 and 2,000 ppm (a.i.).

A further glasshouse experiment with the same cultivar demonstrated that emergence of ears sterilised with 2,000 ppm Ethrel could be substantially improved by applying GA_3 at 100 ppm (a.i.) as a foliar spray between 2 and 5 days after the Ethrel treatment.

Field experiments using Ethrel and GA_3 showed that rates of Ethrel application effective in the glasshouse were also effective in the field.

The chemical induction of male sterility in wheat would make possible the large-scale production of F_1 seed of a hybrid variety without the need for breeding restorer lines and maintaining lines with cytoplasmic male sterility. Such male sterility may also provide a rapid method for screening various parental combinations when seeking F_1 hybrids which show positive heterosis for yield.

Ethrel (2-chloroethylphosphonic acid) can induce full male sterility in wheat if applied at the right concentration and time (Rowell & Miller, 1971; Bennett & Hughes, 1972).

The critical stage of ear development at which Ethrel has to be applied to induce full male sterility coincides with male pre-meiotic archesporial development and terminates at the end of pre-meiotic interphase (Bennett & Hughes, 1972; Hughes *et al.,* 1974). For an ear to be fully sterile therefore, the application has to be made whilst anthers of the basal florets of the oldest spikelets are pre-meiotic. Later applications of Ethrel can be made but progressively higher concentrations are necessary to induce full male sterility (Hughes *et al.,* 1974). When applied at this time, the minimum concentration giving full male sterility in glasshouse grown spring wheat, *T. aestivum* variety Sirius, was 2,000 ppm (a.i.). An aqueous solution of Ethrel was foliar sprayed to the point of run off using 0.02% (v/v) Agral as a wetting agent.

Female fertility of sterilised ears, tested by hand pollinating bagged ears, showed that an average of 70% pollinated florets set grain, indicating that female fertility was high. This concentration however, prevented full ear emergence by shortening the peduncle without a corresponding reduction in length of the flag leaf sheath.

Because the stage of ear development at which the chemical has to be applied to induce male sterility is critical, its effective use at the minimum concentration in the field may be difficult, if for example weather conditions were unsuitable for spraying. In such a situation increasing the rate of Ethrel application may give full male sterility but reduces ear emergence. Investigations were carried out in the glasshouse into the

Table 1. Peduncle lengths (cm) of plants with Ethrel followed two (Time 1) or five (Time 2) days later with GA_3

	GA_3 concentration (ppm, a. i.)				
2,000 ppm Ethrel	0	100	200	300	Mean
Time 1	19.0	28.5	29.9	31.6	27.3
Time 2	16.2*	28.1	29.8	30.7	26.2
Mean	17.6	28.3	29.9	31.1	26.7
No Ethrel					
Time 1	32.7	41.0	43.5	46.4	40.9
Time 2	31.3	43.8	47.7	49.8	43.2
Mean	32.0	42.4	45.6	48.1	42.0
S.E.	Time	Ethrel	GA	Time × Ethrel	Time × GA
	0.95	0.95	1.34	1.34	1.90
		Ethrel × GA		Time × Ethrel × GA	
		1.90		2.69	

* Includes one missing plot.

effectiveness of foliar applications of gibberellic acid GA_3 at improving ear emergence in plants sprayed with Ethrel.

GA_3 was applied at three concentrations, 100, 200 and 300 ppm (a. i.), either two or five days after spraying the variety Sirius with 2,000 ppm Ethrel at the appropriate pre-meiotic stage. The three GA_3 concentrations significantly increased the length of the peduncles in Ethrel treated plants, making them almost as long as those of the control (Table 1). There was no advantage in using the higher GA_3 concentrations and both times of application were equally effective. The GA_3 treatments had no effect on the length of the flag leaf sheath, and therefore resulted in full ear emergence in Ethrel treated plants which otherwise would have had incompletely emerged ears.

There was no effect of the GA_3 treatments on the Ethrel induced male sterility and 78% of the hand pollinated male sterile florets set grain, showing that there was no adverse Ethrel × GA_3 interaction effect on female fertility.

On the basis of these glasshouse results, experiments were carried out during 1972/73 which showed that rates of Ethrel application effective at inducing male sterility in the glasshouse were also effective in the field. Moreover, GA_3 applied at the rate effective in the glasshouse not only increased the degree of ear emergence but also sometimes increased the level of male sterility, thus making lower Ethrel concentrations more effective. It was also found that the proportion of fully emerged sterilised ears was greater in plots sown at 276 plants/m^2 than in those sown at 72 plants/m^2. All sprayings were carried out with a knapsack sprayer, using aqueous solutions applied at a rate of approximately 700 litres/ha, with 0.02% (v/v) Agral as wetting agent.

The level of induced male sterility in the variety Maris Nimrod was assessed by bagging individual ears prior to flowering. At harvest, plants were sampled and the ears classified according to their degree of emergence relative to the ligule of the flag leaf. The number of crossed grain was estimated from the difference between grain set in bagged ears and that on non bagged ears exposed to cross pollination by adjacent drills of the same variety.

Ethrel at 6.4 kg (a.i.)/ha caused 4% selfing, 70% partial or complete emergence of the ears and gave an estimated yield of 7,500 crossed grains per m^2 (Table 2). The

Table 2. Ear emergence and induced male sterility in wheat variety Maris Nimrod treated with Ethrel followed three days later by GA_3 (1.1 kg (a.i.)/ha)

		Rate of Ethrel application (kg [a.i.]/ha)			
		0	6.4	12.8	19.2
Percentage ear emergence	Complete	90.7	54.0	31.3	7.8
	Partial	0.2	15.4	24.1	26.8
	Non emerged	9.1	30.6	44.6	65.4
Percentage male sterility		0	96.4	100.0	100.0
Estimated number of crossed grain/m^2 from fully and partly emerged ears.		–	7,464	4,520	1,675

higher rate of Ethrel application, 12.8 kg(a.i.)/ha gave complete male sterility but only 55% of the tillers were fully or partly emerged. The estimated yield of crossed grain was therefore reduced to 4,500 per m^2.

12.8 kg (a.i.)/ha in the field is approximately equivalent to a glasshouse concentration of 1,500 ppm(a.i.). Even with the added GA_3 treatment this application caused an unacceptable reduction in ear emergence and a rate of Ethrel application between 6.4 and 12.8 kg (a.i.)/ha appears to be desirable for good emergence of sterilised ears. Varenitsa & Zimina (1974) report good results using between 6.0 and 8.0 kg (a.i.)/ha. In another experiment, we obtained good results with 8.6 kg (a.i.)/ha.

More recent investigations using GA_3 to improve ear emergence have indicated that 1.1 kg (a.i.)/ha can cause damage to some varieties in the field by burning the foliage or by promoting too much peduncle elongation. Lower GA_3 concentrations have proved adequate in the glasshouse, and these will be further investigated in the field.

Although the GA_3 treatment could be useful on most varieties, the treatment is not effective when applied to semi-dwarf wheats derived from Norin 10. These varieties have higher endogenous GA levels than conventional varieties but have a block to its utilisation in the stems (Radley, 1970). A similar situation also occurs in the extreme dwarf cultivars Minister Dwarf and Tom Thumb (Gale & Marshall, 1973), and would be expected to occur in dwarf selections derived from them. This would limit the use of the GA_3 treatment with Ethrel as the use of semi-dwarf wheats in breeding programmes is increased.

Because the stage of ear development at which Ethrel has to be applied to induce full male sterility is critical, it would be difficult to use it for large scale production of F_1 hybrid seed, especially if there were large variations in growth stage of the variety to be sprayed. The cost of the chemical would also be prohibitive at current prices. On a small scale however, Ethrel could be used to produce sufficient F_1 seed to give reliable yield data from selected crosses between potential female parents and previously developed restorer lines. The female plants showing the most promising specific combining ability could then be backcrossed to *T. timopheevi* cytoplasm to produce A-lines for further testing and possible commercial exploitation. Crosses which exhibit high general combining ability could also be utilised in pure line breeding programmes.

There are many other applications of Ethrel in a conventional wheat breeding programme, as, for example, in backcrossing selected progeny to the F_1 so creating extensive back cross populations. The development of a technique by which the stem shortening normally associated with its use can be avoided greatly increases the range of circumstances in which it can be employed.

REFERENCES

BENNETT, M. D. & HUGHES, W. G. (1972). Additional mitosis in wheat pollen induced by Ethrel. *Nature* **240,** 566–568.

GALE, M. D. & MARSHALL, G. A. (1973). Insensitivity to Gibberellin in Dwarf Wheats. *Annals of Botany* **37,** 729–35.

HUGHES, W. G., BENNETT, M. D., BODDEN, J. J. & GALANOPOULOU, S. (1974). Effects of time of application of Ethrel on male sterility and ear emergence in wheat *Triticum aestivum. Annals of Applied Biology* **76**, 243–252.

RADLEY, M. (1970). Comparison of endogenous gibberellins and response to applied gibberellin of some dwarf and tall wheat cultivars. *Planta* **92,** 292–300.

ROWELL, P. L. & MILLER, D. G. (1971). Induction of male sterility in wheat with 2-chloroethylphosphonic acid (Ethrel). *Crop Science* **11,** 629–631.

VARENITSA, E. T. & ZIMINA, T. K. (1974). Heterosis in winter bread wheat, *T. aestivum* L., under conditions of the non-chernozem zone of the USSR. *VIIth Congress of Eucarpia, Budapest.* (In this volume.)

Comparative investigations of heterosis effect under laboratory and nursery conditions

F. SÁGI, Z. BARABÁS AND Z. KERTÉSZ
Cereal Research Institute, Szeged, Hungary

SUMMARY

Seven winter wheat varieties were crossed in a diallel system. F_1 seeds of each of the 21 combinations were sown at 100 and 400 seeds/m^2 rates in four replications using the microplot technique. Ten yield components of the hybrids, e.g. grain yield per plot, grain yield per plant, number of spikes per plant, 1000 kernel weight, etc. were determined. Parallel with this, mitochondrial complementation (MC) of the same seed parents were also measured by assaying cytochrome *c* oxidase activity in both parental mitochondria and their 1:1 mixture. Then, all the ten yield components at both seed rates were compared with the results of MC tests. A correlation was found between grain yield per plot and MC values only at low plant density ($r = 0.61$).

These observations show that MC may be a useful technique for predicting yield heterosis under laboratory conditions.

The increasing world need for food and protein on the one hand, and for the so-called "no-chemical" crop plants on the other, unavoidably leads to the establishment of more and more voluminous plant breeding programmes. However, financial and other possibilities necessary for completing of such extended programmes are not increasing proportionally. Therefore, laboratory methods suitable for early diagnosis of various kinds of heterosis would be of great importance.

It is known that the vigour of heterotic hybrids is based upon their more efficient mitochondria having a greater respiratory activity and better energy conservation. This is *mitochondrial heterosis* (MH) described by McDaniel and Sarkissian (1966, 1968), and Sarkissian and Srivastava (1967). When, however, mitochondria prepared from the parents of a heterotic hybrid are mixed in a ratio of 1:1, the mixture will also exhibit increased respiration. This phenomenon is the *mitochondrial complementation* (MC) discovered by the same research group (McDaniel & Sarkissian, 1966; Sarkissian & Srivastava, 1967, 1971). Between MH and MC there is a strong positive correlation as stated in case of maize (Sarkissian & Srivastava, 1967), barley (McDaniel, 1969, 1972), wheat (Sage & Hobson, 1973; Sarkissian & Srivastava, 1969) and sugar beet (Doney *et al.* 1972), respectively. This makes it possible to preselect good-combining parents under laboratory conditions, and by this means breeders can also save considerable time, work and expense.

In order to determine the MC, mitochondria with a good respiratory control should be isolated from extracts of the parental seedlings separately and mixed in 1:1 ratio, using differential ultracentrifugation (McDaniel, 1969; Sarkissian & Srivas-

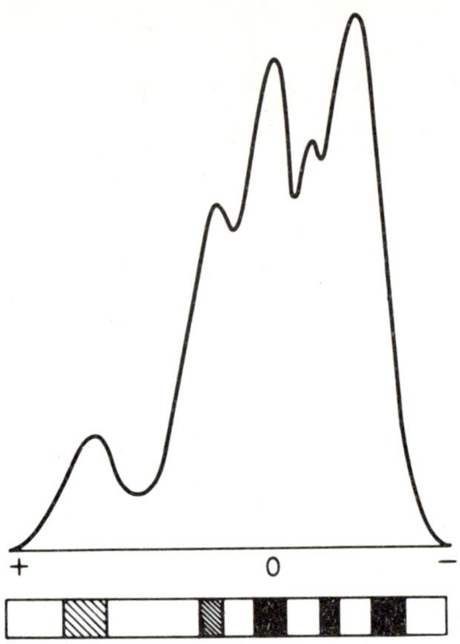

Fig. 1. Mitochondrial cytochrome *c* oxidase isoenzymes; "Tom Pouce".

tava, 1968). Then, the oxygen consumption of both parental and mixed mitochondria should be measured polarographically with a Clark-type O_2-electrode after addition of substrate and ADP. Alternatively, it is possible to measure their cytochrome *c* oxidase activity, since activity of this enzyme also shows MC (Sarkissian & Srivastava, 1971).

Of the authors dealing with MC, Ellis *et al.* (1973) could not find any correlation between MC and hybrid vigour although working with the same barley varieties as McDaniel (1969) and wheat lines as Sarkissian & Srivastava (1969). Negative results were also published by Zobl and Fischbeck (1972).

We did not aim at studying MC from a purely methodological viewpoint, but to obtain more information on the diagnostic value of this method. In the papers which have appeared on MC so far, results described by the authors are confined to a very limited number of lines or varieties. Therefore, in our Institute, seven winter wheat varieties were crossed in a complete diallel system. F_1 seeds of each of the 21 combinations were sown at the rate of 100 and 400 seeds per m^2, in four replications, using the miniplot technique (Barabás *et al.* 1973). As a measure of heterosis, ten yield components, i.e. grain yield per plot, grain yield per plant, number of spikes per plant, number of spikelets per spike, seed number per plant, etc. were determined. Parallel with this, MC exhibited by the same seed parents was also measured. For this purpose, cytochrome *c* oxidase isozymes extracted from parental, and mixed mitochondria were separated by electrophoresis on polyacrylamide gels. An aliquot of the extracts, representing 100 to 200 micrograms of protein was placed on the top of a 6% gel column of $^3/_4$ size. Then, the remaining volume of the glass tube was filled with 6% Cyanogum 41 solution and polymerized again. By this simple technique it was possible to separate the neutral, anionic and cationic isozymes in one run. After

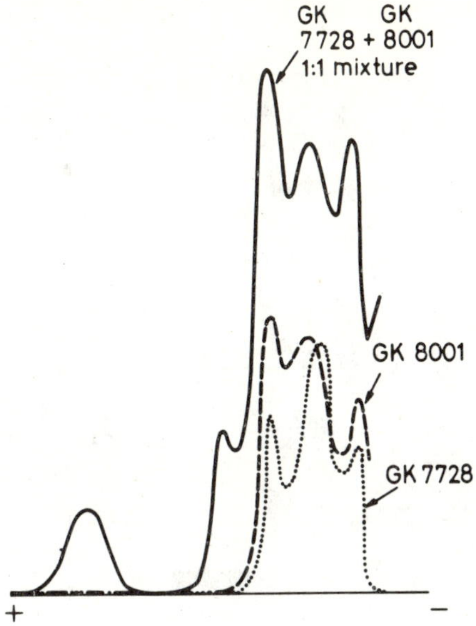

Fig. 2. Cytochrome *c* oxidase isoenzymes.

Table 1. Grain yields in the miniplot test and MC-values as expressed in cytochrome *c* oxidase units (both in per cent of the better parent)

Parents 1	Parents 2	Low seed rate	MC-value	High seed rate
Aurora	GT-76-150	106·1	96·6	79·9
Aurora	GKT-8001	83·4	93 8	91·2
Aurora	ZH-5996	109·9	111·0	146·4
Aurora	Sava	93·1	73·9	122·6
Aurora	Burgas 2	63·2	75·3	125·5
Aurora	Golden Valley	77·1	82·8	117·1
GT-76-150	GKT-8001	91·1	65·5	125·7
GT-76-150	ZG-5996	90·7	129·6*	78·2
GT-76-150	Sava	109·9	106·6	138·8
GT-76-150	Burgas 2	93·0	50·5	156·4
GT-76-150	Golden Valley	170·0	31·4*	256·4
GKT-8001	ZG-5996	117·0	85·7	55.1
GKT-8001	Sava	123·5	99·3	183·4
GKT-8001	Burgas 2	130·2	96·6	126·7
GKT-8001	Golden Valley	163·1	119·2	153·1
ZG-5996	Sava	117·6	117·7	122·2
ZG-5996	Burgas 2	123·9	102·3	130·1
ZG-5996	Golden Valley	136·1	130·8	97·2
Burgas 2	Golden Valley	104·0	121·0	142·7
Sava	Burgas 2	69·2	91·1	114·6
Sava	Golden Valley	155·4	105·9	257·3

* Uncertain MC-values probably due to experimental errors

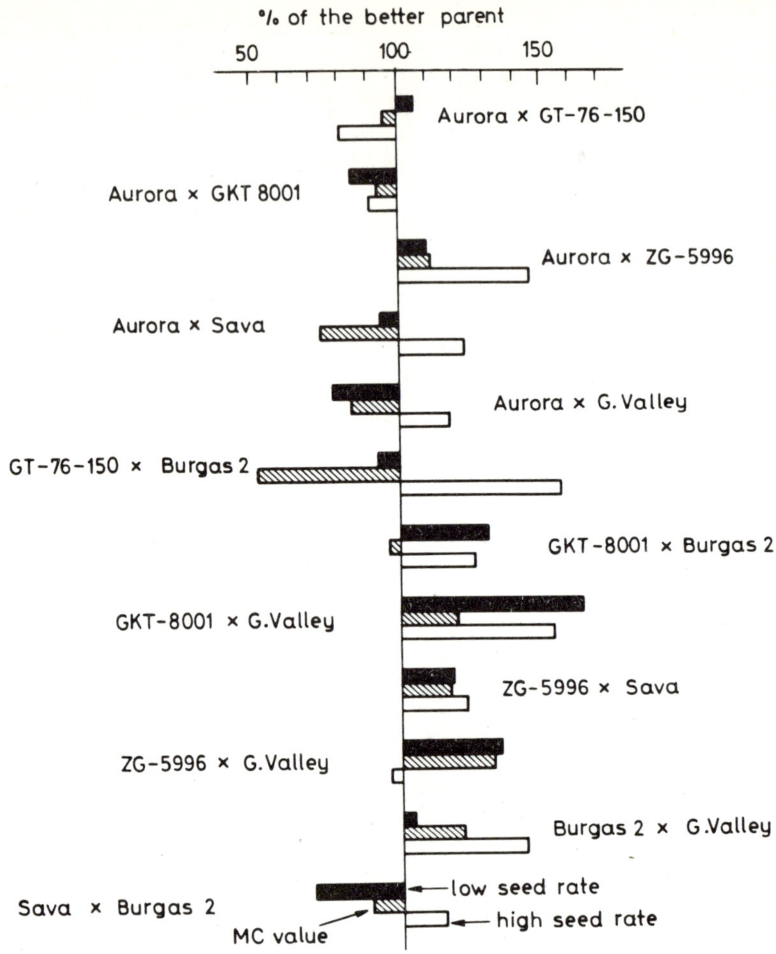

Fig. 3. Grain yields in the mini-plot test and MC-values expressed in cytochrome c oxidase units (both in percentage of the better parent).

developing with the NADI-reaction (Alam & Sandal, 1969), gels were traced by a densitometer. MC was expressed as enzyme units corresponding to the total peak area (cm^2). Figure 1 shows a typical densitometer record of mitochondrial cytochrome *c* oxidase from the wheat variety Tom Pouce (Blé Tome). Figure 2 demonstrates densitometer tracings of cytochrome *c* oxidase isoenzymes extracted from parental mitochondria and their 1:1 mixture showing extremely strong complementation.

In the last step of our investigations, all of the ten selected yield components obtained at both seed rates were compared with the results of MC tests. A correlation was found between grain yield per plot and cytochrome *c* oxidase units only. The related numerical data are given in Table 1. Our results agree with those of Sage & Hobson (1973), who also found that a correlation between the two characters men-

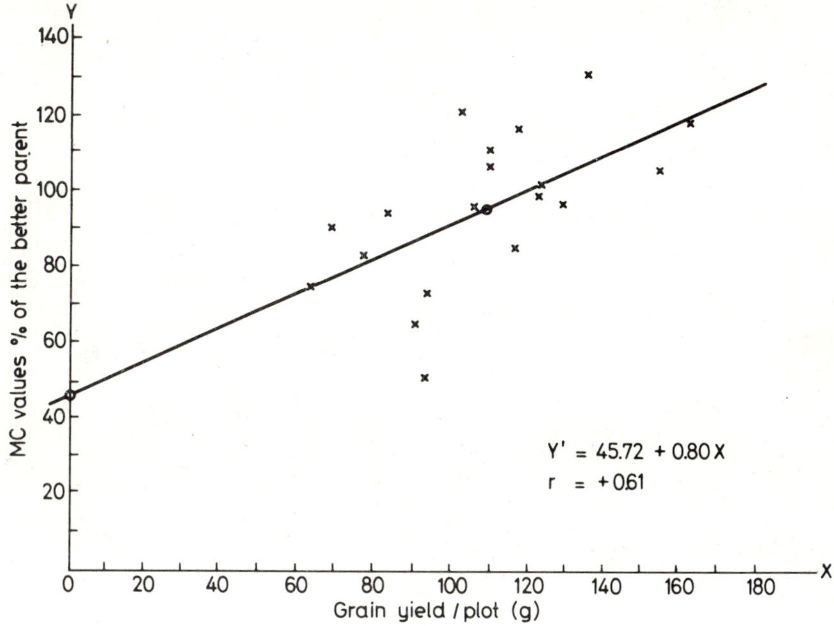

Fig. 4. Correlation between grain yield of real hybrids and cytochrome *c* oxidase of mitochondrial hybrids.

tioned above exists only at low plant density. For sake of a better comparison, the values of some combinations are depicted in Figure 3. It is clear again that the MC data are better correlated with grain yield at low plant density, although some agreements can be found with grain yields at high or both densities. The results were also evaluated with analysis of variance. As can be seen in Figure 4, there is a relatively good positive correlation (r = 0·61) between grain yield per plot of the real hybrids at low plant density and cytochrome *c* oxidase units of the "mitochondrial" hybrids produced in the laboratory. This correlation is almost as good as that found by McDaniel (1972) between MH and grain yield of several heterotic barley hybrids (r = 0·69).

These results are encouraging as far as the future prospects of using the MC test as a tool for predicting of yield heterosis are concerned. With plants raised under optimal nutritive conditions when heterosis is limited exclusively by the mitochondrial energy production, an even better agreement between the degree of MC and grain yield may be expected. Furthermore, an improvement of the analytical technique seems also to be possible.

REFERENCES

ALAM, S. & SANDAL, P.C. (1969). Electrophoretic analyses of anther proteins from male-fertile and male-sterile sudangrass, *Sorghum vulgare* var. *sudanense* (Piper). *Crop Sci.* **9,** 157–159.

BARABÁS, Z., SÁGI, F. & KERTÉSZ, Z. (1973). Special ways for hybrid wheat breeding. *Fourth Int. Wheat Genetics Symp., Columbia,* Mo., USA.

DONEY, D. L., THEURER, J. C. & WYSE, R. E. (1972). Mitochondrial complementation in sugar beet. *Crop. Sci.* **12,** 493–494.

ELLIS, J.R., BRUNTON, C.J. & PALMER, J.M. (1973). Can mitochondrial complementation be used as a tool in breeding hybrid cereals? *Nature* **241,** 45–47.
McDANIEL, R.G. (1969). Relationships of seed weight, seedling vigour and mitochondrial metabolism in barley. *Crop Sci.* **9,** 823–827.
McDANIEL, R.G. (1969). Mitochondrial heterosis in barley. In: *Barley Genetics, Proc. 2nd Int. Barley Genetics Symp.* Washington Sta. Univ. Pr., 323–337.
McDANIEL, R.G. (1972). Mitochondrial heterosis and complementation as biochemical measures of yield. *Nature New Biol.* **236,** 190–191.
McDANIEL, R.G. & SARKISSIAN, I.V. (1966). Heterosis: Complementation by mitochondria. *Science* **152,** 1640–1642.
McDANIEL, R.G. & SARKISSIAN, I.V. (1968). Mitochondrial heterosis in maize. *Genetics* **59,** 465–475.
SAGE, G.C.M. & HOBSON, G.E. (1973). The possible use of mitochondrial complementation as an indicator of yield heterosis in breeding hybrid wheat. *Euphytica* **22,** 61–69.
SARKISSIAN, I.V. & SRIVASTAVA, H.K. (1967). Mitochondrial polymorphism in maize, II. Further evidence of correlation of mitochondrial complementation and heterosis. *Genetics* **57,** 843–850, 1967.
SARKISSIAN, I.V. & SRIVASTAVA, H.K. (1968). On methods of isolation of active, tightly coupled mitochondria of wheat seedlings. *Plant Physiol.* **43,** 1406–1410.
SARKISSIAN, I.V. & SRIVASTAVA, H.K. (1969). High efficiency heterosis and homeostasis in mitochondria of wheat. *Proc. Nat. Acad. Sci. USA* **63,** 302–309.
SARKISSIAN, I.V. & SRIVASTAVA, H.K. (1971). Mitochondrial polymorphism. III. Heterosis, complementation, and spectral properties of purified cytochrome oxidase of wheat. *Biochem. Genet.* **5,** 57–63.
ZOBL, R. & FISCHBECK, G. (1972). Complementation of isolated mitochondria from several wheat varieties. *Plant Physiol.* **50,** 790–791.

Session III

METHODS OF HETEROSIS BREEDING

Chairmen:

DR. J. SNEEP AND PROFESSOR A. BÁLINT

The interaction of restoration with environment in wheat

G.C.M. SAGE
Plant Breeding Institute, Cambridge, England

SUMMARY

Methods for the exploitation of heterosis are well worked out with outbreeding crops where the different outbreeding mechanisms involved facilitate the making of controlled crosses. These methods fall loosely into two categories; those aimed at producing hybrid varieties and those where some form of recurrent selection concentrates genotypes having good combining ability into synthetic strains. Before the exploitation of heterosis can be contemplated with inbreeding crops, such as wheat, some form of male sterility has to be imposed on the crop to provide the outbreeding mechanism which facilitates large scale production of controlled crosses. So far almost all the work with male sterility in wheat has been aimed at producing hybrid varieties. Little consideration has been given to the use of male sterility as a breeding tool to increase the amount of recombinant material available to pure-line breeders or to apply recurrent selection to wheat populations.

For the immediate future the most readily available kind of male sterility in wheat is cytoplasmically inherited. The use of chemical gametocides, as a possible means of switching male fertility on and off at will, is an alternative to the use of cytoplasmic male sterility but the use of gametocides is difficult and, at the moment, likely to be prohibitively expensive on a large scale. Any exploitation of heterosis in wheat is thus likely to be dependent on cytoplasmic male sterility and consequently also dependent on the manipulation of fertility restoring genes. One of the major restrictions to the agricultural use of hybrid varieties of wheat has been that the restorer genes available have been environmentally sensitive and hence the expression of restoration has been unreliable. This unreliability has been looked on, so far, merely as a problem to be overcome. However, if sufficiently understood, there may be aspects of the interaction between restorer genes and the environment which could be utilised to advance heterosis breeding in wheat.

PREVIOUS WORK

Male sterility in wheat has been found to result when the nuclear genomes of hexaploid varieties are introduced into the cytoplasms of some distinct but related species by backcrossing. Recently Tsunewaki & Endo (1973) have shown that male-sterile wheat lines having different cytoplasms each show, in addition to male sterility, a specific spectrum of mostly deleterious plant characteristics that are also cytoplasmically inherited. The additional cytoplasmic effects appear to be developmentally clustered so that, for example, while male-sterile lines having *Ae.umbellulata* cytoplasm show detrimental effects in characters determined in the early vegetative stage of development, isogenic male-sterile lines with *Ae.caudata* cytoplasm show few detrimental effects in vegetative characters but a number of seed characteristics are abnormal. This implies that in male-sterile lines at different stages in plant development, depending on the particular cytoplasm involved, some very fundamental physiological changes from normal are brought about by the imbalance in function between the hexaploid wheat nucleus and the alien cytoplasm which in the course of plant development results in the range of phenotypic effects observed.

Maan & Lucken (1970, 1971, 1972) have shown in a number of reports that the low vigour and poor stunted growth of some male-sterile wheat lines is restored to normal, along with male fertility, when crosses are made to certain restorer lines. The restoring genes involved here must thus be influencing the aberrant physiological development normally associated with the alien cytoplasms. If, as this suggests, restoring genes control basic physiological functions it is, perhaps, not surprising that their expression has been found to vary from location to location (Monteagudo *et al.*, 1966; Lucken & Maan, 1967; Wilson, 1968; Schmidt *et al.*, 1971).

Restoration of male fertility to male-sterile wheat lines having *T.timopheevi* cytoplasm behaves phenotypically as a qualitative or polygenic character. In segregating populations all degrees of partial restoration of male fertility can often be found and a large number of workers have found that restoration is primarily due to up to three major genes but that numerous minor modifier genes are also involved. So far 17 out of the 21 wheat chromosomes have been found to carry genes which influence restoration in *timopheevi* cytoplasm male-sterile lines. It appears that these numerous genes of major and minor effect act in a cumulative manner so that the phenotypic expression observed depends on how many of which genes are present. The interesting controlled environment chamber experiments of Welsh & Klatt (1971) and Johnson & Patterson (1973) have clearly shown that the expressivity of these restoring genes is variable since both sets of workers found that the level of restoration was depressed by high temperatures and to a lesser extent by long photoperiods.

It has often been found by workers attempting to develop hybrid wheat that some *T.timopheevi* male-sterile lines are more easily restored than others (Wilson, 1968; Schmidt *et al.*, 1971; Nettevich & Naumov, 1971). This could well result if such male-sterile lines carried weak restoring genes which on their own were ineffective but which in combination with other restorer genes contributed to the overall fertility of the F_1 generation. The possibility that male-sterile lines might carry latent restoring genes is also suggested by the findings of several workers that some male-sterile lines are able to set some selfed seed in some environments. Maan (1973b), for example, found that male-sterile lines with *Ae.umbellulata* or *Secale cereale* cytoplasm were completely male-sterile in the autumn but partially self fertile if grown in the glasshouse in the spring. Maan (1973a) also found that while *T.timopheevi* male-sterile lines of the varieties Selkirk or Chinese Spring were completely male sterile in the autumn they showed nearly complete male fertility when grown in a growth chamber with a short, 12 hr. photoperiod. All these male-sterile lines were thus switched from the male-sterile condition to partially male-fertile condition by certain combinations of environmental factors.

This paper presents data from crosses involving *T.timopheevi* male-sterile lines which show, 1. that some European varieties carry genes for restoration which by themselves are insufficient to cause any restoration but which segregate in crosses, 2. that certain combinations of these genes may produce genotypes that can be switched from male-sterile to partially male-fertile by different sowing dates in the field, and 3. that these genes are possibly involved in controlling hormonal balances within the plant and hence may have a far reaching influence on plant development.

ENVIRONMENTAL EFFECTS

Nine F_1 hybrid wheats, derived by crossing each of three male-sterile lines to each of three restorer lines were grown in trial in 1971. Six of the F_1 hybrids were grown in trial again in 1972. The parental lines involved and a visual score of the level of restoration in the F_1 plants are given in Table 1. Restoration was lower in 1972 than in 1971 in some hybrids. Crosses involving the male-sterile line TB306 were noticeably more severely affected than the other crosses. In 1971 the hybrid TB306 × Maris Beacon showed a lower level of restoration at a lower seed rate. In both years bagged ears of the F_1 plants were harvested to provide guaranteed selfed seed which was used to sow F_2 populations in 1972 and 1973.

The nine F_2 populations studied in 1972 were grown at two sowing densities of approximately 33 and 133 plants per m^2 from both winter and spring sowings. By observing the anthers in different parts of the ears of main tillers the F_2 plants were classified as fully fertile, that is they had normal dehiscing anthers in all florets, as fully male- sterile, that is those having male-sterile anthers in all florets, and as semi-sterile which were those plants which had normal anthers in florets at the base of the ear but male-sterile anthers in a variable number of florets at the apex of the ear. In 1973 the same classifications were carried out but the lower plant density treatment was omitted and no crosses involving Maris Beacon were included. Of the six parental lines involved in these crosses two, Primepi and Maris Beacon are winter wheats, Maris Ranger is an alternative type, and the remainder are spring wheats. But despite these differences and although the F_2 populations were grown from both winter and spring sowings there was no loss of plants in either year from either winter damage or failure to head from a spring sowing.

The mean percentage of fully fertile or fully male-sterile plants in the crosses involving each of the male-sterile parents are given in Table 2 for the high sowing den-

Table 1. Visual score of level of restoration in F_1 hybrid wheats 0–9 (9=fully restored)

Male-sterile Female parent	Male parent	1971 High seed rate	1971 Low seed rate	1972 High seed rate
Kloka	× Primepi	9	9	9
Kloka	× TW 74 (Minister × Els)	9	9	9
Kloka	× Maris Beacon	9	9	–
Maris Ranger	× Primepi	9	9	9
Maris Ranger	× TW 74	9	9	8
Maris Ranger	× Maris Beacon	9	9	–
TB 306	× Primepi	9	9	6
TB 306	× TW 74	9	9	5
TB 306	× Maris Beacon	8	6	–

TB 306 = Jufy × Svenno

sity in each year. In both years the TB306 crosses produced markedly fewer fully fertile plants than the crosses of the other two male-sterile parents, in the TB306 crosses apparently more fully fertile plants were produced in 1973 than in 1972, and in the crosses involving male-sterile Kloka or Maris Ranger fewer fully fertile plants were produced in the spring sowing than in the winter sowing in both years.

Thus it appeared that the segregation of fully fertile plants in TB306 crosses was different from that of Kloka or Maris Ranger crosses and that the TB306 type of segregation was sensitive to year differences while that of Kloka and Maris Ranger crosses was sensitive to sowing date differences. In contrast to this the number of fully male-sterile plants produced was similar in all the crosses and apparently did not differ with year and sowing date to the same extent as the number of fully fertile plants.

In Table 3 the mean percentage of fully fertile or fully male-sterile plants found at the different sowing densities in the 1972 F_2 populations are given for TB306 crosses and jointly for Kloka and Maris Ranger crosses. The analysis of variance of angular transformed data showed that in Kloka and Maris Ranger crosses the mean percentage of fully fertile plants was significantly lower in the spring sowing than in the winter sowing and that in the winter sowing it was significantly lower at the lower plant density than at the higher plant density. In contrast the mean percentage of fully fertile plants produced by TB306 crosses was not significantly affected by sowing date or plant density.

With Kloka and Maris Ranger crosses the mean percentage of fully male-sterile plants produced behaved in an opposite manner to the mean percentage of fully fertile plants. At 30% it was significantly higher in the spring sowing than from the winter sowings where only 26% of the plants were fully male-sterile. There were, however, no significant differences in the mean percentage of fully male-sterile plants produced in the TB306 crosses and in no case was this percentage affected by sowing density differences. Thus, in the one year, it appeared that the proportion of fully male-fertile or fully male-sterile plants produced was less affected by environmental differences in TB306 crosses than in Kloka and Maris Ranger crosses and that within Kloka and Maris Ranger crosses the proportion of fully male-sterile plants was less affected than the proportion of fully fertile plants.

The overall mean percentages of the two plant types given by crosses involving each of the three restorer lines showed a significant effect of restorer line genotype in

Table 2. Mean percentages of fully male-fertile or fully male-sterile plants in F_2 populations sown at 133 plant per m^2
(1972 figures are means of crosses involving Primepi, TW 74/34/3 and Maris Beacon as male parents. 1973 figures are means of Primepi and TW 74/34/3 crosses only)

Plant type	*Female parent*	*Year*	*Winter sown*	*Spring sown*
Fully male-fertile	Male-sterile TB306	1972	27.49	21.82
		1973	32.52	34.20
	Male-sterile Kloka	1972	62.48	57.51
		1973	63.64	54.55
	Male-sterile Maris Ranger	1972	61.71	40.04
		1973	62.79	56,36
Fully male-sterile	Male-sterile TB306	1972	28.35	30.66
		1973	28.67	25.63
	Male-sterile Kloka	1972	24.76	27.90
		1973	24.55	24.68
	Male-sterile Maris Ranger	1972	26.45	31.16
		1973	26.36	24.40

Table 3. Mean percentage of fully male-fertile and fully male-sterile plants in the F_2 of crosses involving the restorer lines Primepi, TW 74/34/3 and Maris Beacon grown in 1972 in four field environments

Plant type	Restorer line involved	Winter sown 133 plants per m²	Winter sown 33 plants per m²	mean	Spring sown 133 plants per m²	Spring sown 33 plants per m²	mean	overall mean
Fully male-fertile	Primepi	66.24	61.52	63.88	54.97	55.47	55.22	59.55*a*
	TW 74/34/3	62.64	56.56	59.60	53.16	53.82	53.49	56.55*a*
	Maris Beacon	59.94	48.92	54.43	35.97	43.54	39.76	47.09*b*
	mean	62.94*a*	55.67*c*	59.30*b*	48.03*d*	50.94*d*	49.49*d*	
Fully male-sterile	Primepi	24.37	24.24	24.31	25.78	27.17	26.48	25.39*a*
	TW 74/34/3	23.28	24.06	23.67	28.16	28.51	28.34	26.00*b*
	Maris Beacon	29.44	30.22	29.83	35.31	34.93	35.12	32.48*b*
	mean	25.70*b*	26.17*b*	25.94*b*	29.75*a*	30.20*a*	29.98*a*	

Within either plant type environmental or genotypic means having the same letter are not significantly different ($P = 0.05$) as shown by analysis of variance of angular transformed data.

Kloka and Maris Ranger crosses but not in TB306 crosses. In Kloka and Maris Ranger crosses those involving Maris Beacon produced significantly fewer fully fertile, but significantly more fully male-sterile plants than those involving Primepi or TW 74 which suggests that Maris Beacon carries minor restoration modifying genes not carried by the other two lines. Thus of our two different kinds of F_2 segregation one, that shown by Kloka and Maris Ranger crosses, appeared to be influenced by sowing date, sowing density and genotypic background while the other, that shown by TB306 crosses appeared to be much less affected not only be environmental differences but by the genotypic background differences as well.

SEGREGATION RATIOS

The actual numerical segregations observed in both years in Kloka and Maris Ranger crosses were very similar to those found in a previous study (Sage, 1972). In this study the F_2 segregation from a winter sowing was 75 fully fertile plants to 21 semi-sterile plants to 32 fully male-sterile plants and was accounted for by a hypothesis of three major genes each of which made a different contribution to restoration but which acted cumulatively to produce the phenotypic expression of restoration observed.

Table 4 gives the overall F_2 segregation observed in 1972 in Kloka and Maris Ranger crosses. Overall from the winter sowing or overall crosses involving Primepi or TW 74 as the male parent, the segregations were not significantly different from the 75:21:32 ratio. There was considerable divergence from this ratio in the overall segregation from the spring sowing and overall in the crosses involving Maris Beacon. However in the environment most favourable to restoration, that is at the high plant density winter sown, where the highest proportion of fully fertile plants were produced, the segregation of the Maris Beacon crosses also fitted the 75:21:32 ratio. Thus it is possible that the basic segregation in all the Kloka and Maris Ranger crosses was 75:21:32 and that the observed deviations from this segregation were principally due to the modifying genes carried by Maris Beacon which may have conferred a greater sensitivity to the effects of spring sowing to Maris Beacon derived material.

Table 4. Overall F_2 segregation of fertility plant type in different environments or in crosses involving different genotypes

Genotype involved and/or environment	Number of Plants: Fully male-fertile	Semi-sterile	Fully male-sterile	Total	χ^2 75:21:32
Winter sowing	1204	302	524	2030	3.623 NS
Spring sowing	843	349	504	1696	55.901 ***
Primepi crosses	788	192	331	1311	3.036 NS
TW 74/34/3	718	220	327	1265	1.367 NS
Maris Beacon crosses	541	239	370	1150	63.240 ***
Maris Beacon crosses at high density winter sown	205	49	106	360	4.730 NS

*** = P 0.001 NS = not significant

The hypothesis put forward previously to explain the 75:21:32 segregation is presented diagramatically in Table 5. The 27 possible genotypes of a three gene segregation are set out from left to right in descending order of genotypic value with respect to restoration of male fertility. In the hypothesis two sets of genotypes, numbers 13 and 14 and numbers 18 and 19, are assumed to fall at threshold genotypic levels such that genotypes 1–12 are fully fertile, genotypes 15–17 are semi-sterile and genotypes 20–27 are fully male-sterile. Since they fall at the upper threshold level, genotypes 13 and 14 are assumed to have a 50% change of classification as either fully fertile or semi-sterile. Similarly genotypes 18 and 19 are assumed to have a 50% chance of classification as either semi-sterile or fully male-sterile. Given these assumptions the segregation of the triple heterozygote, genotype number 8, would be in whole numbers 75:21:32. It is probable that the F_1 hybrids involving male-sterile Kloka or Maris Ranger segregated in this manner in the F_2 generation and hence were probably triple gene heterozygotes.

The controlled environment work referred to earlier would lead us to expect that the higher temperatures and longer day lenghts that would be experienced by plants developing from spring sowings would depress the expressivity of the restoring genes. In a segregation of the kind illustrated in Table 5 this would reduce the value of each genotype in terms of restoration. The phenotypic expression of each genotype in a spring sowing might thus be expected to shift to the right of its expression from a winter sowing. Thus genotype 12 might be expected to produce semi-sterile phenotypes and genotype 17 might be expected to produce fully sterile phenotypes. In other words the phenotype of genotypes close to the threshold levels would slide to that of a lower category of restoration. Since in this F_2 there is a higher proportion of genotype 12 than genotype 17 spring sowing might change more plants from fully fertile to semi-sterile than from semi-sterile to fully male sterile. This is exactly what was observed with the F_2 populations involving Kloka and Maris Ranger as female parents. Spring sowing increased the proportion of fully male-sterile plants from about 25 to 30%, but decreased the proportion of fully fertile plants to an even greater extent from 59 to about 49%.

The numerical segregations found in both years in TB 306 crosses under the environmental conditions most favourable to restoration, that is at a high plant density winter sown, are given in Table 6. The proportion of fully male-sterile plants found was consistently about 25 % so that each of the crosses fitted a 3:1 segregation of some restoration to none. However the proportion of fully fertile plants varied so that some crosses fitted a ratio of 3 fully fertile: 3 semi-sterile: 2 fully male-sterile plants and some fitted a ratio of 1 fully fertile: 2 semi-sterile: 1 fully male-sterile plant. The 3:3:2 ratio was found in F_3 progenies in the previous study and probably arose as follows.

Genotype 14 (Table 5) would on selfing produce one genotype 13 to two genotype 14 to one genotype 25. Since genotypes 13 and 14 would be classified 50/50 as either fully fertile or semi-sterile this would result in a 3:3:2 segregation. In some genetic backgrounds or in some environments the threshold nature of genotypes 13 and 14 might mean that something nearer the basic 1:2:1 segregation ratio would be observed. Thus it is possible that the TB 306 crosses were derived from F_1 plants possessing genotype 14. This possibility is strengthened by several of the observations already mentioned. First the fact that the level of restoration of the F_1 plants involving TB 306 was observed to be more variable than in the other F_1 plants might be

Table 5. Hypothetical model of the inheritance of restoration

	Fully male fertile (FF)												FF or SS		Semi-sterile (SS)			SS or FS		Fully male-sterile (FS)							
Genotype number	1	2	3	4	5	6	7	8	9	10	11	12	13	14	15	16	17	18	19	20	21	22	23	24	25	26	27
F_2 Genotype	*AA*	*Aa*	*AA*	*Aa*	*AA*	*Aa*	*AA*	*Aa*	*AA*	*Aa*	*AA*	*Aa*	*AA*	*Aa*	*AA*	*Aa*	*aa*	*aa*	*aa*	*aa*	*AA*	*Aa*	*aa*	*aa*	*aa*	*aa*	*aa*
	BB	*BB*	*BB*	*BB*	*Bb*	*Bb*	*Bb*	*Bb*	*BB*	*BB*	*Bb*	*Bb*	*bb*	*bb*	*bb*	*bb*	*BB*	*BB*	*Bb*	*Bb*	*bb*	*bb*	*BB*	*Bb*	*bb*	*bb*	*bb*
	CC	*CC*	*Cc*	*Cc*	*CC*	*CC*	*Cc*	*Cc*	*cc*	*cc*	*cc*	*cc*	*CC*	*CC*	*Cc*	*Cc*	*CC*	*Cc*	*CC*	*Cc*	*cc*	*cc*	*cc*	*cc*	*CC*	*Cc*	*cc*
Expected frequency	2	4	4	8	4	8	8	16	2	4	4	8		6	4	8	2		8	8	2	4	2	4	2	4	2
							72						3	3		14		4	4					28			
									75 FF							21 SS							32 FS				

Winter sowing: 59% FF = 75/128
Spring sowing: 49% FF = 63/128

25% FS = 32/128
30% FS = 38/128

expected if they possessed a threshold genotype, number 14, while the others possessed a genotype, number 8, which not only had a higher genotypic value but also had one well above the upper threshold value. Secondly the consistency of the proportion of fully male-sterile plants produced in F_2 populations of TB 306 crosses and the fact that in general such F_2 populations were less environmentally variable than in Kloka or Maris Ranger crosses might be expected if fewer genotypes were segregating and the single fully male sterile genotype had a value that was well below the lower threshold level.

Table 6. Numerical segregations of TB 306 crosses sown in the winter at approximately 133 plants per m^2

Year		Fully fertile	Semi-sterile	Fully male-sterile	Total	χ^2 1:2:1	χ^2 3:3:2	χ^2 3:1
1972	Primepi	60	68	57	185	13.076*	3.793 NS	3.332 NS
	TW 74	35	66	36	137	0.196 NS	9.471 **	0.119 NS
	Maris Beacon	32	70	38	140	0.514 NS	14.095 ***	0.343 NS
1973	Primepi	34	32	28	94	10.340 **	1.206 NS	1.149 NS
	TW 74	59	79	54	192	6.281 *	3.778 NS	1.000 NS

*, **, *** = P 0.05, 0.01, 0.001 respectively NS = not significant

All this evidence points to the conclusion that TB 306 crosses were derived from F_1 plants possessing genotype 14 and Kloka and Maris Ranger were derived from F_1 plants possessing genotype 8. Since the same three male parents were involved in all the crosses this implies that the genes A and C were contributed by the restorer lines, Primepi, TW 74 and Maris Beacon and that Kloka and Maris Ranger contributed the B gene and TB 306 contributed the C gene to their respective F_1 plants. Each of the male sterile lines investigated thus carried a latent restoring gene which on its own was unable to cause any restoration. The implication that Primepi contributed two major genes for restoration to its F_1 plants is supported by the work of several other investigators all of which concluded that the restoring ability of Primepi in due principally to two major genes (Nettevich & Naumov, 1970; Miller & Schmidt, 1970; Shebeski, 1971).

It is possible that the effect of spring sowing on genotypes with genotypic values for restoration close to a threshold level is to cause them to be classified in a lower restoration category than they would be in a winter sowing. If this were so, as I have suggested, for genotype 17 then a homozygous genotype consisting of a combination of the Kloka or Maris Ranger gene with the TB 306 gene would be semi-sterile, and hence set selfed seed, in winter sowings but would be male sterile in spring sowings. This genotype would thus behave like those male sterile lines found by Maan in which male fertility could be partially switched on by suitable environments.

In the F_3 generation F_2 plants having genotype number 2 (Table 5) would, according to the hypothesis, segregate one genotype 1 to two genotype 2 to one genotype 17 and hence would segregate 3 fully fertile plants for every one semi-sterile plant. Such segregations have been found in winter sown F_3 progenies (Sage, 1972) and since, according to the hypothesis, genotype 2 is the only one that could segregate in this manner the semi-sterile plants produced must carry the homozygous genotype number 17.

The possibility of finding for a given location a simply segregating genotype in which the expression of restoration is switched by simply applied environmental differences such as sowing date might have two consequences for the exploitation of heterosis in wheat. Firstly the switchable genotype might be used as the female parent in the production of hybrid wheat. In spring sowings it would function as a normal male sterile A line but in winter sowings, because it would be at least partially self fertile, it could be grown in pure stand for multiplication purposes thus reducing the amount of land required and the absolute dependence on wind-pollination that occurs when normal A lines are multiplied by B-line wind-pollination.

An alternative use for such a switchable genotype might be its use as the basis of a recurrent selection programme in wheat populations. A cross between plants of genotype 17 (Table 5) and genotype 1 would have genotype 2 in the F_1 generation and would segregate 3 fully fertile to 1 semi-sterile in winter sown F_2 populations. Continued pedigree selection from winter sowings of segregating progenies would result in later generations in elite lines which could be easily intercrossed on a large scale merely by growing a blend of the seed from a spring sowing and harvesting the open-pollinated seed set on the male sterile segregates.

SPACING EFFECTS

In both cases where differences in the level of restoration were found in different density plantings sown at the same time in the same year, that is with the F_1 plants of TB 306 × Maris Beacon in 1971 and the winter sown Kloka and Maris Ranger crosses in F_2 in 1972, higher levels of restoration were found to be associated with higher plant densities. To examine the apparent response of fertility to plant spacing further a four replication trial was grown in the glasshouse during the winter. Seed from bagged ears of five semi-sterile plants, having the possibly switchable genotype 17, was sown, together with the spring variety Sirius as control, at two densities each slightly greater than those used in the field. All the ears produced were bagged and the level of male fertility was assessed by calculating the percentage of spikelets per ear in which some grain was found to be set at harvest. Both for this estimate of the level of male fertility and for mean tiller height there were significant plant density and variety effects. Varietal and plant density treatment means are given in Table 7. The level of male fertility was not complete in the control variety Sirius but was significantly higher than that found in the semi-sterile plant progenies. Both a higher level of male fertility and greater height were found at the high density. Despite the significant variety differences the plant density × variety interactions were not significant for either character so that the semi-sterile progenies and Sirius responded to the density treatments in the same way.

DISCUSSION

Given the evidence which suggests that restorer genes affect fundamental physiological functions it is tempting to speculate that the increased male fertility and the increased height observed at the high density in the glasshouse trial were due to a common cause. Wheat plants, other than those carrying dwarfing genes none of

Table 7. Varietal and spacing treatment means of glasshouse trial 1973–74

Varietal means	Percentage spikelets with grain per ear	Mean tiller height cm
Sirius	83.25 *a*	75.83 *d*
Semi-sterile progeny A	74.28 *b*	91.53 *ab*
Semi-sterile progeny B	73.70 *b*	90.02 *b*
Semi-sterile progeny C	72.68 *b*	93.22 *a*
Semi-sterile progeny D	72.58 *b*	91.40 *ab*
Semi-sterile progeny E	67.50 *b*	86.95 *c*
L.S.D. (P=0.005)	6.986	3.020
Spacing means		
High density	79.81 *a*	91.66 *a*
Low density	68.19 *b*	84.64 *b*
L.S.D. (P=0.005)	6.915	5.597

Within either character or treatment means having the same letter subscript are not significantly different

which were present in the material I have been discussing, respond to exogenously applied gibberellins by growing taller. That the plants growing at the high density were taller suggests that the endogenous lèvels of gibberellins at this density was higher than in plants growing at the low density. Kirby & Farris (1970) have suggested that many morphological and physiological changes that occur in barley when grown at high densities are the result of high endogenous levels of gibberellins. If a higher endogenous level of gibberellins was responsible for both the increased height and the increased male-fertility observed then it is reasonable to suppose that the level of male fertility and hence of restoration in the semi-sterile progenies was determined by gibberellin levels in the plants, or, at least, by its interactions with other development controlling hormones. Restorer genes may thus function by controlling the relative balance of hormonal levels. Puckeridge (1968) suggested that the differences in leaf growth that he observed in different densities were brought about by the destruction of gibberellin in low densities by the higher light intensities penetrating the sparser canopy. If gibberellin levels are critical for the expression of restoration this might also be the explanation for both the depression of restoration observed at low densities and for the depression of restoration caused by long day lengths.

The interaction of restoration with spacing may thus provide further circumstantial evidence that restorer genes affect basic metabolic processes. But, since it is the basic metabolic processes that determine plant development and ultimately yield we may have to realise that restoring genes are not only the means by which we can produce heterotic crosses in inbred crops but also components of the heterotic yield we are attempting to exploit. This may well place a constraint on the use of the kind of environmentally switchable genotypes discussed in this paper. If the particular hormonal situation that produces the delicately balanced level of male fertility in switchable genotypes is not also compatable with the hormonal situation required by the highest expression of yield then it will be a waste of time first of all trying to find such genotypes and secondly trying to use them to exploit heterosis.

The number of environmentally sensitive genes that have to be rationally manipulated in order to exploit heterosis using cytoplasmic male sterility in a crop like wheat makes handling the material an extremely complex and time consuming process. By comparison the alternative, the use of chemical gametocides, appears extremely attractive despite the difficulties. But even some of the successful chemicals so far investigated probably work by affecting the plant's hormonal balances. It may be that it will be possible to find genotypes, possibly those that have a genotypic value for restoration at or close to some internal threshold level, that can be sterilised by gametocides more easily or more reliably than normal varieties. If this were the case heterosis breeding in wheat might most successfully be pursued by combining the cytoplasm and gametocide approaches. We should then be looking not so much for male sterile lines that are easy to restore but for male-fertile lines that are easy to sterilise.

REFERENCES

JOHNSON, J.W. & PATTERSON, F.L. (1973). Pollen Production of Fertility Restored Lines of Soft Red Winter Wheats. *Crop Science* **13**(1), 92–95.

KIRBY, E.J.M. & FARIS, D.G. (1970). Plant Population induced Growth Correlations in the Barley Plant Main Shoot and Possible Hormonal Mechanisms. *J. Exp. Bot.* **21,** 787–798.

LUCKEN, K.A. & MAAN, S.S. (1967). Effect of Genotype and Environment on Fertility Restoration of Cytoplasmically Male-Sterile Spring Wheat (*Triticum aestivum* L.). *Agronomy Abstracts, Madison* 14.

MAAN, S.S. (1973a). Cytoplasmic Male Sterility and Male Fertility Restoration Systems in Wheat. Paper given to *Eucarpia Cereals Section Meeting, University of Italy* May 14–18.

MAAN, S.S. (1973b). Cytoplasmic Variability in Triticinae. *Proceeding of the Fourth International Wheat Genetics Symposium, Missouri* 367.

MAAN, S.S. & LUCKEN, K.A. (1970). Interaction of *Triticum boeoticum* Cytoplasm and Genomes of *T. aestivum* and *T. durum:* Restoration of Male Fertility and Plant Vigour. *Euphytica* **19,** 498–508.

MAAN, S.S. & LUCKEN, K.A. (1971). Male-Sterile Wheat with Rye Cytoplasm. Restoration of Male Fertility and Plant Vigour. *Journal of Heredity* **62,** 353–355.

MAAN, S.S. & LUCKEN, K.A. (1972). Interacting Male Sterility – Male Fertility Restoration System for Hybrid Wheat Research. *Crop Science* **12,** 360–364.

MILLER, J.F. & SCHMIDT, J.W. (1970). Inheritance of Genes Controlling Male Fertility Restoration in Two Different Lines of Wheat. *Agronomy Abstracts, Madison* 23.

MONTEAGUDO, A., SANCHEZ-MONGE, E. & LACADENA, J.R. (1967). Trigo Hibrido, V. Ensayos de restauracion de la fertilidad en 1966. *Bol. Inst. Invest. Argon., Madrid* **27** (56), 79–81.

NETTEVICH, E.D. & NAUMOV, A.A. (1971). Influence of the Genotype of the Sterile Line on the degree of restoration of fertility in bread wheat forms with cytoplasmic male sterility (Russian). *From Plant Breeding Abstracts* **42,** No. 4531.

PUCKERIDGE, D.W. (1968). Competition for Light and its effect on Leaf and Spikelet Development of Wheat Plants. *Australian J. Agr. Res.* **19,** 191–201.

SAGE, G.C.M. (1972). The Inheritance of Fertility Restoration in Male Sterile Wheat Carrying Cytoplasm derived from *Triticum timopheevi. Theor. Appl. Gen.* **42** (6), 233–243.

SCHMIDT, J.W., JOHNSON, V.A., MORRIS, M.R. & MATTERN, P.J. (1971). Cytoplasmic Male Sterility and Fertility Restoration. *Seiken Ziho,* **22,** 113–118.

SHEBESKI, L.H. (1971). Hybrid Wheat Breeding.*Information Bulletin on the Near East Cereal Improvement and Production Project* **8,** No. 1, 1–7.

TSUNEWAKI, K. & ENDO, T. (1973). Genetic Relatedness among five cytoplasms in *Triticum* and *Aegilops. Proceedings of the Fourth International Wheat Genetic Symposium, Missouri* 391.

WELSH, J.R. & KLATT, A.R. (1971). Effects of temperature and photoperiod on spring wheat pollen viability. *Crop Science* **11** (6), 864–865.

WILSON, J.A. (1968) Hybrid Wheat Developments with *T. timopheevi* Zhuk. Derivatives. *Proceedings of the Third International Wheat Genetics Symposium, Canberra* 423–430.

Breeding of F_1-hybrids in flowers

R. REIMANN-PHILIPP

Institute for Horticultural Plant Breeding, Ahrensburg, FRG

SUMMARY

Flower breeders had already exploited "heterosis" before the phenomenon became of great interest in corn. When Shull in 1914 introduced the term "heterosis" the first true F_1-hybrid variety in flowers had already been released (*Begonia semperflorens* "Primadonna" by Benary [Erfurt]). However, heterosis breeding remained limited essentially to *Begonia semperflorens* for a long time. About 1950, the "know-how" of breeding F_1-hybrids in flowers became widely distributed. The expansion of F_1-hybrid production to other species (*Petunia, Tagetes, Antirrhinum*) depended upon the potential for artificial cross-pollination on a larger scale which only worked in such cases where emasculation could be replaced by any type of inherited male-sterility. Self-incompatibility, pollen sterility and male-sterility by deformation of the male sex have been applied and are discussed with examples of the flower species concerned and the practical results achieved. Besides the economic profit offered, F_1-hybrid production in flowers resulted also in some important findings on genetic principles of cross-pollination. For instance, from the F_1-hybrid production of self-pollinators such as *Tagetes* we learned that significant performance of F_1-hybrids does not necessarily presuppose inbreeding-depression of the parental lines or real "hybrid vigour" in the F_1 generation. Thus, the heterosis model deduced from corn and the explanation scheme based on quantitative genetics does not cover the type of F_1-hybrid performance which we have actually got in self-pollinating crops. A rather widely occurring "superdominance" of single "major-genes", which are composed of closely linked characters with both positive and negative influence on yield may present an explanation, demonstrated also by experimental results.

Recently the high expenses for manual labour have restricted to some extent F_1-hybrid production in flowers, and may lead to production methods which provide the simplest way of harvesting plentiful, good seed conceding some compromise with respect to uniformity in favour of lower seed prices. This aspect also is demonstrated by some experimental results.

INTRODUCTION

When Shull in 1914 introduced the term "heterosis" and started to propound F_1-hybrid-production in corn the first true F_1-hybrid variety in flowers had already been released, i.e. *Begonia semperflorens* "Primadonna", 1909, by Benary (Erfurt). Though subsequently Frimmel (Eisgrub) was rather successful in elaborating special techniques for applying "heterosis"-breeding to some further flower species, the procedure remained limited to *Begonia semperflorens* for a long time, because no other subject could offer the same facilities of large number of seeds per capsule and ease of emasculation. In 1928, the term "heterosis" in direct connection with the name of a variety was protected in Germany as a trade-mark for Benary (Erfurt). Since a great number of seeds per capsule as a precondition for F_1-hybrid-seed production occurs also in some other flower species as for example *Antirrhinum* or *Petunia,* the latter became the next subjects of breeding F_1-hybrid varieties. About 1950, the "know-how" of producing F_1-hybrids in flowers was widely distributed and was promoted chiefly by Japanese and North-American breeders, at the present time, but it does not nearly reach the limits which it might on principle achieve.

FACILITIES FOR ARTIFICIAL CROSS-POLLINATION ON A LARGE SCALE

Strangely enough those flower species, which seem obviously to be predestined for F_1-hybrid production because of cross-pollination systems such as heterostyly in *Primula,* were only recently subjected to heterosis. In *Petunia,* F_1-hybrid varieties are predominant, it is true, but the very well experienced system of gametophytically controlled self-incompatibility has not yet been used for practical purposes. On the other hand, there is in the seed-world a supply of F_1-hybrids even in some of those flower species which do not show any inherent facility for F_1-hybrid production, as for example in *Viola tricolor.* Formerly, those inherent principles which favoured cross-pollination were eliminated involuntarily by the breeders who consequently selected for self-fertility to simplify the breeding method. But nowadays, F_1-hybrid production is extended to a remarkably large number of flower species, particularly those, such as *Pelargonium,* which are usually propagated vegetatively. Vegetative propagation, however, in some cases ought to be replaced by generative propagation for phytosanitary reasons. F_1-hybrid production was also applied to some of those flower species, such as *Tagetes,* which are partly self-pollinating. In the same way several genetic systems providing self-sterility were introduced and made practical use of, as illustrated in the following sections.

1. Genes producing "pure females" by morphological deformation of the male parts of the flowers

1.1 Double flowers

Such a "feminisation" may be ascribed to the "double flower" character which transforms anthers into petals more or less completely. Problems may arise if a reduction of the female fertility (by partial transformation of the stigmas and styles into petals) occurs at the same time or if environmental influences reduce double-flowering. I have been able, however, to make use of the double-flower character in the production of rather good F_1-hybrids in carnation *(Dianthus caryophyllus)* types which are propagated by seeds. The male sterile, homozygous double-flowered parent is propagated by growing the material under poor nutritional conditions which reduce double flowering and repair fertility to some extent. This can also be done with other double-flowered ornamentals such as *Chrysanthemum.* As double flowering as a means of producing male-sterility will probably not be valuable for agricultural crops, however, I need not go into more details here.

1.2 *Mutants like "Femina" in Tagetes (Bolz, 1961) or "Cinderella" in Begonia (Reimann-Philipp & Seidel, 1963)*

Among several mutants of similar function, "Femina" and "Cinderella" represent a type of sex transformation which results in a complete absence of the male parts of the flowers ("Femina") or which produces sterile, misshapen style- or stigma-like anthers ("Cinderella"). Both mutants are easily used as the female parent in F_1-hybrid production and may encourage breeders of other crops to search for similar genotypes. Since these mutant genes are recessive, and the male parent provides the normal, dominant allele, the F_1-hybrid shows the normal fertile phenotype of the inflorescence, which is required. Because of its male-sterility the seed parent (=homozygous recessive) can only be propagated by backcross-lines of the type ♀ *cc* × ♂ *Cc* (or ♀ *cccc* × ♂ *Cccc* in autotetraploids). This requires an elimination of the 50% males which appear within the seed-parent in the production field.

2. *Self-incompatibility* (by "*S*"-alleles)

Although self-incompatibility is widely used today in vegetable-breeding (even in the more complicated form of sporophytic control, as in Brussels sprouts), the system has not been practised in *Petunia,* as has already been mentioned, though it operates there in the rather simpler form of gametophytic control. There is, however, an expanding application of self-incompatibility in flower crops such as *Ageratum houstonianum* and *Bellis perennis* of which the single plants can easily be propagated vegetatively (Reimann-Philipp, 1963). It is then rather simple to select self-incompatible individuals which show a good combining ability after free, reciprocal pollination. After vegetative multiplication the parts of the two clones which are all self-incompatible are planted in a mixed stand. Then, every seed harvested from any plant in the field represents an F_1-hybrid seed.

Good results can be achieved by application of this method. On the occasion of "IGA 1973" (= International Horticultural Exhibition Hamburg, 1973) our Hungarian colleagues won a gold medal for excellent *Ageratum*-F_1-hybrids which had most probably been produced by application of self-incompatibility. We, ourselves, won at the same competition a gold medal for an F_1-hybrid of *Heliotropium peruvianum* which had also been bred in this way.

3. Pollen sterility

In *Petunia* pollen sterility has been of great interest both in its practical application in the production of F_1-hybrids and also with respect to basic research on the genetics of the system. There was for a long time uncertainty in the literature, whether a cytoplasmic factor alone or an interaction of genetic and cytoplasmic influences determine pollen sterility. There was also doubt, whether it was possible to transmit sterility by grafting – which would lead to the postulation of a virus-like nature of sterility, – and concerning the extent to which pollen sterility could be influenced by environmental factors (Edwardson & Warmke, 1967; van Marrenwijk, 1969 and 1970; Izahar & Frankel, 1971, etc.). We may now start from the conception that there is a cytoplasmic-genetic base of pollen sterility and that a "restorer" gene exists which repairs fertility. However, because of an unfavorable dependence on environmental influences the pollen sterility in *Petunia* can break down at times. We may observe, therefore, that some well-known *Petunia*-F_1-hybrids which had formerly been bred from male-sterile seed parents now show pollen fertility, thus indicating that they are no longer produced from male-sterile mother-plants. Breeders obviously abandoned the use of male-sterility because it did not work satisfactorily.

Here, I should like to add that in flowers like *Petunia* the F_1-generation need not be fertile, as would be the case in crops of which the fruits are wanted. Since the pollen-sterility in *Petunia* is dominant in crosses with normal fertile varieties, presumably following the scheme ♀ *(S) ms ms* × ♂ *(N) ms ms* the introduction of a "restorer" gene is not necessary (Reimann-Philipp, 1964). On the other hand, we must find such a repair of fertility and introduce it into the breeding cycle if we wish to use a type of pollen sterility available in *Ageratum.* Here, the F_1-hybrids from crosses of the pollen sterile seed parent with normal fertile varieties are fertile as would result from the formula

♀ *(S) ms ms* × ♂ *(N) Ms Ms.*

In this case, we have to search for a genotype *(N) ms ms* (as we actually do) in order to breed "all-females" from

♀ *(S) ms ms* × ♂ *(N) ms ms*

and to use them as the seed parent.

Alternatively the pollen sterility in *Ageratum* may be inherited by a single recessive gene, thus:

♀ *ms ms* × ♂ *Ms ms.*

If the pollen parent is heterozygous, as expressed in the formula, the generative propagation of the pollen-sterile material takes place as indicated. Of course, for the production of F_1-hybrids a homozygous seed parent would have to be chosen. If this hypothesis was true, 50% of the plants of the seed parents (i. e. the male-fertile ones) would have to be removed, and we are at present actually proceeding so in the production of F_1-hybrids. However, if we identify the genotype *(N) ms ms* (mentioned above), thus demonstrating that a cytoplasmic factor is also operating, we would be able to provide a pure female seed parent and to produce the F_1-hybrids much more economically (Fig. 1/a–b).

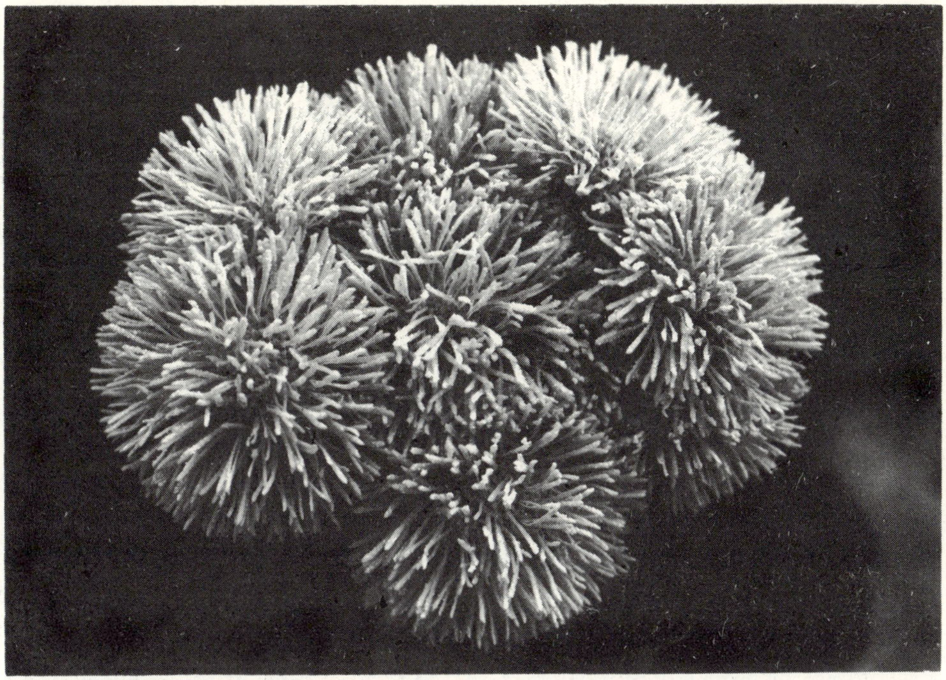

Fig. 1/a. Inflorescence of *Ageratum houstonianum* with normal fertile flowers (anthers densely covered with pollen).

Fig. 1/b. Flower of male-sterile genotype of *Ageratum houstonianum* (pollen entirely lacking).

A comparison of the frequency relations in *Petunia* and in *Ageratum* suggests that *Ms* is the normal (wild-type) allele in *Ageratum* but that *ms* should be regarded as the normal allele in *Petunia.* As it is difficult to believe that this is really so, some other new principles may have to be postulated determining pollen sterility in *Petunia,* in *Ageratum* or in both.

According to Edwardson and Warmke (1967), the cytoplasmic-genetic pollen sterility in *Petunia* may have originated from species-crosses. For *Begonia* (and possibly also to some extent for *Cyclamen*) Matouš (1969) demonstrated the derivation of various types of cytoplasmic-genetic male-sterility from species crosses. This may mean that Lamprecht's idea of "intraspecific genes" which cannot be transferred by crossing to another species (Lamprecht, 1964), or Schwemmle's (1941) similar results, which showed that there are specific genes which cannot express themselves in a foreign cytoplasm, could serve as a guide in selecting types of male-sterility in any other crop.

4. *Different genetic principles available to secure cross-pollination*

It is difficult to understand why in flower breeding the application of genetic principles guaranteeing cross-pollination is not more widely used considering that such excellent facilities are available, including a choice of different systems already mentioned for *Petunia* or *Ageratum.* Trying to explain this I would imagine that the genetic "know-how" in the application of such matters as genetic male-sterility is needed, but not available in the seed producing industry. Because of the relatively small sales of most of the flowers compared to agricultural crops, it is not economic in most cases to employ a graduate flower breeder. So, in ornamentals today many a fine facility is not utilized, and the simplest way of producing an F_1-hybrid is necessarily considered as the best one. For many a flower seed company this implies that not too much applied genetics can be undertaken even if some more manual labour has to be employed. Of course, this is not necessarily so in other crops! In flowers, however, there will not be a strong tendency for a further expansion of true, real F_1-hybrids. In my opinion, the further development of breeding methods will aim at varieties which can be produced as simply and as cheaply as possible. To achieve this, plentiful seed of high germination capacity and high genotypic value are expected to be produced from the seed parent, whereas some concessions with respect to uniformity would have to be admitted. Some ways to solve this problem are suggested in the following paragraphs.

SIMPLIFICATIONS IN THE PRODUCTION OF HYBRID VARIETIES

1. *F_2-hybrids*

As just mentioned, F_1-hybrid production in new species of flowers will presumably not show increasing progress. My statement should refer, however, only to "true" F_1-hybrids, that is to say to those hybrid varieties which are entirely composed of F_1-individuals derived from 100% of crosses between two inbred parents. Because of good facilities for vegetative propagation, there are rather good chances, in flowers, for breeding "synthetic" varieties, which result from more than two parents by polycross techniques. Up to date, however, little use has been made of this possibility in flowers. Another special method of a similar kind which has been successfully used in flower breeding and which could surely find interest also in agricultural crops, is "F_2-hybrids". While F_2-performance will not reach that of F_1, of course, it may in many cross-combinations exceed that of the parents and of normal inbred varieties. A second, rather important aspect of F_2-hybrid flower-breeding is that uniformity of the F_2 may not be so bad, provided that special care has been taken in selecting two inbred parents of reasonably similar growth habit, etc. or – particularly in flowers – of similar segregation for flower colours. In flowers F_2-generations sometimes proved to be so good that the F_1-hybrid did not obtain a protection against illegal reproduction of the variety in question, to obtain which was properly the intention of creating the F_1-hybrid.

It is unnecessary to say that the production of an F_2-hybrid (which represents a "by-product" of the F_1-hybrid concerned) is much less expensive than the production of masses of F_1-hybrid seed for sale, because only a few cross-pollinations to raise some few F_1-individuals must be performed.

2. *Improvements of the production techniques where "true" F_1-hybrids still have to be produced*

Begonia semperflorens (Link & Otto) may be regarded as a subject of the best imaginable fitness for breeding F_1-hybrids. This has been known for nearly 65 years, ever since the introduction of "Primadonna" (and proves true particularly in cases where heterosis and triploidy are – as usually – combined). Nevertheless, an improvement in the production technique saving manual labour such as emasculation and pollination will be possible, and in my opinion urgently necessary. This could be done by incorporating male-sterility (e.g. the "Cinderella" – character) into the 4n inbred lines (which serve as the seed parent) and selecting 1:1 segregating backcross-lines (of the type ♀ *cccc* × ♂ *Cccc*) in order to maintain and propagate the material. As was shown by Preil & Lorenz (1973), there is good evidence for the production of sufficient F_1-hybrid seed by artificial ventilation. Using a hairdrier, one person could easily blow the pollen from the diploid parent to the styles of the 4n male-sterile females. Further experiments are required to find the best proportion of paternal and maternal plants in the pollination group, the best spacing of the pots, the frequency of ventilation and the possibility of installing a fixed, permanent ventilation conduit. I believe, however, that such simplifications can already be regarded as quite useful.

3. *Selection for high fitness of heterozygotes*

Horticulturists study and handle single plants (particularly those in pots) to a much greater extent than agriculturists, and consequently become more familiar with the proportion of homozygous and heterozygous genotypes. Such observations, particularly in autotetraploids as *Cyclamen, Petunia* (*superbissima*-types) or *Viola tricolor,* often show that homozygotes for the characters in question occur less frequently than would be expected at random, or even show sublethal performance. The worst thing breeders can do – and sometimes actually do – is to exaggerate inbreeding in order to improve uniformity. If heterozygotes have high fitness values, inbreeding will not only lead to reduced vigour, but also fail to improve uniformity. Instead of inbreeding, those heterozygotes should be selected, which yield as few homozygotes as possible, because the homozygotes concerned are in any way sublethal: both vigour and a rather good uniformity would then be combined. For example in *Petunia* the *grandiflora*-locus exhibited a strong selective advantage in the heterozygous condition and this tendency was greatly increased by selection in diploids and in autotetraploids (Reimann-Philipp, 1962, 1963, 1964, 1968). As Table 1 demonstrates, nearly all homozygous genotypes eliminate themselves because of linkage to a sublethal gene (c = a chlorophyll-factor linked to G) or a lethal (l linked to g). So, I kept available for demonstration some material which was rather uniform for the *grandiflora* characters and highly vigorous because homozygotes were lacking (except those few which originated from crossing-over).

Originally, the linkage-group was composed of cGl/CgL in the *grandiflora*-varieties of *Petunia.* The uniform but heterozygous material could not therefore be produced before a cross-over attaching g to l (and vice versa G to L) had been selected.

As Table 1 also demonstrates, chlorophyll-mutants of a reduced vigour may serve to advance the fitness values of the respective heterozygote, whether they had arisen spontaneously or from artificial induction. If both of the inbred parents carried a chlorophyll-mutant in the genetic block which gives rise to heterosis as a heterozygote, then the chlorophyll-homozygotes would possibly eliminate themselves under normal spacing conditions in the field, so leaving a crop which is uniformly composed of heterozygotes. At the present time, we are experiencing such a balance in the production of F_1-hybrid monoecious spinach (Handke, 1973). Even if those chlorophyll-mutants were not directly linked to "combining ability" (i.e. segregating freely from "superdominance") they could be useful as "marker-genes" which allow an easy separation of offspring from inter- and intra-line pollination. I remember a discussion at the XIth International Congress of Genetics, Scheveningen 1963, where the application of chlorophyll-mutants as marker-genes was recommended for Brussels sprout F_1-hybrids.

4. *Fixation of heterosis*

When the model of breeding F_1-hybrids was transferred from maize to a good number of other crops (some of them self-pollinating) it became obvious that the theory of heterosis based on quantitative genetics, which might have explained rather well the situation in maize did not equally well cover the experimental results in those new fields of application. There is for instance in some cases no inbreeding effect or only a minimal depression even in extremely open-pollinated crops like dioecious spinach or insect-pollinated *Tagetes* though heterosis performance in the F_1-hybrids concerned is excellent. The phenomen of "combining-ability" was found to behave like a segregating Mendelian character, which apparently is composed of

Table 1. Fitness of "*grandiflora*" heterozygotes in diploid *Petunia*. Offspring from a selfed "*grandiflora*"-heterozygote *cGL/CgL* exhibiting strong selective disadvantage of the homozygotes due to linked lethals

		Normal gametes		Cross-over gametes	
		cGL	*Cgl*	*cGl*	*CgL*
Normal gametes	*cGL*	*cGL/cGL* large flowers bright green sublethal	*cGL/Cgl* large flowers green vigorous	*cGL/cGl* large flowers bright green sublethal	*cGL/CgL* large flowers green vigorous
	Cgl	*Cgl/cGL* large flowers green vigorous	*Cgl/Cgl* (small flowers) (dark green) lethal	*Cgl/cGl* (large flowers) (green) lethal	*Cgl/CgL* small flowers dark green vigorous
Cross-over gametes	*cGl*	*cGl/cGL* large flowers bright green sublethal	*cGl/Cgl* (large flowers) (green) lethal	*cGl/cGl* (large flowers) (bright green) lethal	*cGl/CgL* large flowers green vigorous
	CgL	*CgL/cGL* large flowers green vigorous	*CgL/Cgl* small flowers dark green vigorous	*CgL/cGl* large flowers green vigorous	*CgL/CgL* small flowers dark green vigorous

only a few "major genes" or even a single one (see e.g. Schnell & Hepting, 1971). As is well known, the substantiation and practical introduction of "early testing" resulted from this perception. Furthermore, if those major-genes were able to confer "heterosis" to the genotype concerned, they would no doubt have to express some degree of dominance. As to those "major genes", I have just demonstrated a quite typical example in the "*grandiflora*-locus" in *Petunia* (Table 1), where not only is the heterozygous condition extremely heterotic, but also where the homozygotes are to some extent not viable. While at first sight the *grandiflora*-locus seems to be monogenic, and, therefore, to represent a case of "superdominance", an intensive study revealed it to be composed of at least 3 different genes. If superdominance was understood as an operation of only two alleles, and heterosis as a result of superdominance, then we would have to conclude that there is no superdominance in the *grandiflora*-locus of *Petunia* because several closely linked genes showing normal dominance expression simulate a phenomenon of "monogenic heterosis". Such major-genes which appear to represent a complex of linked factors, however, are no exceptions. Many examples showing an extremely high heterozygote fitness value are known in the textbooks as systems of "balanced lethals". In a less superior but still predominant performance, the heterozygotes of such loci may produce the heterosis-effect of those cases which do not correspond to the postulations of quantitative genetics.

If this were true (as I believe), then several conclusions of practical interest would have to be drawn. Not only would the application of "early testing" find a further justification, but it would lead to the recommendation that fixation of the heterosis effect should be attempted! Returning once more to the content of Table 1, I would like to draw attention to the crossing-over events which are capable of explaining "superdominance": while the linkage of *C* and *l* or *g* and *L* has already been broken, a separation of *G* and *c* has not yet been realised. Of course, as long as there is no crossing-over we must take into consideration that it may never occur but that there is a real pleiotropic effect of the alleles concerned. I believe, however, that this will not prove true; for there are so many cases of "major genes" known in the literature which proved to be complex-characters, the linkage relations of which became apparent only after late crossing-over events occurred. Therefore, one day I expect to find separated too, the "large flower" and "bright green leaf" characters in *Petunia*. For another very close linkage in *Petunia,* which from a logical understanding was assumed to represent a case of pleiotropic function, such a rare crossing-over event revealed the complex nature of the factors concerned. This example was connected with the linkage of the width of the flower petals with width of the flower sepals. While normally a large flower is associated with a large calyx, in the course of breeding work individuals suddenly appeared with small flowers and broad sepals, and vice versa (Fig. 2).

Why, then – it may be asked, – have only very few examples of successful fixation of heterosis become known in recent years, and has the idea of "fixing" hybrid vigour been almost entirely abandoned? The explanation may be that "superdominance" in the sense of "monogenic heterosis" which had to be regarded as an essential part of "hybrid vigour" would render unimaginable any idea of a fixation. The conclusion,

Fig. 2. *Petunia hybrida* Vilm. Left: small flower with a calyx of small sepals of a normal "multiflora" variety. Middle and right: small flower with relatively very broad sepals. This genotype apparently originated from crossing-over.

however, that there is in many cases no real difference between "superdominance" and dominant genes located in linkage groups may bring forth new aspects, and consequently new breeding techniques for the selection of appropriate F_2-genotypes.

CONCLUSIONS

Flower-breeding has been concerned with "heterosis" from the very beginning of F_1-hybrid-production. A scientific evaluation of the wide practical work has at present scarcely been tried, however, as the relatively small field of flower-breeding did not allow much basical research-work to be done. If we tried to draw some retrospective conclusions, we would find some interesting results, which would possibly be worth imitation in agricultural plant breeding, particularly with respect to the following points:

The production "F_2-hybrids" might be considered – particularly in cases where the essential heterozygotes show strong fitness values.

They could possibly be advanced by selection of the respective homozygotes in a sublethal condition (e.g. by chlorophyll defects), based on the assumption that "superdominance" will prove in many cases not to be "monogenic" but to be determined by "major genes" which are composed of several closely linked factors.

Rare crossing-over events might then have a chance to be detected by an appropriate technique and, therefore, a fixation of F_1-hybrid vigour might also be attempted.

Several different genetical principles securing cross-pollination have been tested in flowers which perhaps could also be successfully applied in agricultural plant-breeding.

REFERENCES

BOLZ, G. (1961). Genetisch-züchterische Untersuchungen bei Tagetes. II. Herstellung, Genetik und Verwendung röntgeninduzierter Mutanten in der Züchtung. *Zeitschrift für Pflanzenzüchtung* **45,** 121–142.

EDWARDSON, I. R. & WARMKE, H. E. (1967). Fertility restoration in cytoplasmic male-sterile Petunia. *Heredity* **58,** 195–196.

FRIMMEL, F. von (1941). Die Bedeutung der Züchtung von Heterosissorten. *Zeitschrift für Pflanzenzüchtung* **23,** 638–660.

HANDKE, S. (1973). Jahresbericht 1973 der Bundesforschungsanstalt für gartenbauliche Pflanzenzüchtung, Ahrensburg. *Forschung im Geschäftsbereich des Bundesministeriums für Ernährung, Landwirtschaft und Forsten, Teil D.*

IZHAR, Sh. & FRANKEL, R. (1971). Mechanism of male sterility in *Petunia:* The Relationship between pH, Callase Activity in the Anthers, and the Breakdown of the Microsporogenesis. *Theor. and Appl. Genetics* **41,** 104–108.

LAMPRECHT, H. (1964). Species concept and the origin of species. The two categories of genes, intra- and interspecific ones. A simplified presentation. *Agri Hortique Genetica* **XXII,** 272–280.

LAMPRECHT, H. (1968). Wesen und Werdegang naturbedingter Arten. *Phyton (Austria)* **13,** 1–14.

MARREWIJK v., G. A. M. (1969). Cytoplasmic male sterility in Petunia: I. Restoration of fertility with special reference to the influence of environment. *Euphytica* **18,** 1–20.

MARREWIJK v., G. A. M. (1970). Cytoplasmic male sterility in Petunia: II. Discussion on male sterility transmission by means of grafting. *Euphytica* **19,** 25–32.

MATOUŠ, J. (1962). Über die Möglichkeit der Ausnutzung von Heterosissorten und die Frage der Pollensterilität bei *Cyclamen persicum* Mill. *Acta Pruhoniciana* **4,** 13–51.

MATOUŠ, J. (1969). Die cytoplasmatische männliche Sterilität bei *Begonia semperflorens* hort. und Möglichkeiten ihrer Ausnutzung in der Züchtung. *Acta Pruhoniciana* **19,** 1–45.

PREIL, W. & LORENZ, A. (1973). Herstellung von triploidem F_1-Hybrid-Saatgut durch „künstliche" Windbestäubung bei Verwendung von männlich sterilen, tetraploiden Inzuchtlinien als mütterlichem Kreuzungspartner bei *Begonia semperflorens. Vortrag anläßlich des Treffens der Eucarpia, Sekt.* „Zierpflanzen", Arbeitsgruppe *Begonia semperflorens* und *Ageratum houstonianum,* in der BFA in Ahrensburg am 19.7.1973.

REIMANN-PHILIPP, R. (1962). Untersuchungen über die Vererbung des grandiflora-Merkmales bei *Petunia hybrida Vilm. Zeitschrift für Pflanzenzüchtung* **48,** 143–176.

REIMANN-PHILIPP, R. (1963). The application of incompatibility in plant breeding. *Genetics Today,* **3,** 649–656.

REIMANN-PHILIPP, R. (1964). Entwicklungsarbeiten zur Züchtung von F_1-Hybriden bei Blumen. *Zeitschrift für Pflanzenzüchtung* **51,** 249–314.

REIMANN-PHILIPP, R. (1968). Untersuchungen über die Vererbung des grandiflora-Merkmals bei Petunia hybrida Vilm. II. Die Verwendung tetraploider Linien der grandiflora-Klasse für die Züchtung neuer Formen der superbissima-Klasse, insbesondere solcher mit gelber Blütenfarbe. *TAG,* **38,** 58–65.

REIMANN-PHILIPP, R. & SEIDEL, H. (1963). Untersuchungen bei *Begonia semperflorens* Link et Otto über die Eignung des Merkmals der Blütenfüllung und des Merkmals der Antherendeformation der „Cinderella"-Sorten für ihre Verwendbarkeit als männlich steriler Kreuzungspartner in der Züchtung von F_1-Hybriden. *Zeitschrift für Pflanzenzüchtung* **50,** 59–70.

SCHNELL, F. W. & HEPTING, L. (1971). Segregation for combining ability in early generations of maize. *Proceedings of the sixth Meeting of the Maize and Sorghum Section of EUCARPIA, Freising-Weihenstephan 14.–17.9.1971.*

SCHWEMMLE (1941). Weitere Untersuchungen an Euoenotheren über die Bedeutung des Plasmas und der Plastiden. *Zeitschrift Abst. u. Vererbgl.* **79,** 321.

SHULL, G. H. (1914). The genotype of maize. *Amer. Naturalist* **45,** 234–252.

Heterosis after diversifying selection

P. M. A. TIGERSTEDT

Department of Plant Breeding, University of Helsinki, Helsinki, Finland

SUMMARY

In the reported study some breeding strategies were simulated using *Drosophila melanogaster* populations. The aim of this study was to follow changes in the genetic structure of populations after diversifying selection and to evaluate its effect on population variability under variable ecological conditions. Changes in pheno- and genotypic variance of pronounced fitness traits, including productivity and competitive ability, were carefully scrutinized and some conclusions regarding heterosis were drawn on the basis of studies on adaptive polymorphisms by Dobzhansky. Presented results indicate clearly that disruptive selection has improved population fitness.

Results are discussed in the light of observations on ecological tolerance reported by Beardmore. It is concluded that disruptive selection in plant breeding programs is advantageous from two different aspects: 1. disruptive selection increases ecological tolerance; 2. disruptive selection increases euheterosis due to chromosomal heterozygosity.

The use of diversifying selection in plant breeding programs is discussed and some breeding strategies are suggested for predominantly inbreeding and predominantly outbreeding species. Particularly the CIMMYT programs on small grain are scrutinized.

A program called "divergent-convergent-selection" (Lonnquist, pers. comm.) is presented for predominantly outbreeding species. This appears to be the first plant breeding program that deliberately aims at higher ecological tolerance.

DIVERSIFYING SELECTION UNDER NATURAL CONDITIONS

The evolutionary significance of disruptive selection has been subject to much investigation and argument during the last ten years. The theory of diversifying or disruptive selection actually contains one of the most controversial questions of modern evolutionary theory, namely that of conditions for speciation. Essentially the argument centres on the question of whether incipient speciation can occur during sympatric conditions or whether allopatry is a prerequisite for speciation.

Experiments on plants have been largely neglected in the discussions but recent studies have shown that genetic divergence in adjacent plant populations can occur even in populations which are practically sympatric. In *Anthoxanthum* and *Agrostis* for example tolerance to heavy metal contamination in soils can take place even if patches of normal and heavy metal soils are no more than some tens of centimeters apart. Genetically such plant populations are essentially sympatric with gene flow exceeding 50% between the patches (see e.g. Jain & Bradshaw, 1966).

The heterogeneous or patchy environment that a species encounters in nature must lead to disruptive selection. The pollen of a given plant, at least in all outbreeding species, is distributed around its source of release in a leptocurtic fashion. Essen-

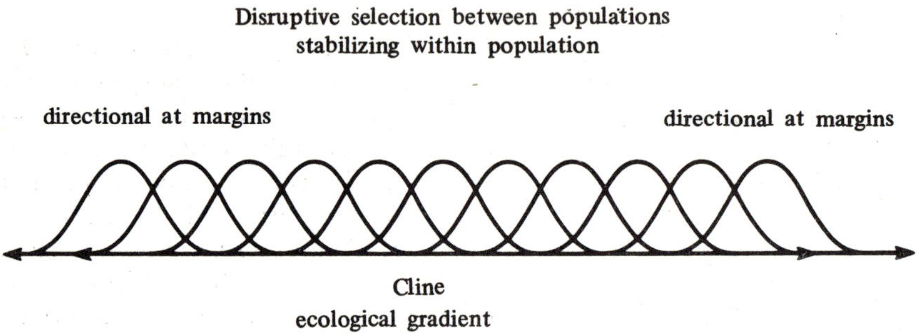

Fig. 1. Selection along clines.

tially the same must apply to predominantly inbreeding species which show occasional bursts of outbreeding. This causes "neighbouring effects" and subsequent strong local population divergence. Along an ecological cline this would lead to disruptive selection between adjacent populations and stabilizing selection within any population on the cline. Only at the extreme ends of a species distribution (under marginal conditions) would one expect to find directional selection (Fig. 1).

Diversifying selection may or may not cause incipient speciation. However, there is abundant confirmation that it causes gradual development of polymorphism (Ford, 1964). Diversifying selection has been shown both empirically and theoretically to cause chromosomal polymorphism. Thus it seems paradoxical to conclude that diversifying as well as its counterpart stabilizing selection both cause heterogeneity and polymorphism in populations. Stabilizing selection causes population polymorphism on a genic level through overdominance or "one-gene-heterosis". Disruptive selection acts on larger gene complexes and causes polymorphism on a higher, supergene level. This too may often be advantageous to the population and so balanced polymorphism is maintained at both levels of gene organization. It must be emphasized that one mode of selection does not necessarily exclude the other; both operate simultaneously under natural conditions but the latter, diversifying or disruptive selection, has been largely neglected in artificial plant selection programmes. The reader is referred to the classical works on adaptive polymorphism by Dobzhansky (1951) for further information.

Ecological tolerance of cultivars has been given little attention in plant breeding. However, recent large-scale disease outbreaks and also, the establishment of wide international breeding programmes have focused interest in breeding for widely adapted cultivars. Very little experience is however available from agricultural experiments and breeding programmes.

From a theoretical point of view, small and unpredictable variations in environmental conditions (fine grained) tend to select for individual tolerance through balanced heterozygosity on the genic level. Large regular environmental variations (coarse grained) tend to select for population tolerance. This is disruptive selection. What is the effect of this mode of selection on population tolerance, fitness, etc? The author has undertaken some selection experiments using *Drosophila melanogaster* as a model organism (Tigerstedt, 1969). A situation was simulated where there was random mating between disrupted parts of the population in each generation. Three such disrupted lines were selected in this manner, starting from a highly heterogeneous base population. To be sure that no inbreeding would affect the results, populations and selected fractions were kept sufficiently large in each generation. Populations averaged 6,000 individuals per generation and of these 4% of the slowest and fastest developing respectively were selected in each generation and allowed to mate at random. The development of fitness, given as progeny per female *($\bar{x}$)*, mean rate of development *($\bar{y}$)* and population phenotypic variance (σ^2_p) are given in Fig . 2.

The results of this experiment show that fitness almost trebled over 50–60 generations of disruptive selection. Evidently all three populations were also attaining higher developmental stability as the phenotypic variance kept decreasing. A slight shortening of the average rate of development (faster development, faster growth) must here be a sign of heterotic effects. Some pilot experiments on the exponential growth of populations (Fig. 3) showed that disruptively selected populations were generally superior to unselected or directionally selected populations. Experiments on competitive ability which were extensively carried out in De Wit's (1960) arrangements showed superiority of disruptive lines. Results clearly indicated that disruptively selected lines showed general superiority of fitness and growth; the lines were euheterotic sensu stricto. Later experiments with the same lines (Namvar, 1972) have confirmed these conclusions. Analysis of controlled crosses within and between slow and fast population fractions indeed indicated that two different (slow and fast) chromosome complexes were building up. Increased genetic variance in disruptive lines did not, however, increase phenotypic variance in the populations; on the contrary, disruptively selected lines showed decreased phenotypic variance probably due to better population buffering,i.e. higher tolerance!

The existence of two kinds of heterozygosity, genic and chromosomal, is of course well known. Both kinds of genetic heterogeneity are sustained in populations but the underlying selection pressure is different in each case. In particular recent investigations in predominantly self-pollinating species have shown that much heterozygosity can be preserved even after many generations of selfing (Allard & Kahler, 1972). If bulk selection is applied under heterogeneous environmental conditions, this means

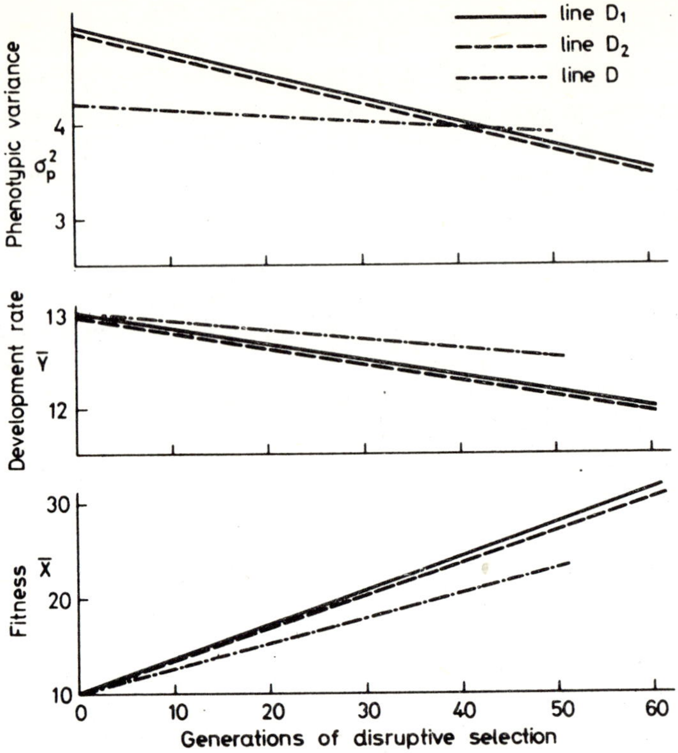

Fig. 2. Outcome of disruptive selection in *Drosophila melanogaster* (Tigerstedt, 1969).

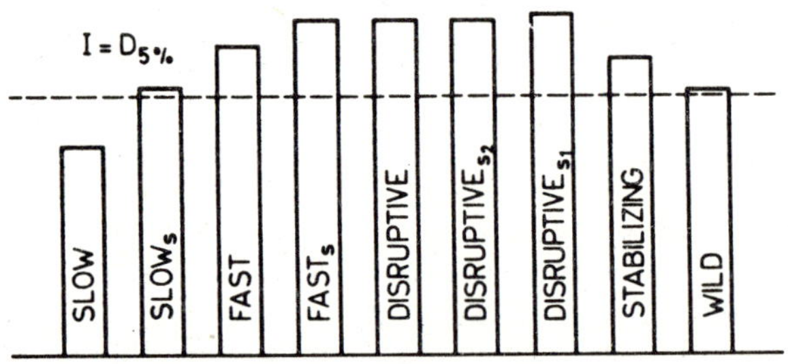

Fig. 3. Population expansion under unrestricted conditions relative to "white" control (Tigerstedt, 1969).

that in the long run, segmental heterozygotes can arise – a situation very similar to balanced polymorphism in animals or predominantly outbreeding plants. It must be emphasized again, that the difference between outbreeding and inbreeding species is one of degree, both genetic systems tending to maintain heterozygosity and genetic diversity under heterogeneous (patchy) environmental conditions. This situation has not previously been realized and breeders have thought of the two genetic systems as basically different. In predominantly inbreeding species genetic diversity is of course mainly expressed as differences between lines (population buffering) although even here investigations on isoenzymes indicate higher than expected individual heterozygosity (individual buffering).

In discussing relations between ecological heterogeneity, genetic variability and fitness in *Drosophila* populations, Beardmore (1970) has found that under variable temperature conditions populations tend to retain high amounts of genic or chromosomal heterogeneity. I quote: "Under all temperature conditions the more productive populations tend also to be the least variable in this component of fitness. As more productive populations are also genetically richer, an inverse relationship between the homeostatically significant property of productivity variability and that of genetic variability is indicated." Again, this is euheterosis *sensu stricto*.

DIVERSIFYING SELECTION IN PLANT BREEDING PROGRAMMES

Orthodox plant breeding methods take advantage of stabilizing selection during bulk selection periods subsequent to F_2 transgression. It is a phase of breeding very much akin to the negative part of Waddington's normalizing selection – sieving the population for phenodeviants. Ultimately bulk selection aims at higher fitness of the population under prevailing conditions. If the conditions under which bulk selection is practised are rather uniform from year to year and also from site to site this then becomes a simple example of selection for intermediate fitness optima. This includes fitness components like plant hardiness, gross disease resistance and competitive ability. If this bulk phase is continued for a number of generations, selection may become "canalizing" aiming at greater phenotypic stability. However, under very uniform modern agricultural conditions this might cause rather low population tolerance! This must be the case particularly if climatic conditions are also quite regular from year to year. Geneticists would here rather talk about "laboratory environments" and no doubt modern agricultural conditions resemble such environments to some extent. This must be a reason for the drastic loss of genetic variation during the initial phases of breeding. An effective method for increasing population tolerance and euheterosis generally would evidently be that bulk populations be replicated over different sites intended for cultivation of the crop in question. Unfortunately this is seldom the case, and much bulk selection is carried out within narrow limits of ecological variability. In addition, modern agricultural methods, including heavy fertilizing, must cause bulk populations to adapt to extremely favourable nutritional conditions. Such adapted populations would then become sensitive to small variations in nutritional levels. In the long run this must be harmful as farmers cannot maintain such uniformity over long periods. It seems then, that bulk selection should

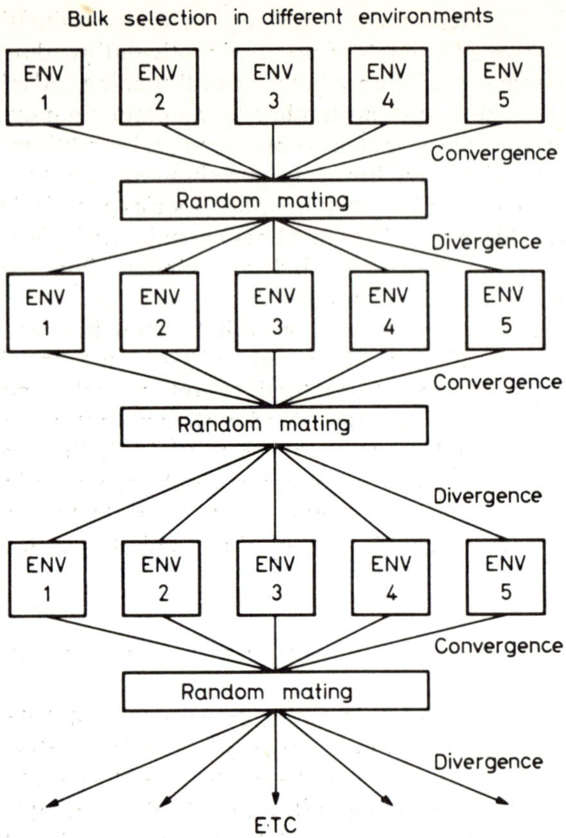

Fig. 4. Diversifying selection in *Zea mays*. "Convergence – divergence selection" (after Lonnquist, 1972, personal communication).

be practised over rather diverse ecological conditions to ensure that natural selection renders a high amount of genetic diversity and subsequent tolerance to the populations. Genetically, it means that as well as normalizing selection we utilize diversifying selection to increase the ecological tolerance of the cultivar, to increase euheterosis!

Evidently the highly tolerant Mexican wheat cultivars that came out of multiline mixes or crosses bulk selected under variable environmental conditions (CIMMYT bulk fields in Obregon and Toluca) are the only practical examples of true disruptive selection in plant breeding so far. There is much interest in developing cultivars with good phenotypic buffering properties at the present time, due to the disastrous consequences of poor tolerance recorded in highly specialized and genetically uniform plant populations. The most warning example was the corn leafblight epidemic in the U.S. in 1970, which evidently was caused by the use of a single source of cytoplasm in developing a major portion of the corn hybrids (*T* male sterility-inducing cytoplasm) (Nas, 1972).

Recently Lonnquist (personal communication,1972) has started a selection program on corn in Wisconsin which he calls "convergence-divergence selection" (Fig. 4). This is in essence a programme of diversifying selection and random mating between fractions. Fractions are selected naturally (or artificially) in different locations and random mating is arranged by constructing a central "pollination field" with equal parts from each location. Actually Lonnquist's programme is more sophisticated than described here because he deliberately restricts stabilizing selection. This is a programme very similar to the one described for *Drosophila* above with the difference that there are 5 fractions instead of two. As a result one should end up with a line with high adaption to all five locations.

CONCLUSIONS

Euheterosis can depend on genic or chromosomal polymorphism. Genic heterozygosity and overdominance depend on stabilizing selection for intermediate optima and single gene heterosis. Chromosomal heterozygosity depends on diversifying selection and this process is a major factor in regulating ecological tolerance. Plant breeding has largely neglected this mode of selection. By using diversifying selection in plant breeding programmes true euheterosis can be achieved. Examples from *Drosophila* experiments have shown:

- that disruptively selected lines show better fertility
- that disruptively selected lines have higher competitive abilities
- that there is better phenotypic stability in disruptively selected lines
- that there is an inverse relationship between developmental homeostasis and genetic variability as a result of disruptive selection.

Plant breeders should recognize the importance of diversifying selection and introduce it into breeding programmes in order to obtain higher ecological tolerance and true euheterosis.

REFERENCES

ALLARD, R. W. & KAHLER, A. L. (1972). Patterns of molecular variation in plant populations. In: Proc. 6th Berkeley Symp. Math. Stat. and Prob. Vol. V. Darwinian, neo-darwinian and non-darwinian evolution. Univ. Cal. Press. 237–254.

BEARDMORE, J. (1970). Ecological factors and the variability of gene-pools in *Drosophila.* In: Essays in Evolution and Genetics in Honor of Theodosius Dobzhansky (M. K. Hecht & W. C. Steere eds). North Holland Publ. Co. Amsterdam. 299–314.

DE WIT, C.T. (1960). On competition. Versl. landbouwk. Onderz. Ned. 66, 8.

DOBZHANSKY, Th. (1951). Genetics and the Origin of Species. Columbia Univ. Press.

FORD, E.B. (1964). Ecological Genetics. Methuen, London.

JAIN, S. K. & BRADSHAW, A. D. (1966). Evolutionary divergence in adjacent plant populations L. The evidence and its theoretical analysis. *Heredity,* **21,** 407–441.

LONNQUIST, J. H. (1972). Area improvement in corn. Stenciled mimeogr. Univ. of Wisc.

NAMVAR, K. (1971). Versuche über den Aufbau einer genetischen Isolierungsbarriere durch disruptive Auslese bei *Drosophila melanogaster.* Diss. Forstl. Fak. Univ., Göttingen. 68.

NAS, (1972). National Academy of Science, Committee on Genetic Vulnerability of Major Crops. Working group report. Wash. D.C.

TIGERSTEDT, P. M. A. (1969). Experiments on selection for developmental rate in *Drosophila melanogaster. Ann. Acad. Sci. Fenn.* Ser. A. **58.**

Some progress in breeding for restoring ability in timopheevi-cytoplasmic-genic interaction systems of male-sterility-fertility restoration*

W. ODENBACH

Institut für Vererbungs- und Züchtungsforschung
Berlin–Dahlem, FRG

SUMMARY

1.There were some indications for defining our location in Berlin as a "sterile environment" from an analysis of F_2 segregants of the cross tcms Bison × Nebraska-restorer and from a restoration ability test of the restorer genes from the varieties Primepi, Palmaress and Minister, which was carried out at several locations in different regions of Germany.

2.There was no restorer source, which we could accept as a well acting male sterile-restorer system.

3.In order to test for good F_1 restoration the information about the localization of seed set in the ear is very valuable, i. e. the fertility in the upper quarter of the ear.

4.Apart from genotype-environmental fertility interactions the genetical background in a restored genotype plays a role in restoration expressivity. Reduced restoration was found in crosses with tcms Moisson, tcms Novi Sàd 171 – 1 – 3, tcms Novi Sad 737, tcms Novi Sad 440, tcms Ranaja 12 and tcms Tenor as the female parents.

5.A good restoration ability was observed in selections from crosses between the variety Primépi and CIMMYT restorers. F_1 hybrids from tcms *aestivum* × F_2-Primépi/CIMMYT combinations showed a high proportion of plants with anther dehiscense up to the top of the ear. The segregation analysis suggests the hypothesis that in this material the restorer genes on chromosomes 6 B or 7 D and on 5 D possibly do not influence male fertility in the heterozygotes. A trifactorial model of segregation is discussed including restorer genes on chromosomes 1 A, 1 B and 6 B or 7 D.

Since the late sixties, when we started hybrid wheat research in Berlin, it has been known that a different fertility behaviour of restored genotypes is found when they are cultivated in variable environments. Based on this information Wilson (1967) proposed a hypothetical scheme of genetic-environmental fertility interactions. From this one can define a particular environment as highly sterile, sterile, or partially sterile by analyzing the restoration ability of a set of genotypes containing restorer genes in increasing dosage. Monogenic restoration is reported from the

* This research was supported by a grant from GFP.

varieties Primépi, Palmaress and Minister (Oehler & Ingold, 1966; Goujon & Ingold 1967; Zeven, 1967).

Our objectives inbreeding for restoring ability were to test the environmental fertility interactions in Berlin and other places in order to find a location with a high selection pressure against restoration ability and at the same time to obtain some information about the restoration capability of different known restorer gene sources and selections from crosses between them.

TEST FOR ENVIRONMENTAL FERTILITY INTERACTIONS

There were some indications for defining our location in Berlin as a "sterile environment" from an analysis of F_2 segregants of the cross tcms Bison × Nebraska-restorer in 1969. Expecting a 15:1 ratio we found a higher proportion of male-sterile segregants (Odenbach, 1970). An F_3-generation analysis in 1970 of selected F_2-plants showing different patterns of ear fertility confirmed our classification of Berlin as sterile environment (Table 1). Although in 1970 the proportion of male sterile segregants from the double heterozygous genotypes was less than in 1969 we did not find the expected 15:1 ratio. Monogenic heterozygous F_2-plants showed the expected 1:3 and 3:1 segregations. These data are in agreement with the results of Schmidt (1964) in showing that restoration ability is not only a question of gene dosage but also of gene action. The gene with the lower restoration capability (B) could not induce male fertility in our test in the heterozygous stage (Table 2).

Table 1. Comparison of ear fertility in F_2- and F_3-populations from the cross tcms Bison × Nebraska-restorer (Berlin 1969, 1970)

Segregating genotype [%/o fertility]*	Year/ generation	Segregation				P
		sterile	partially fertile	fertile	ratio	
AaBb [47]	1969/F_2	25	49	4	5:10:1	0·90
AaBb [67]	1970/F_3	10	52	7	3:12:1	0·31
Aabb [42]	1970/F_3	8	28		1:3	0·70
AABb [109]	1970/F_3		57	18	3:1	0·84

* 100%/o = 2 kernels/spikelet.

Table 2. Classification of F_2 genotypes from the cross tcms Bison × Nebraska-restorer (Berlin, 1969, 1970)

Fertile	Partially fertile		Sterile	
1 *AABB*	2 *AABb*	1 *AAbb*		2 *aaBb*
	2 *AaBB*		2 *Aabb*	1 *aabb*
			1 *aaBB*	

Critical chromosomes and expressivity of Rf-genes (after Maan, 1973) $A = 1A > B = 7D$.

In 1969 apparently quite a high proportion of *Aabb* and/or *aaBB* F_2-genotypes were male sterile. In both years only the double dominant homozygous genotype (*AABB*) showed complete fertility. The analysis of restorer genes and their expressivity (Bahl & Maan, 1973) showed that the genes were located on chromosomes 1 *A* and 7 *D*. Our gene (A) is identical with the one on chromosome 1 A.

In 1972 we tested the restoration ability of restorer genes from the varieties Primépi, Palmaress and Minister in the heterozygous stage. These genotypes were crossed with two male-sterile populations, tcms Format and tcms Habicht, respectively. The test was carried out at several locations in different regions of Germany. In Table 3 the data are given for six locations, summarized for all F_1-combinations. There are significant differences between several locations in mean fertility as well as in the proportion of ears showing male fertility in the upper quarter of the ear. Only 24.8% of the analyzed ears from Berlin showed seed set in this region of the ears. In Cologne and in Schöningen we observed less F_1 tip sterility. These data confirmed our opinion that the genetic-environmental fertility interactions at our station favour the search for restorer genes and gene combinations in which fertility restoration has to be independent of variation in environments: one of the conditions which should be fulfilled for the practical use of male sterile-restorer systems (Ingold, 1967).

Table 3. F_1-restoration ability test at 6 locations (1972) (tcms Format resp. tcms Habicht crossed with Primépi, Primépi-Rest. Sel., Palmaress and Minister

	F_1 fertility [%]	F_1 Kernels/ear	F_1 (**) 0.75–1.0[%]	Isolated ears
Berlin	54.9	23.6	24.8	105
Seligenstadt	57.7	26.0	22.3	202
Cologne	82.1	37.6	44.4	144
Herford	55.5	24.3	20.7	140
Schöningen	73.3	29.4	46.6	176
Schladen	55.7	24.1	29.6	206

F_1-fertility L.S.D. $P_{5\%} = 12.8$ F_1 0.75–1.0 L.S.D. $P_{5\%} = 18.2$.
(**) Proportion of ears fertile in the upper quarter of the ear.

RESTORATION ABILITY OF DIFFERENT RESTORER SOURCES

One of the main problems in restorer breeding is the search for genotypes which induce complete male fertility in the F_1-hybrids. There was no restorer source, tested in the last four years, which we could accept as a well acting tcms male sterile-restorer system. Even such genotypes carrying restorer genes on two or three chromosomes showed a high proportion of tip sterility. There were remarkable differences in F_1 fertility. Some data from these tests are given in Table 4.

In segregation tests with the restorer gene from the French variety Primépi we found in most of the analyzed backcross generations a 1:1 ratio ($P = 0.40$). Only in the crosses including the German variety Habicht was the 1:1 ratio not confirmed. The mean fertility of the F_1's had not differed significantly. From these observations one can guess that in addition to the restorer gene action there are also interactions with the genetical background which influence the expressivity of restoration.

Table 4. Restoration ability of different restorer genes
(Berlin, 1973)

♂	♀ male-sterile	Critical chromosomes*	F_1-fertility [%]	F_1 Kernels/ear	Isolated ears
Primépi	Champlein	1B, 5D	69.9	15–36	9
T.tim/Goldcoin-Sel. 2	Bezostaja 1	(1A, 7D ~)	75.3	19–40	18
R_2/Son 64/Wis. 255	St. 223	1A, 6B, 7D	68,6	18–43	20
T. zhuk/Justin[3]	Maris Widgeon	1A, 7B, 7D	50.8	6–32	22
Cimmyt-2	Farino	1A, 6B, 7D	73.0	20–55	15
T. spelta duham.	Maris Widgeon	1B	56	13–32	23
Maris Beacon	Pantus	–	72	20–41	18

*After Maan, 1973
100% fertility = 2 kernels/spikelet

Table 5. F_1-restoration ability test of the varieties Primépi, Palmaress, Minister and one Primépi-restorer selection*
(1972, 6 locations)

♂	♀ male-sterile	F_1fertility [%]	F_1 Kernels/ear	F_1(**) 0.75–1.0 [%]	Isolated ears
Primépi	tcms Format	74.2	31.3	48.0	123
Primépi-Rest. Sel.	tcms Format	75.2	31.0	44.6	157
Palmaress	tcms Format	37.8	16.8	2.0	153
Minister	tcms Format	60.7	20.8	10.7	150
Primépi-Rest. Sel.	tcms Habicht	75.2	31.4	74.3	136
Palmaress	tcms Habicht	54.9	25.3	7.9	131
Minister	tcms Habicht	68.4	30.6	35.0	123

F_1-fertility L.S.D. $P_{5\%} = 13.8\%$
$F_1$0.75–1.0 L.S.D $P_{5\%} = 19.6$

*Selected by Oehler (67/418–6)
(**)proportion of ears fertile in the upper quarter

In the 1972 F_1-restoration ability test we found a significant difference in mean F_1 fertility in crosses between two different male sterile genotypes and the variety Palmaress (Table 5). No differences were found in crosses with a Primépi-restorer selection and Minister respectively. In classification of restored genotypes we examined the fertile regions of the spikes, the mean fertility of the spikelets and the mean ear fertility. In order to test for good F_1 restoration the information about the localization of seed set in the ear is very valuable. If one compares the fertile regions of the seven F_1 hybrids in Table 5 more marked differences are to be found than by analyzing the mean F_1 fertility. The character, fertility in the upper quarter of the ear, showed significant differences in the F_1 hybrids of Primépi-restorer selection and Minister with the two male sterile genotypes tcms Format and tcms Habicht respectively. In general these observations are strengthened by data obtained from the further eight locations.

Apart from genotype-environmental fertility interactions the genetical background in a restored genotype plays a role in restoration expressivity. The case of reduced restoration after crossing tcms Moisson with Primépi compared with other F_1 hybrids carrying the Primépi restorer gene indicates the influence of genetic interactions as already mentioned by Goujon & Ingold (1967). From the results of a series of testcrosses with different restorer sources we could classify some of our male sterile lines into two groups. We found better restoration effects in crosses with tcms Bezostaja 1, tcms Carstens VIII, tcms Champlein, tcms Dunav, tcms Heine VII and tcms Skorospelka as the male sterile parent. Combinations with tcms Moisson, tcms Novi Sad 171 – 1 – 3, tcms Novi Sad 737, tcms Novi Sad 440, tcms Ranaja 12 and tcms Tenor showed less restoration. The F_1 hybrids with tcms Format as the female parent lie somewhere in the middle of these two groups.

These facts indicated that selection work for a universal restorer has to take these genetical interactions into account, i.e. for testcrosses one should use not easily restorable male-sterile lines, testing the F_1 generation in a steril environment.

RESTORATION ABILITY OF DIFFERENT RESTORER GENE COMBINATIONS

Since no testcross with known restorer sources showed complete F_1 fertility we tried to get better results from combinations of several restorer sources. In 1971 we observed some apparently successful populations in which we had combined on the basis of *T. timopheevi* cytoplasm the restorer genes from the varieties Primépi and Palmaress with restorers we received from CIMMYT, Stillwater (Oklahoma R 92 – 25) and Fargo (69 B, 11-). Especially in the tcms Primépi – CIMMYT combinations, a high proportion of F_1 plants showed anther dehiscense up to the top of the ear. For a preliminary test we produced F_1 hybrids by crossing tcms Tenor and tcms Ranaja

Table 6. F_1-restoration ability test of different F_2-restorer combinations (1973, 13 locations).

♂	♀	F_1-fertility [%]	F_1 0.75/1.0 [%]*	Isolated ears
F_2-Prim/zhuk/Just^3	tcms Tenor/Dunav	50.2	55.1	279
F_2-Prim/CIMMYT-R	tcms Ranaja 12	88.8	74.2	310
F_2-Prim/Oklahoma R	tcms Ranaja 12	66.4	61.2	350
F_2-Prim/Oklahoma R	tcms Skorospelka	81.2	70.8	386

100% fertility = 2 kernels/spikelet
*proportion of ears fertile in the upper quarter of the ear

Table 7. F_1-restoration ability test. Proportion of male sterile segregants (1973, data from 6 locations)

Cross	Expressivity of critical chromosomes (after Maan, 73)	Segregation model	*n*	Male sterile obs.	Male sterile exp.	χ^2	*P*
tcms × F_2-Prim /zhuk/Just. 3	1A = 7B > 7D 1B (5D)	trifactorial*	126	35	31	0.68	0.43
tcms × F_2-Prim /CIMMYT-R.	1A > 6B = 7D 1B (5D)	trifactorial*	136	29	33	0.64	0.44

*factors on 5 D and 7 D do not induce ♂-fertility in the heterozygous stage.

12 which seemed to be less restorable, and tcms Dunav and tcms Skorospelka with F_2 populations of these restorer combinations (Table 6). The F_1 test was carried out at 13 locations spread over Germany. As expected the F_1 tcms Ranaja 12 × F_2 Primépi / Oklahoma restorer showed less restoration than the cross based on tcms Skorospelka. The highest mean F_1 fertility as well as the highest proportion of plants with fertile ears in the upper quarter were observed in the F_1 tcms Ranaja 12 × F_2 – Primépi/CIMMYT restorer. In accordance with the results from Bahl & Maan (1973) we tried to get an idea about the segregation of restorer genes in this material. The compiled data from six locations indicate a trifactorial segregation (Table 7). It seems that the restorer genes on chromosomes 7 D and 5 D do not induce male fertility in the heterozygous stage in F_1 tcms × F_2-Primépi/zhuk. Justin[3], the combination with the restorer from Stillwater. The segregation analyses of the cross tcms × F_2-Primépi/CIMMYT-restorer suggest the hypothesis that in this material the restorer genes on chromosomes 6 B or 7 D and on 5 D possibly do not influence male fertility in the heterozygotes. If this is true we have to take into account a three-gene restoration in offspring of these crosses.

Selection work is going on in this material and this year we are producing new testcross material in a larger scale with both groups of male-sterile lines mentioned above. These F_1 testcrosses will be grown under different environmental conditions.

REFERENCES

BAHL, P.N. & MAAN, S.S. (1973). Chromosomal location of male fertility restoring genes in six lines of common wheat. *Crop Sci.* **13,** 317–320.

GOUJON, C. & INGOLD, M. (1967). Restauration de fertilité de quelques lignées mâle-stériles de blés hexaploides. *C.R. Acad. Sci. (Paris) Ser. D.* **264,** 334–336.

INGOLD, M. (1967). Male sterility and restorer systems in wheat. *Euphytica* **17,** Suppl. 1, 69–74.

MAAN, S.S. (1973). Cytoplasmic male-sterility and male-fertility restoration systems in wheat. *EUCARPIA, Cereal Sect. Bari (Italy),* 1–54.

ODENBACH, W. (1970). Versuche zur Wiederherstellung der Fertilität pollensteriler hexaploider Weizen. I. Timopheevi-Cytoplasma induzierte Pollensterilität sowie deren Überwindung unter Freilandbedingungen in Berlin-Dahlem. *Z. Pflanzenzüchtung* **64,** 73–89.

OEHLER, E. & INGOLD, M. (1966). Männliche Sterilität und Wiederherstellung der Fertilität bei den Triticinae. *26. Jahresber. Schweiz. Ges. Vererbungsforschung* **41,** 26–37.

SCHMIDT, J.W. (1964). The mechanism of sterility and restoration in an experimental wheat. In: *Hybrid Wheat, Crop Quality Council, Minneapolis, Minn.* 20–21.

WILSON, J.A. (1967). Problems in hybrid wheat breeding.*Euphytica* **17,** Suppl. 1, 13–33.

ZEVEN, A.C. (1967). Cross pollination and sources of restorer genes in wheat and a semi-hybrid variety. *Euphytica* **17,** Suppl. 75–81.

EUCARPIA VII (1976)

Microtrials in Austria concerning heterosis in winter wheat

R. HRON
Federal Institute for Plant Production and Seed Testing, Vienna, Austria

SUMMARY

23 F_1's from crosses between Austrian and foreign winter wheat varieties were tested in micro-yield-trials from 1971 to 1973 in five locations and several replications, seeding rates ranging from medium to high. Seed shortage necessitated non-orthogonal testing.

Only about half of the F_1's tested showed a more or less distinct yield heterosis as compared with the better parent. So far no F_1 with consistent heterosis could be found; F_1 showing positive heterosis reached a mean yield of only 2% over the better parent, as an average of all trials. A clear tendency towards a decrease in yield heterosis with increased seeding rates was revealed.

Baking qualities were inherited in an intermediary to incomplete dominant manner.

INTRODUCTION

Since 1971 the extent of F_1-heterosis in winter wheat in crosses between conventional varieties has been tested in performance trials at several sites. The trials were carried out within the framework of the Austrian Working Group for Research on Plant Breeding (formerly Working Group for Hybrid Wheat Breeding), in which the Austrian cereal breeders cooperate with public institutes under the leadership of the Federal Institute for Plant Production and Seed Testing (Hron, 1971). The Working Group was set up by the Austrian Plant Breeders' Association.

The trials in question do not contribute to breeding methods of heterosis varieties as such; they are, however, destined to answer the main preliminary question for the breeding of hybrid varieties in wheat, namely the possible amount of yield heterosis. We aim at comprehending the extent of yield heterosis under conditions which are in agreement with the usual field conditions and seeding rates, as much as the available seeds and the technical circumstances of trials on microplots permit. The heterotic performance should not be examined in individual plants but in a plant population. According to the use of wheat not only the quantity but also quality and processing value of the crop were recorded. The results gained under the specific growing conditions of our own country should enable forecasts for technical and economical possibility of hybrid breeding in wheat.

MATERIAL AND METHODS

Over a period of three years (1971 to 1973) a total of 23 different F_1's from 25 different winter wheat varieties were tested in comparison with the parent varieties used; occasionally, also other high-yielding standard varieties were tested. All F_1's tested were crosses between conventional self-fertile varieties.

The F_1 seed was produced by castration of the parent varieties and successive free wind-pollination in the crossing blocks. The series of trials has not yet been concluded, it is planned to be continued for another 2 or 3 years.

The varieties used for the crosses come from the following countries:

Austria: Danubius, Lentia, Multiweiss, Neuhof Nr. 1, Probstdorfer Extrem (Bundesanstalt 1971, 1972, 1974); furthermore, among others the Austrian stocks (unnamed varieties) NR 70/1094, NR 5150/6/71, P 645/4/70, P 8040-70, SL 9/63–65 and two others were used for crosses;

FRG: Caribo, Jubilar, Kormoran (Bundessortenamt 1973);

France: Moisson;

Great Britain: TL 363/30 (stock from Plant Breeding Institute, Cambridge);

Netherlands: Manella;

USA: Blueboy, Lancer, Scout;

USSR: Bezostaja 1, Kavkaz, Mironovskaja 808, Rannaja 12.

The trials were carried out in several replications on plots of 1 sq.m. size with a row width of 120 to 150 mm. The seeding rates were 150 or 300 kernels per sq.m. The tests were performed in the following 5 places:

a) Fuchsenbigl, district Gänserndorf, Lower Austria, Experiment Station of the Federal Institute for Plant Production and Seed Testing, Vienna;

b) Probstdorf, district Gänserndorf, Lower Austria, Breeding Station of Probstdorfer Saatzucht GmbH;

c) Neuhof-Rohrau, district Bruck an der Leitha, Lower Austria, Agricultural Enterprise Neuhof-Rohrau, Plant Breeding;

d) Marchtrenk, district Wels-Land, Upper Austria, Breeding Station of Probstdorfer Saatzucht GmbH;

e) Reichersberg, district Schärding, Upper Austria, Plant Breeding Enterprise of the Upper Austrian Provincial Seed Growers' Cooperative, Linz.

The sites Fuchsenbigl, Probstdorf and Neuhof-Rohrau are situated in the semiarid cereal production area of Austria with a mean annual precipitation between 500 and 600 mm. Marchtrenk and Reichersberg have higher precipitation between 800 and 900 mm per annum.

RESULTS

Table 1 demonstrates the yield heterosis so far achieved in the trials, and Table 2 shows the yield performances of F_1's and parents in the individual trials, as percentage of the respective trial means. All results so far gained are included in the tables, except 2 F_1 from crosses derived from Austrian stocks, whose yields were considerably below that of the better parent. A total of 23 F_1's were tested in the different years, locations, and seeding rates, giving altogether 73 experimental units in two to four replications.

Table 1. Relative yields of 21 F_1's of winter wheat (in percentage of the better parent)

Seeding rate	150 kernels per sq. m.							300 kernels per sq. m.							
Year	1971				1972			1972			1973				
Location	*a)*	*b)*	*c)*	*e)*	*a)*	*b)*	*e)*	*a)*	*b)*	*e)*	*a)*	*b)*	*c)*	*d)*	*e)*
F_1 (Bezostaja 1 × Probstdorfer Extrem)	90	111	—	—	72	111	—	—	106	—	—	—	—	—	—
F_1 (Bezostaja 1 × Mironovskaja 808)	82	111	109	—	—	—	—	—	—	—	—	—	—	—	—
F_1 (Bezostaja 1 × Neuhof Nr. 1)	90	98	85	—	77	124	—	—	—	—	—	86	—	—	—
F_1 (Blueboy × Probstdorfer Extrem)	—	—	—	—	134	109	—	121	109	—	—	—	—	—	—
F_1 (Caribo × Jubilar)	—	—	—	103	—	—	—	—	—	—	—	—	—	—	—
F_1 (Danubius × Caribo)	—	—	—	—	—	—	—	—	—	—	97	—	—	—	102
F_1 (Jubilar × Caribo)	—	—	—	—	—	—	—	—	—	—	99	91	—	105	99
F_1 (Kavkaz × Probstdorfer Extrem)	—	—	—	—	—	—	—	—	—	—	96	93	105	—	112
F_1 (Kormoran × Caribo)	—	—	—	—	—	—	—	—	—	—	111	103	—	95	—
F_1 (Lancer × Probstdorfer Extrem)	—	—	—	—	113	107	—	—	—	—	—	—	—	—	—
F_1 (Lentia × Jubilar)	—	—	—	102	—	—	95	—	—	—	—	—	—	—	—
F_1 (Manella × Multiweiß)	—	—	—	116	93	—	93	—	—	81	—	—	—	—	—
F_1 (Moisson × Multiweiß)	120	99	116	106	—	—	99	—	—	105	—	—	—	—	—
F_1 (Rannaja 12 × Probstdorfer Extrem)	91	—	94	—	—	—	—	—	—	—	—	—	—	—	—
F_1 (Scout 66 × Probstdorfer Extrem)	—	—	—	—	—	—	—	—	—	—	79	66	99	—	—
F_1 (NR 70/1094 × Multiweiß)	—	—	—	—	89	—	—	77	—	—	—	—	—	—	—
F_1 (NR 5150/6/71 × Probstdorfer Extrem)	—	—	—	—	—	—	—	—	—	—	104	—	112	—	—
F_1 (P 645/4/70 × Neuhof Nr. 1)	—	—	—	—	95	98	—	100	90	—	85	—	—	—	—
F_1 (P 8040-70 × Caribo)	—	—	—	—	—	—	—	—	—	—	95	103	—	85	—
F_1 (SL 9/63-65 × Multiweiß)	—	—	—	—	91	—	97	—	—	91	—	—	—	—	—
F_1 (TL 363/30 × Caribo)	—	—	—	—	—	—	—	—	—	—	108	91	101	—	91

Table 2. Relative yields of 21 F_1's and their parent varieties (percentage of the trial means)

Seeding rate	150 kernels per sq. m.							300 kernels per sq. m.							
Year	1971				1972			1972			1973				
Location	*a)*	*b)*	*c)*	*e)*	*a)*	*b)*	*e)*	*a)*	*b)*	*e)*	*a)*	*b)*	*c)*	*d)*	*e)*
F_1 (Bezostaja 1 × Probstdorfer Extrem)	102	105	—	—	87	113	—	—	109	—	—	—	—	—	—
F_1 (Bezostaja 1 × Mironovskaja 808)	99	111	105	—	—	—	—	—	—	—	—	—	—	—	—
F_1 (Bezostaja 1 × Neuhof Nr. 1)	102	96	99	—	91	123	—	—	—	—	—	98	—	—	—
F_1 (Blueboy × Probstdorfer Extrem)	—	—	—	—	161	105	—	149	118	—	—	—	—	—	—
F_1 (Caribo × Jubilar)	—	—	—	114	—	—	—	—	—	—	—	—	—	—	—
F_1 (Danubius × Caribo)	—	—	—	—	—	—	—	—	—	—	92	—	—	—	111
F_1 (Jubilar × Caribo)	—	—	—	—	—	—	—	—	—	—	90	87	—	107	107
F_1 (Kavkaz × Probstdorfer Extrem)	—	—	—	—	—	—	—	—	—	—	111	108	106	—	103
F_1 (Kormoran × Caribo)	—	—	—	—	—	—	—	—	—	—	111	100	—	97	—
F_1 (Lancer × Probstdorfer Extrem)	—	—	—	—	136	110	—	—	—	—	—	—	—	—	—
F_1 (Lentia × Jubilar)	—	—	—	97	—	—	100	—	—	—	—	—	—	—	—
F_1 (Manella × Multiweiß)	—	—	—	109	70	—	104	—	—	93	—	—	—	—	—
F_1 (Moisson × Multiweiß)	111	103	109	100	—	—	111	—	—	120	—	—	—	—	—
F_1 (Rannaja 12 × Probstdorfer Extrem)	103	—	99	—	—	—	—	—	—	—	—	—	—	—	—
F_1 (Scout 66 × Probstdorfer Extrem)	—	—	—	—	—	—	—	—	—	—	92	76	91	—	—
F_1 (NR 70/1094 × Multiweiß)	—	—	—	—	95	—	—	93	—	—	—	—	—	—	—
F_1 (NR 5150/6/71 × Probstdorfer Extrem)	—	—	—	—	—	—	—	—	—	—	120	—	103	—	—
F_1 (P 645/4/70 × Neuhof Nr. 1)	—	—	—	—	112	87	—	114	89	—	96	—	—	—	—
F_1 (P 8040-70 × Caribo)	—	—	—	—	—	—	—	—	—	—	103	106	—	91	—
F_1 (SL 9/63-65 × Multiweiß)	—	—	—	—	62	—	108	—	—	104	—	—	—	—	—
F_1 (TL 363/30 × Caribo)	—	—	—	—	—	—	—	—	—	—	95	87	91	—	99
Bezostaja 1	81	94	61	—	71	89	—	86	70	—	—	108	93	—	—
Blueboy	—	—	—	—	102	111	—	111	109	—	—	—	—	—	—

Table 2. (continued)

Caribo	—	—	—	111	—	—	—	—	—	94	88	95	90	—	109
Danubius	—	—	—	—	—	—	—	—	—	—	95	—	—	101	105
Jubilar	—	—	—	96	61	—	91	—	—	92	91	95	—	97	103
Kavkaz	—	—	—	—	—	—	—	—	—	—	103	116	101	—	92
Kormoran	—	—	—	—	—	—	—	—	—	—	100	97	—	100	—
Lancer	—	—	—	—	106	53	—	—	—	—	—	—	—	—	—
Lentia	—	—	—	89	—	—	105	—	—	—	—	—	—	—	—
Manella	—	—	—	93	76	—	92	—	—	85	—	—	—	—	—
Mironovskaja 808	121	100	97	—	—	—	—	—	—	—	—	—	—	—	—
Moisson	93	105	88	95	113	—	104	—	—	95	—	—	—	—	—
Multiweiß	77	81	94	94	68	—	112	64	—	114	—	—	—	—	—
Neuhof Nr. 1	113	98	116	—	118	100	—	114	100	—	113	115	96	—	—
Probstdorfer Extrem	113	98	105	—	120	102	—	124	103	—	116	115	92	—	—
Rannaja 12	84	—	96	—	120	102	—	124	103	—	—	—	—	—	—
Scout 66	—	—	—	—	—	—	—	—	—	—	104	108	86	—	—
NR 70/1094	—	—	—	—	107	—	—	121	—	—	—	—	—	—	—
P 645/4/70	—	—	—	—	88	76	—	86	72	—	78	—	—	—	—
P 8040-70	—	—	—	—	—	—	—	—	—	—	108	103	—	106	—
SL 9/63-65	—	—	—	—	64	—	97	—	—	97	—	—	—	—	—
TL 363/30	—	—	—	—	—	—	—	—	—	—	85	93	84	—	71
L.S.D. 5% relative scale	—	—	—	11	19	—	11	see lower			9	9	—	18	10
Trial mean, q/ha	80	66	56	89	47	55	76	seeding rate			82	65	71	66	86

Locations of the trials:

a) Fuchsenbigl, district Gänserndorf, Lower Austria.
b) Probtsdorf, district Gänserndorf, Lower Austria.
c) Neuhof-Rohrau, district Bruck an der Leitha, Lower Austria.
d) Marchtrenk, district Wels-Land, Upper Austria.
e) Reichersberg, district Schärding, Upper Austria.

No sufficient yield heterosis could be proved in the great majority of the 73 units tested; however, it must be said that in 1973 the quality of the F_1 seed and the field germination were worse than those of the parent varieties. Only in 18 cases (25% of the units) was yield heterosis observed or at least a non-significant tendency towards increases in yields as compared with the better parent; from these only 3 cases (18%) showed a heterosis of more than 10% over the better parent, and only 6 cases a yield heterosis of more than 15%.

Considering F_1 performance in several sites and over several years, substantial fluctuations in the relative yields are found generally. Thus no consistent yield heterosis could be proved in any of the F_1's so far examined. Only the F_1 (Blueboy ×Probstdorfer Extrem) showed sufficient yield heterosis, but the results of a single year do not suffice to give evidence of a stable heterosis; the trials are being continued.

Only 12 of the total of 23 F_1's tested showed a more or less distinct yield heterosis as against the better parent. From 47 individual results of these 12 F_1's responding positively as to heterosis, an average yield heterosis of only 2% above the better parent was obtained.

In order to achieve maximum yields, seeding rates of 350 to 450 kernels per sq.m. are recommended for winter wheat cultivation in Austria. In 1972 from a total of 9 experimental units of 7 different F_1's results of two different seeding rates were obtained. In our trials a seeding rate of 300 kernels/sq.m. showed already a tendency towards a diminution of yield heterosis. On the average the higher seeding rate resulted in 6% less heterosis than the lower seeding rate of 150 kernels/sq.m.

The following baking characteristics were determined: protein, wet gluten, swelling number, and sedimentation value. These characters showed intermediary or incompletely dominant inheritance. This fact gives rise to the hope that – should an F_1 with sufficient and stable yield heterosis be found – the plant breeder might succeed in maintaining a high quality level as requested in Austrian wheat production, in spite of an increase in yield by heterosis. Keydel found similar results (1973).

Though our investigation was limited to a relatively small number of F_1's and trials, we have the impression, that the probability of finding wheat hybrids of superior performance when cultivated in dense stands is not very high. For the present we are not yet able to judge on the chance of wheat hybrids (cf. Keydel, 1973).

ACKNOWLEDGEMENTS

The following gentlemen have cooperated in the trials within the framework of the Austrian Working Group for Research on Plant Breeding: Prof. Dr. H. Hänsel, Probstdorfer Saatzucht; Dipl. Ing. J. Adam, Agricultural Enterprise Neuhof-Rohrau; Ing. J. Körber and Dipl. Ing. E. Kirsch, Upper Austrian Provincial Seed Growers' Cooperative, Plant Breeding Enterprise Reichersberg/Inn; and colleagues of the Federal Institute for Plant Production and Seed Testing, Vienna, in particular Dr. K. Waltl in determining characteristics decisive for baking quality.

The Working Group established within the Association of Austrian Plant Breeders is supported by the Austrian Federal Ministry of Agricultural and Forestry from research funds of the Green Plan since 1970.

REFERENCES

Bundesanstalt für Pflanzenbau und Samenprüfung in Wien (1971). Sortenbeschreibung landwirtschaftlicher Kulturpflanzen, 2. Auflage, Wien. 1972, 1. Nachtrag, Wien, 1974. 3. Nachtrag, Wien

Bundessortenamt (1973). Beschreibende Sortenliste für Getreide, Mais und Ölfrüchte. Hannover

HRON, R. (1971). Tätigkeitsbericht 1970. Züchtungsforschung: Hybridweizenzüchtung. *Die Bodenkultur* **22,** Sonderheft, 32–34.

KEYDEL, F. (1973). Stand und Aussichten der Hybridzüchtung bei Weizen. *Der Förderungsdienst* **21,** 186–193.

A possible method for the production of F_1 hybrid forage grasses

F.J.W. ENGLAND

Scottish Plant Breeding Station, Pentlandfield, Roslin, Midlothian, UK

SUMMARY

A possible method for the production of F_1 hybrid seed of forage grasses in commercial quantities is described. The method exploits the incompatibility system in many *Gramineae* which is controlled by two or more loci.

Inbred lines, derived from a single genotype, are expected to have a high degree of within-lines incompatibility so that when two, or more, such lines are interplanted the seed produced will be largely of hybrid origin. It is shown that it is possible to derive the transition matrix for the two-locus system or for similar incompatibility systems depending on any specified number of loci. All such systems eventually lead to populations with stable frequencies of the incompatibility genotypes and hence it is possible to calculate the expected proportion of outcrosses when such populations are sown together. The two-locus system results in populations which may be expected to produce 83% of F_1 hybrids when two lines are allowed to interpollinate. The proportion of hybrids decreases as the number of loci controlling the incompatibility mechanism increases.

The merits, or otherwise, of F_1 hybrids in normally out-breeding species have been discussed in other sections of this congress. This paper will be confined to demonstrating a possible basis for the production, in out-breeding members of the *Festucaceae* and some other families, of populations containing a high proportion of F_1 hybrids.

The proposed method depends on the incompatibility system described by Hayman (1956) and by Lundqvist (1961). These authors postulate a gametophytic system controlled by two independently segregating loci S and Z. For a pollination to be compatible at a minimum of one of the loci the allele in the pollen must not match either allele in the stigma. Some recent work (Hayward & Wright, 1971) suggest that in *Lolium perenne* the control may depend on more than two loci but is otherwise as described by Lundqvist. The incompatibility mechanism is not perfect; it is usually possible to obtain a few viable seeds on selfing and some genotypes are much more self-fertile than others.

On selfing a plant of constitution $S_1S_2Z_1Z_2$ we can obtain of the progeny genotypes in the first column of Table 1 with expected frequencies as shown in column two. Such an inbred (I_1) population is expected to produce a pollen cloud containing equal proportions of the four possible pollen genotypes shown at the top of the table, and will produce a first advance (A_1) population of the expected composition shown on

Table 1. Frequency of genotypes in the I_1 generation and in its A_1 progeny

I_1 Female genotype	I_1 Frequency	Male gamete compatibility				Frequency of progeny genotypes in the A_1 within each female genotype in the I_1				
		1.1	1.2	2.1	2.2	11.12	12.11	12.12	12.22	22.12
11.11	1/16	−	+	+	+	1/3	1/3	1/3	−	−
11.12	1/8	−	−	+	+	−	1/4	1/2	1/4	−
11.22	1/16	+	−	+	+	1/3	−	1/3	1/3	−
12.11	1/8	−	+	−	+	1/4	−	1/2	−	1/4
12.12	1/4	−	−	−	−	−	−	−	−	−
12.22	1/8	+	−	+	−	1/4	−	1/2	−	1/4
22.11	1/16	+	+	−	+	−	1/3	1/3	−	1/3
22.12	1/8	+	+	−	−	−	1/4	1/2	1/4	−
22.22	1/16	+	+	+	−	−	−	1/3	1/3	1/3
		Overall frequency of progeny genotypes in A_1 weighted for parental frequencies in I_1				5/36	5/36	16/36	5/36	5/36

Note that the completely homozygous offspring do not occur in the A_1 generation and that the completely heterozygous genotypes do not produce offspring.

the right of the table. At the bottom of the table are shown the overall expected frequencies in the A_1 generation after weighting for the maternal genotype frequencies in the I_1. If we wish to find the frequencies in the A_2 and subsequent generations, we can repeat the above process indefinitely by substituting the appropriate maternal frequencies in the second column. This however becomes rather cumbersome, especially when we come to consider the cases of three or more loci. Fortunately, we can classify the various genotypes into a small number of groups according to their degree of homozygosity. With two loci the nine possible genotypes can be divided into three groups according to whether they are heterozygous at both loci (hom^0), homozygous at one locus (hom^1) and homozygous at both loci (hom^2). The number of such groups is obviously one more than the number of loci involved. If we enter the frequencies with which each of these genotypes is expected to occur in the progeny of each of them as the elements of a matrix then we can use this matrix as a transition or generation matrix. If **A** is this matrix (shown in Table 2 for two loci) and **f** is a normalized column vector of the frequencies of the genotype group in any one generation, then $\mathbf{Af}_n$ gives the frequencies in the next generation. This is:

$$\mathbf{f}_{n+1} = \mathbf{Af}_n$$

If we start with the expected I_1 frequencies for two loci we can repeat this process as shown in Table 2.

In practice, the frequencies stabilize by the A_2 generation, a 50:50 mixture of hom^0 and hom^1 genotypes. The general rules for constructing the matrix **A** for any number of loci will be described later in this paper, but at this point it should be mentioned

Table 2. The matrix **A** for two loci and the frequencies of the different genotype groups in successive generations

		$\mathbf{f}_n$ genotypes			Frequency of genotypes in I_1 and subsequent generations		
		hom^2	hom^1	hom^0	I_1	A_1	A_2
	hom^2	0	0	0	$^1/_4$	0	0
$\mathbf{f}_{n+1}$ genotypes	hom^1	$^2/_3$	$^1/_2$	0	$^1/_2$	$^5/_{12}$	$^5/_{24}$
	hom^0	$^1/_3$	$^1/_2$	0	$^1/_4$	$^4/_{12}$	$^5/_{24}$

The I_1 generation frequencies are assumed to be in binomial proportions.
The frequencies in the A_1 and A_2 generations have not been normalized, the ratio between hom^1 and hom^0 frequencies in the total A_1 population is of course $^5/_9$:$^4/_9$ and similarly in the A_2 $^1/_2$:$^1/_2$.

that with three- and four-locus control the equilibrium proportions of the possible genotypes are:

	three loci	four loci
hom^4	–	0.00
hom^3	0.00	0.16
hom^2	0.28	0.40
hom^1	0.50	0.34
hom^0	0.22	0.10

and this position is reached by the A_3 generation.

Each of the genotype groups is compatible with some proportion (always less than 1) of the pollen from its own population. I have called this proportion the "within population compatibility" or *WIC* and its value can be calculated from the expression:

$$WIC = 1 - (^1/_2)^p$$

where p is an integer representing the degree of homozygosity of the genotype group. If a population of the kind described is exposed to alien pollen from a similarly inbred population each of its genotype groups will produce a mixture of seed derived from within population fertilization and fertilization by alien pollen. With no incompatibility, we would expect the proportion of alien fertilization to be 0.5. If we know the *WIC* of a genotype group we can calculate the effect of incompatibility on the proportion of outbred seed produced by that group from the expression

$$1/(1+WIC)$$

and the proportion of outbred seed produced by each population by summing over all the genotype groups after weighting for their frequencies. For two loci for example we have 0.5 of the population of type hom^0 and 0.5 type hom^1. Their respective *WIC*'s are $1-(^1/_2)^0 = 0$ and $1-(^1/_2)^1 = 0.5$ and the proportion of outbred seed produced by such a population is:

$$0.5\ (1.0/1.5) + 0.5(1.0/1.0) = 0.83$$

For three and four loci the corresponding values are 0.71 and 0.64. For eight loci the value is 0.53, effectively no greater than with no incompatibility at all.

It is reasonable to assume that seed produced as a result of fertilization within any one population would give rise to relatively weak seedlings which would be eliminated in the early stages of sward establishment.

The method described depends to some extent on a contradiction. It assumes that the incompatibility mechanism is less than perfect since the process begins by selfing. The prediction of the composition of the A_1 and later generations assumes that the mechanism is perfect. However, during the multiplication phase, incompatible pollen will normally be in competition with that which is compatible and "illegitimate" fertilization will be rare.

A serious practical objection lies in the difficulty of finding adequate isolation for the multiplication of the inbred lines. The fact that hom^0 genotypes are incompatible with all pollen from their own population renders them particularly liable to fertilization by foreign pollen. Apart from this maintenance of the lines should not be an insuperable problem, since their coefficient of inbreeding will be 0.5 and while the plants will undoubtedly be less vigorous than the outbred material they should be reasonably viable.

CONSTRUCTION OF THE MATRIX **A**

This matrix consists essentially of the matrix of frequencies of the genotype groups to be expected with no incompatibility minus the frequency of those which do not occur because of the incompatibility mechanism. Let k be the number of loci involved and **X** be a $(k+1) \times (k+1)$ matrix such that each of its columns contains the coefficients of the binomial $(a+b)^k$. Let **Y** be a $(k+1) \times (k+1)$ matrix such that each of its columns contains the coefficients of the binomial $(a+b)^{j-1}$ (where j is the column) in its first j rows and zeros elsewhere. Let $\mathbf{Z} = \mathbf{X} - \mathbf{Y}$. Then by normalizing the columns of **Z** (*i.e.* by dividing each entry by the sum of the column in which it occurs) we obtain the required matrix **A**. The process for two loci is illustrated below

$$k = 2$$

$$\mathbf{X} = \begin{matrix} 1 & 1 & 1 \\ 2 & 2 & 2 \\ 1 & 1 & 1 \end{matrix} \qquad \mathbf{Y} = \begin{matrix} 1 & 1 & 1 \\ 0 & 1 & 2 \\ 0 & 0 & 1 \end{matrix}$$

$$\mathbf{Z} = \mathbf{X} - \mathbf{Y} = \begin{matrix} 0 & 0 & 0 \\ 2 & 1 & 0 \\ 1 & 1 & 0 \end{matrix} \qquad \mathbf{A} = \begin{matrix} 0 & 0 & 0 \\ 2/3 & 1/2 & 0 \\ 1/3 & 1/2 & 0 \end{matrix}$$

$$a_{ij} = z_{ij}/z_{.j}$$

REFERENCES

HAYMAN, D.L. (1956). The genetical control of incompatibility in *Phalaris coerulescens.* Desf. *Austr. J. Biol. Sci* **9,** 321–331.

HAYWARD, M.D. & WRIGHT, A.J. (1971). The genetic control of incompatibility in *Lolium perenne* L. *Genetica* **42,** 414–421.

LUNDQVIST, A. (1961). Self incompatibility in *Festuca pratensis* Huds. *Hereditas* **47,** 542–562.

EUCARPIA VII (1976)

Prediction of heterosis levels from parental information

R.L. THOMAS
Twyford Plant Breeders, Banbury, UK
and
J. E. GRAFIUS
Michigan State University, USA

SUMMARY

The prediction from parental information of the expected degree of yield heterosis of F_1 hybrids constitutes a particular instance of one of the main problems of plant breeding – namely advance prediction of the performance of any possible F_1 combination. Central to one approach to the problem is a knowledge of the parental yield components which we take initially to be those measured attributes which when multiplied together give yield exactly. It is demonstrated that prediction methods based on parental yield components can be more accurate and meaningful than those based on yield alone. The prediction methods rely on a geometric approach involving the bio-mathematical inter-relationship of the components and crop yield.

INTRODUCTION

We initially define and justify the type of parental information we wish to utilize and thereafter proceed to the prediction method. The information, which should be obtained on a set of potential parental populations, varieties or inbred lines in roughly the environments under which the putative offspring might be grown, are simply the principal yield components. These however need some definition since the term is often used loosely to include both components of and contributors to yield. The former we define here to be strictly those characters which when multiplied together give yield exactly. To consider two examples which are later more fully utilized the principal components of yield in wheat we take to be heads/unit area, spikelets/head, seeds/spikelet, and weight/grain; and in maize are cobs/plant, kernel rows/cob, kernels/row and weight/grain. In some cases the main components, to which we incidentally confine ourselves below, may be further and logically subdivided, retaining the multiplicative principle, e.g. component one in wheat into tillers per unit area times the percentage of fertile tillers. Most crops are amenable to the subdivision of yield into comparable main components. What we term contributor characters and then exclude from further consideration are generally less amenable to practical measurement and these are mainly included in the domain of crop physiologists and pathologists rather than breeders; examples: effective photosynthetic area, flag leaf size and duration, nitrate reductase activity, etc. These have one common feature and disadvantage namely the incompleteness of such measures in accounting for yield – at least this is so at the present state of the science – and thus in contrast with the yield component situation. Ultimately it is to be hoped that such contributor characters may achieve the status of major or minor yield components, even to the extent of replacing them. In the meanwhile let us inquire to what extent consideration of present and simple main yield components may help in predicting offspring yield – surely the central problem of plant breeding. We suggest below that manipulation of the information provided by components may constitute a useful predictive tool and that this is certainly preferable to considering only parental yield.

The method of yield prediction we propose to outline may conveniently be regarded as being in three stages. The first involves removing the correlations between the sequentially developed component traits to render them orthogonal to one another. The second is a form of genetic analysis aimed at measuring the loss in variation of the second and subsequent traits in the sequence caused by the influence of previous traits and obtaining a measure of "true relative genetic variance" – (Thomas *et al.*, 1971*a, b, c*). The third step is to indicate how this information may be utilized in prediction itself. As implied earlier we take two crops, maize and wheat, to illustrate the technique. Although breeding systems and methods differ, the prediction of F_1 yield levels, for F_1 hybrids in the first instance and nowadays in wheat for either F_1 hybrids and or pure line selections from segregating material, is of equivalent importance. "Heterosis" was present in both sets of data and although we concentrate here on prediction of the absolute yield of the offspring it should be understood that this embraces prediction of heterosis.

TRANSFORMATION OF YIELD COMPONENT DATA

Thomas *et al.* (l. c.) have pointed out that yield component development is generally a particular instance of sequential trait formation. The problem of any consideration, be it physiological or genetic, in such sequences is that the expression of the later characters in the sequence are influenced to varying extents, negatively and positively, by the earlier – this is measurable by inter-traits correlations. Although we would favour a physiological rather than a genetic e.g. pleiotropic or linkage etc., interpretation of the nature of the causative mechanism of such correlations this does not concern us here so much as do the genetic consequences. Undoubtedly the consideration of n traits and all possible $([n^2 - n] + 2)$ correlations between them rapidly becomes frustrating as n become large. In any case it is advantageous for the present purposes to examine the n traits separately from the previous developmental events and to this end several techniques were evolved. The first depended on vector mathematical methods, and the second on a statistical transformation. These were found to be identic in effect, the former being more easily understandable by visual methods but since the latter is more concise this only is considered here (see Thomas *et al.* 1971*a*). We also explored another approach to the problem by considering an analogy of the yield component situation with the mathematics of oscillatory waves and this may provide an alternative insight (Grafius & Thomas, 1971), this is also not further considered here. The statistical solution may be briefly summarized. With

$x_1 \ldots. x_p$ observed yield components we obtain $y_1 \ldots.. y_p$ transformed and uncorrelated components thus: –

$$1.\ Y_1 = X_1$$
$$2.\ Y_2 = X_2 - a_{21}Y_1$$
$$3.\ Y_3 = X_3 - a_{32}Y_2 - a_{31}Y_1$$

.....

.....

$$P/\ Y_p = X_p - a_{p,p-1}\,Y_{p-1} \quad \ldots - a_{p1}\,Y_1$$

Where

$$a_{21} = \frac{\lambda_{21}}{\lambda_{11}} = \frac{\text{covariance } Y_2\, Y_1}{\text{variance } X_1}$$

$$a_{31} = \frac{\lambda_{31}}{\lambda_{11}}$$

$$a_{32} = \frac{\lambda_{32} - a_{21}\,\lambda_{31}}{\lambda_{11}}$$

A more comprehensive coverage of the method is given by *e.g.* Rao (1952), and we proceed here to the next step.

GENETIC ANALYSIS OF CORRELATED SEQUENTIAL YIELD COMPONENTS APPLIED TO WHEAT DATA

The original data was published by Whitehouse *et al.* (1958) and constituted the F_1 and F_2 of a 4 × 4 diallel, with 4 measured yield components. The above transformation was then applied giving uncorrelated values for the 3 yield components spikelet/head, seeds/spikelet and weight/seed; the first component, heads/unit, remains, of course, untransformed – these data are given by Thomas *et al.* (1971*c*) for both parents and all possible F_1 and F_2 combinations (the generations separately transformed). Genetic analysis is however concerned with the more concise consideration of variances rather than means and it is this that we now undertake. We present in Table 1 a method for arriving at what we termed the "true relative genetic variance". This may be defined as the percentage of observed variations of the later yield components in the developmental sequence which operates independently of the influence of previous characters. This measure, as may be seen from Table 1, was obtained by comparing, as a ratio, the variation of transformed parental data with that of the untransformed; the measure of variation used in both cases was the variance divided by the mean. Only the parental data rather than F_1 was so utilized since in conventional diallel analysis the statistic variance of parents *(V_p)* has a particular interpretation.

It is very obvious from the last column of Table 1 that when we examine the latter characters in a sequence with the influence of previous characters removed the "true relative genetic variance" differs considerably from the apparent variance of the untransformed character. For the F_1 diallel the amount of variance present in the second character in the sequence is only *c.* 30% of the variance which was observed phenotypically i.e. *c.* 70% is a reflection of the effect of the first character. This drops off to only *c.* 2% and less than 1% true relative genetic variance for the third and fourth character respectively – i.e. the variance of these latter are very strongly determined by the variance of previous characters to the tune of respectively *c.* 98% and *c.* 100%. The situation is similar in the F_2 which could more conveniently – see Whitehouse *et al.* (l. c.) – be regarded as the same parental set grown in a different environment. However the details differ here, the drop off in the second character being non existent – in fact the sequential independence of the second character is due to the lack of connecting stresses (i.e. correlations) caused or imposed by the first. However by the time the last character is produced there is little or no true relative genetic variance left – all the variation is again accounted for by the influence of previous characters. Removal of correlations also revealed a difference in the quality as well as in the quantity of genetic variance remaining in those transformed variables – generally they appeared less additive than in the analysis of straight phenotypic values. However since the quantity was so substantially reduced in these sets, the situation may presently be considered in a simplified form taking true relative genetic variance without conversion to a true *absolute* genetic variance.

The central point we wish to make about these results is that they lead to postulate the existence of two general types of yield component – "influencing", e.g. heads/plant in the F_1 and spikelets per head in the F_2, and "influenced", e.g. all subsequent traits to these. To these categories we would add "neutral" component traits – ones

Table 1. Estimation of "true relative" genetic variance of sequential characters in wheat (after Thomas et al. 1971c)

	Untransformed data			Transformed data			"True relative" genetic variance
	Parental mean	Parental variance	Weighted parental variance	Parental mean	Parental variance	Weighted parental variance	
F_1	A_1	B_1	$(B_1/A_1) \times 100$	A_2	B_2	$(B_2/A_2) \times 100$	$\frac{(B_2/A_2)}{(B_1/A_1)} \times 100$
Heads/plant	22·85	1·04	—	—	—	—	100·00
Spikelets/head	20·70	1·19	5·748	26·60	0·487	1·83	31·83
Seeds/spikelet	2·24	0·18	8·036	16·89	0·032	0·19	2·35
Weight/grain	37·15	108·99	293·405	476·98	0·003	<0·01	<0·01
F_2							
Heads/plant	3·83	0·53	—	—	—	—	100·00
Spikelets/head	16·30	1·93	11·85	14·72	1·936	12·47	105·32
Seeds/spikelet	1·29	0·03	2·32	1·52	0·020	1·32	56·89
Weight/grain	42·05	116·86	277·91	165·91	0·001	<0·01	<0·01

which are neither influencing nor influenced; an obvious example here would be heads/plant in the F_2. We would expect to find the influencing/influenced balance in a high interplant competition situation. The resultant high competition between component traits would be expressed by high correlations and revealed by transformation in a rapid drop off rate in true relative genetic variance as we proceed along the developmental sequence. The corollary would also seem a reasonable hypothesis, where interplant competition is low, so also are the correlations; the variance of transformed components will correspond more closely with that of the untransformed and the degree of neutrality of traits would be higher. This is borne out by the present results and by Grafius & Okoli (1974), who have presented evidence that such inter-trait correlations are very strong in barley grown under competitive conditions. On the other hand Lee & Kaltsikes (1972) using "our transformation" on wheat grown under spaced plant conditions found for the same yield components as are used here, and moving along the same developmental sequence, true relative genetic variances 100%, 24%, 106% and 38% respectively. Thus in these non competitive conditions we could classify the sequence as "influencing", "influenced", "non-influenced", "possibly influencing and influenced", but the degree of influencing (or influenced) is substantially less here, and significantly there was no heterosis.

We shall indicate how the more realistic competitive findings may be used in prediction later but now turn to the consideration of the maize data.

GENETIC ANALYSIS OF CORRELATED SEQUENTIAL TRAITS APPLIED TO MAIZE DATA

This data was also obtained in a diallel arrangement, a 14 × 14, grown at two separate semesters at Cali, Colombia. Four yield components were estimated. The results have been previously described by Grafius *et al.* (1974), by Grafius *et al.* (in preparation) and by Thomas & Grafius (also in preparation). The last paper concerns us presently and we shall briefly present a preview of our findings. We applied

Table 2. Estimation of true relative genetic variance in maize

	True relative genetic variance in	
	a) Semester 1	*b)* Semester 2
Cobs/plant	100%	100%
Kernel rows/cob	85%	51%
Kernels/kernel row	58%	50%
Weight/grain	41%	23%

the transformation and analysis to obtain the true relative genetic variance for the 14 parents in the same manner as for the wheat data. A summary of findings are given in Table 2. Patently in this case (as a consequence of the lower inter-trait correlations) the distinction between influencing and influenced characters is less marked, and the drop off along the sequence, whilst discernable is of a considerably lesser magnitude than in the wheat data.

This situation is thus in contrast with the wheat data of Table 1. It must be remembered that both situations represent normal i.e. highly competitive growing conditions and we should look beyond a possibility of differential competition as a causative agent of the difference in findings. The two crops are however very different in growth habit and physiology etc. and we for the present simply hypothesize that the maize varieties here exhibit a lesser inter-trait competitive situation than wheat due to a greater total supply to photosynthates and proceed to the implications of the contrasting findings for prediction purposes.

PREDICTION OF F_1 YIELD IN WHEAT

The finding from our genetic analysis that there were two very different types of trait in wheat led us to the following hypothesis. It is suggested that strongly influencing traits not only affect substantially the subsequent traits but also contribute disproportionately to the determination of variance of the complex trait – yield. Thus midparent values of these strongly influencing yield components might be utilized more or less exclusively to predict F_1 yield. The correlations *(r)* and coefficients of determination *(r^2)* were calculated between F_1 yield and midparent values for each of the components in turn (Table 3). Midparent yields were calculated indirectly from the multiplicand of midparent values of the 4 components and directly from midparental yield. Obviously F_1 yield is not accounted for when prediction is attempted from the midparental values of all four components, although the situation is improved if midparental yield alone is used for prediction. The situation is still comparatively unfavourable for the F_2 set although the percentage prediction is improved. The best and only significant prediction of yield is obtained in both diallel generations from the most strongly influential character – 94% determination in the

Table 3. Determination coefficients of observed F_1 and F_2 diallel progeny yield with calculated midparent values of individual yield components in wheat. Yield 1 is calculated from the product of midparent component values and yield 2 directly from midparental yield

	Heads/plant	Spikelets/ear	Grains/spikelet	Weight/grain	Yield 1	Yield 2
F_1	94%**	63%	0%	12%	1%	63%
F_2	20%	80%*	10%	8%	66%	47%

F_1 from component 1 and 80% determination in the F_2 from component 2. These are substantial improvements over the determination of both the remaining components taken singly, or in combination. We hypothesize that the improvement over taking midparent values of all 4 components – 93% in the F_1 and 14% in the F_2 – is due to the "static" created by the inclusion of all four by the latter method. We recall obtaining a greater improvement percentage-wise over even the main influencing component by obtaining midparent values for all four *transformed* component values but did not present this originally because it is impossible statistically to prove an improvement over 94% and 80% determination respectively with the degrees of freedom available. Nevertheless the percentage of determination is, by any criterion, encouraging when we utilize strongly influencing yield components and our method of prediction was shown to have merit and statistical respectability.

We would anticipate that the predictive methods employed for wheat would apply wherever the genetic analysis performed here revealed a comparable situation. Nevertheless two general questions are posed by the results. First, can strongly influenced characters be selected and secondly can the method be improved on? If it is necessary to improve mean seed weight for example this may be accomplished by "normal" selection methods but were this to be done on a set such as the wheat considered here it would follow that such improvement would be at the expense of previous traits, which in fact we are indirectly selecting phenotypically but directly in the sequential sense. The method of prediction could conceivably be improved if data were available to allow the transformation to uncorrelated yield components on each putative parent independently and separately. This procedure would not therefore include the intrinsic assumption made here, that all varieties have similar stresses between traits. Midparent values for all the transformed components could then be calculated and back transformed by the mean stresses. The multiplicand of these new components could then be taken as the expected F_1 yield. We have no suitable data to test this suggestion.

PREDICTION OF OFFSPRING YIELD IN MAIZE

It is tempting to suggest a very opposite hypothesis for the facility of yield prediction in maize, stemming from the contrasting picture for the true relative genetic variance of the sequential traits to the wheat data. Namely, in the absence of any evidence for extremely influential traits we might suggest no great advantage of utilizing the midparent values of any particular component, but might do as well by using all yield components, or indeed merely straight midparental yield. We present in Table 4, for only one semester so far analyzed the r^2 values for the various midparental values and F_1 yield. In general these appear disappointing for the absolute degree of prediction attained although it is the relative values for the different methods which primarily concern us. Although the marginally most influential character (number of kernel rows) does give a marginally better prediction compared with any other component or with the method based on using all 4 yield components or indeed parental yield, the differences are not significant. We could conclude from this test, in the absence of strongly influential characters, that the scope for improvement in the relative level of yield prediction is less promising and although the results might support such an hypothesis it is hardly comforting to maize breeders.

Table 4. Determination coefficients for semester 1 between observed F_1 diallel progeny yield with calculated midparent values in maize. Yield 1 is calculated from the midparent values for all components and yield 2 directly from midparent yield

Cobs/plant	Kernel rows/cob	Kernels/kernel row	Weight/kernel	Yield 1	Yield 2
2.8%*	14.9%***	12.3%***	9.4%**	9.4%**	13.0%***

Nonetheless, more optimistically the high values for true relative genetic variance do indicate that the yield components are not bound tightly together either physiologically or genetically in this set. This would seem to allow the possibility for construction of hybrids with high values for each component without worrying too much about inter-component compensation. Such an approach would necessitate a more logical construction of inbred lines with a view to the ultimate combination of complementary types, which have high expression of different yield components. The ramifications of this are naturally extendable to three and four way cross systems.

We conclude with some explanations for the low absolute prediction levels for this data set although again emphasizing that for present purposes the relative values for the different prediction methods are of primary importance. Grafius *et al.* (1974) have indicated that, whether midparental yield components or midparent yield itself are considered, a substantial improvement in the absolute level of prediction is achieved by utilizing the diallel array means for each parent to obtain midparental values. They further suggest that this improvement is mainly due to a reduction in intrinsic technical error of parental yield component estimates rather than the involvement of systematic test crosses *per se* and that the same ends would have been achieved by an increase in replication of the parental set. Finally we should mention that an improvement over the present method of prediction was obtained, using only the parental sets by imposing a new concept – that of an optimum distribution of yield components in varieties or as we prefer to consider it – varietal shape. This method however lies outside the scope of the present paper and is further considered by Grafius & Okoli (1974) and Grafius *et al.* (in preparation).

REFERENCES

Grafius, J.E. & Thomas, R.L. (1971). A case for indirect genetic control of sequential traits and the strategy of deployment of environmental resources by the plant. *Heredity,* **26,** 433–442.

Grafius, J.E., Harpstead, D.D., Sarria, D. & Barnard, J. (1974). Specific combining ability and the components of yield in a 14 × 14 diallel cross of Corn, *Zea mays* L. *Crop Science* (in press).

Grafius, J.E. & Okoli, L. (1974). Dimensional balance among yield components and maximum yield in an 8 × 8 diallel of barley, *Hordeum vulgare. Crop Science* (in press).

Lee, J. & Kaltsikes, P.J. (1972). Diallel analysis of correlated sequential characters in Durum wheat. *Crop Science,* **12,** 770–772.

Rao, C.R. (1952). *Advanced Statistical Methods in Biometric Research.* John Wiley & Sons, New York, 390.

Thomas, R.L., Grafius, J.E. & Hahn, S.K. (1971*a*). Transformation of sequential quantitative characters. *Heredity,* **26,** 189–193.

Thomas, R.L., Grafius, J.E. & Hahn, S.K. (1971*b*). Stress: an analysis of its source and influence. *Heredity,* **26**; 423–432.

Thomas, R.L., Grafius, J.E. & Hahn, S.K. (1971*c*). Genetic analysis of correlated sequential characters. *Heredity,* **26,** 177–188.

Thomas, R.L. & Grafius, J.E. (in preparation) Sequential trait analysis applied to yield components in maize

Whitehouse, R.N.H., Thompson, J.B. & Ribeiro, M.A.M. de V. (1958). Studies on the breeding of self pollinated cereals. 2. The use of the diallel cross analysis in yield prediction. *Euphytica,* **7,** 147–169.

Session IV

HETEROSIS IN OUT-POLLINATING SPECIES

Chairmen

PROFESSOR G. RÖBBELEN AND PROFESSOR E. KURNIK

Heterosis in out-pollinating species

Y. DEMARLY

Université de Paris-Sud,
Laboratoire d'Amélioration des Plantes,
Orsay, France

SUMMARY

Homozygote or heterozygote structures as well as ploidy level seem to be connected with various systems of reproduction. These biological characteristics act as a whole from the point of view of selective value. They define the breeding system of a species.

Whenever the heterozygote structure has been selected by a plant, it seems obviously that each possibility of its genetic information evolved to maximization of efficiency; this covers dominance, overdominance, epistatic effects and regulation, interactions between nucleus and cytoplasm. These various aspects are illustrated in the genetic structure of maize and lucerne.

These selective values of a breeding system are suited to natural conditions. For agricultural purposes we need a reorganization of the qualities contained in the genetic stock of the species.

The breeder has to find the most efficient programme of improvement maintaining both additive and interaction effects. Various methods are discussed regarding out-pollinating field crops.

Heterosis is a comprehensive phenomenon. The plant behaves as a whole, each of its genetic and physiologic resources is involved in a strategy as outlined by J. F. Mac Key and B. G. Konarev. For these reasons, it is obvious that we must not look for one single explanation of heterosis but for an integrated set of coordinate elements such as:

- interaction between homologues;
- cohesion of genes in blocks known as linkats, possessing their global selective value (Demarly, 1968);
- relational balance between linkats whether homologous (dominance) or not (epistasy);
- chromosome number, ploidy level (allo-or autoploidy);
- mating system, population structure and genetic load in connection with the flexibility to environmental conditions.

The amount of heterosis may be interpreted as one element of this set which defines a genetic strategy of a plant species. To analyze this as a whole two ways are offered to the plant breeder: (a) Statistical and quantitative genetics; and (b) Genetical aspects at the cellular level. The purpose of this paper is to afford some personal ideas on both points.

The mean performance of an individual plant in a progeny can be expressed as: the sum of the mean gametic value of its parents and of the interaction in the mating of parental gametes. The first part represents the general combining ability of each parent (GCA), the second the specific combining ability (SCA) interaction. In terms of variances we write:

Genetic variance = Variance (GCA) + Variance (SCA)

The genetic supports of these effects are drawn from Fig. 1 where F is the inbreeding coefficient.

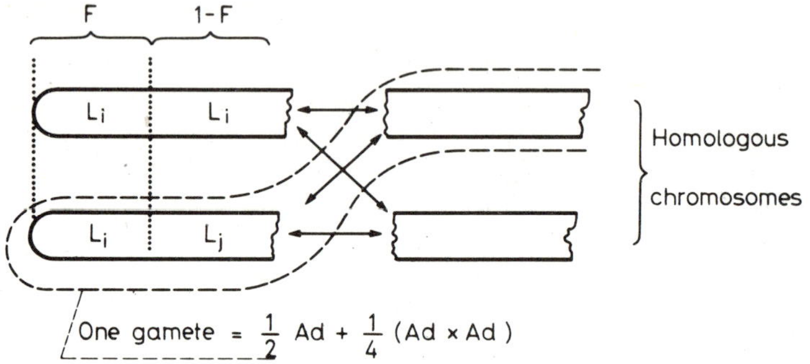

Fig. 1.

The general combining ability is for one linkat l_i/l_j

$$\frac{1}{2}(l_i + l_i)F + \frac{1}{2}(l_i + l_j)(1 - F) = Fl_i + \frac{1}{2}(l_i + l_j)(1 - F)$$

For a polygenic trait we have then:

$$Var(GCA) = \frac{1+F}{4} \; Var\,Additive + \sum_n \left(\frac{1+F}{4}\right)^n Var \underbrace{Ad \cdot Ad \cdot \ldots}_{n}$$

where n is the number of interacting loci we derive from the total genetic value.

So, the amount of heterosis when combining two parents depends upon additive effects, dominance and overdominance, as well as upon various epistatic interactions (modifiers, complementary effects, regulatory systems, etc). In recent years plasmatic actions which are not easily drawn from classical analysis have been discovered (it is easy to see from Fig. 2 that for tetraploid species, the mean gametic value (GCA) implies not only additive effects, but also dominance and additive × dominance interactions. For one locus GCA = $\frac{1}{2}$ Additive + $\frac{1}{6}$ Dominance).

According to the concepts of inbreeding and heterosis in quantitative genetics inbreeding arises:

1. from a negative interaction between two identical homologous allels (or linkats);
2. from a loss in the diversity of epistatic effects.

Heterosis depends on interactions between homologous or non-homologous loci. During the last five years, various quantitative analyses gave evidence on the great importance of multiple interactions. Nowadays, we cannot regard the value of an allel as a unit, but as giving one of the oscillatory effects which interact in the cytoplasm (as pointed out by Goodwin, 1963).

A comparison between the yield of single and double hybrids in lucerne, an autotetraploid species, shows the superiority of the tetragenic diversity and gives an idea of the importance of complex interactions.

If we deal now with the breeder's point of view, the question arises: how to maximize heterosis, or rather, how to get the best GCA and SCA?

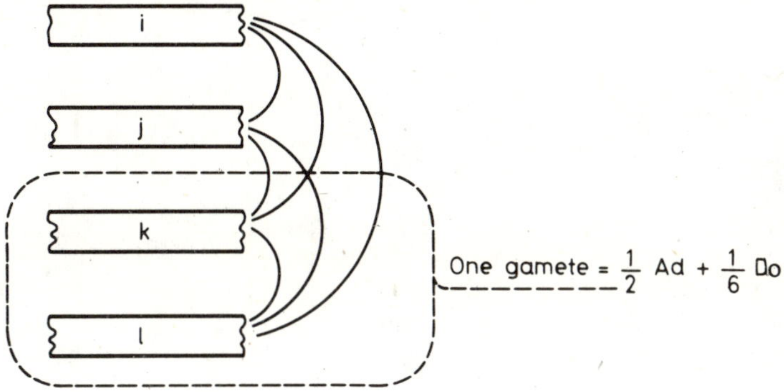

Fig. 2.

The most common methods to improve cross pollinated species lead to first generation hybrids or to synthetic varieties. A first generation hybrid or progeny may be obtained from: controlled pollination (avoiding any inbreeding); between lines or homogenous structures (homozygous lines, clones, F_1 hybrids). Synthetics imply: a fixed generation derived from free pollination of a given number of parental structures previously improved for their combining abilities, and each of them identically reproducible (Fig. 3).

The four main steps to improve these plants are as follows:

1. An inbreeding period to draw lines from populations, during which line selection is undertaken.
2. Choice of the best combinating parents.
3. Crossing parental structure.
4. Multiplication and flexibility tests.

This approach is summarized in Fig. 4.

During the phase of inbreeding, the choice between lines only concerns additive and cis additive × additive interactions. In fact, the breeder selects upon aggregative blocks of interacting genes or linkats (see the works of Mukai, Kimura, Franklin and Lewontin). At each generation of selfing, the epistatic relations rapidly vanish. For a character which is governed by 10 loci, one pure line retains only one within one thousand homozygous formulas.

Thus inbreeding appears as a drastic restrictive process and a loss of linkats having superior dominance but bad additivity. (Theory of junctions (Fisher, 1949) shows the main importance of a choice at self 1 generation.) At the end of such a process,

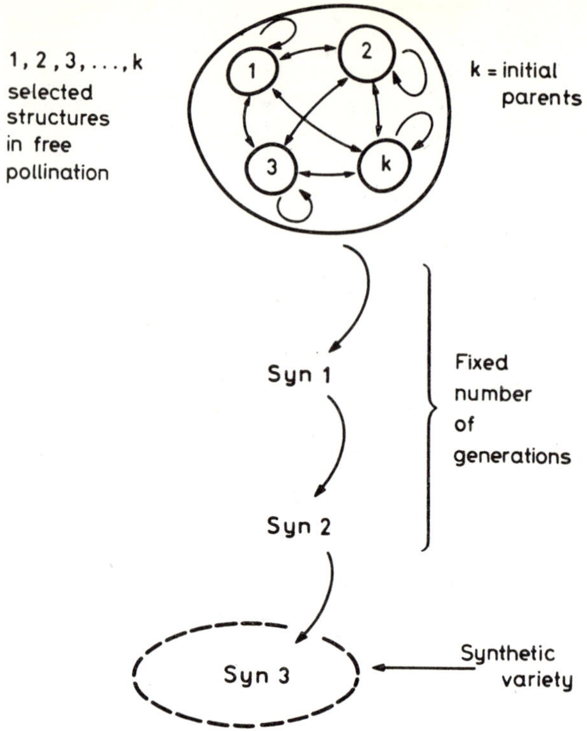

Fig. 3.

when trying to restore hybrid vigour must we be surprised at finding only predominant additive variance and general combining ability? The probability to get back the best specific interactions is then near to zero because of the poverty in allelic polymorphism. This is probably the reason why the gains in corn breeding have been so low during the last five years.

Methods using recurrent selection were proposed to avoid such a narrow genetic basis. The principles of these methods are:
– to keep a larger variability through cyclic generations of gene mixing;
– to enrich the pool with favourable genes (Fig. 4).

The value of recurrent selection lies in the method used to increase favourable genetic combinations in the improved population pool. Various processes can be worked with; such as recurrence following:

1. Mass selection; 2. Ear to row selection; 3. Top cross selection; 4. Selfed progenies selection; 5. Reciprocal general combining ability test; 6. Full-sib selection.

It is obvious that the first three methods only deal with general combining ability, i.e. with additive effects and additive interactions between linkats. Yet successful results were obtained (cf. Sprague, Millet & Brimhall, Webel & Lonquist).

The choice based on selfing leads to improved populations with higher frequency of good additive effects and a lower rate in recessive defects. Yet allelic polymor-

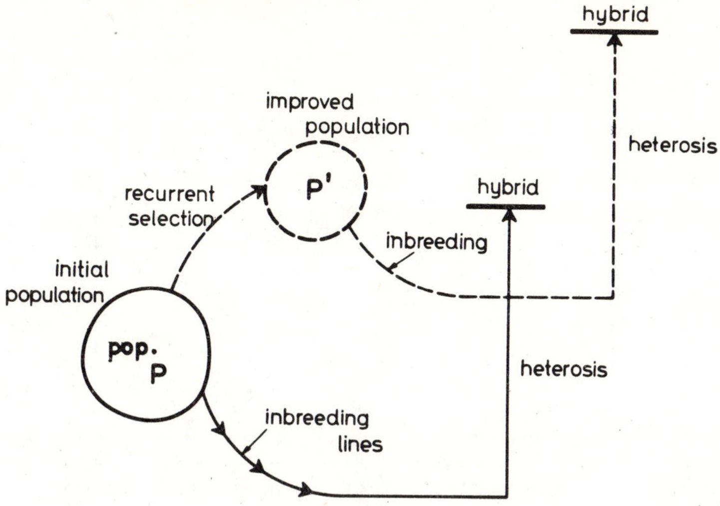

Fig. 4 (after J. L. Marchand).

phism and diversity of allelic interactions are strongly reduced in this choice. It seems that full-sib recurrent selection (Eberhart, Hallauer, Moll) and reciprocal recurrent selection based on specific combining ability would be the most efficient. They maximize various quantitative effects, additive as well as dominance and epistasy. More sophisticated methods can be proposed in various particular conditions (for tropical maize see Le Conte; for lucerne see diallel-genealogic method).

For each of these programmes an estimation of combining abilities must be obtained. Since the diallel technique is the most commonly used it seems necessary to study its limits. We shall not come again to the hypothesis in the various models of diallel-cross. Our purpose is to comment on the availability of such a test from a biologist's point of view.

In our laboratory, H. Feyt uses simulated genotypes with the following parameters: diploid plants with only one chromosomal pair; quantitative characters controlled by 36 loci; rate of additivity, dominance, overdominance; interactions between loci. Sharing between cis and trans epistasy; aggregative blocks of 3 loci (linkats); random design for diallel.

The main results we got from many replications of such a diallel are:

1. very poor precision on estimated variances for GCA and SCA; when a 10 × 10 diallel is used variation coefficient is about 50%;
2. large bias between genetic variance and table variance (due to covariance between crossed parents): genetic variance is estimated by 2 variance GCA + variance SCA; the diallel table variance comes from covariance Full Sib;
3. important changes connected with genic effects which strengthen the part played by additive effects on GCA and of dominance and overdominance on SCA;

Table 1. Variances *(V)* under various genetic structures (extracted from H. Feyt)

Additivity	Dominance	Overdominance	Epistasy	Ratio cis/trans	Homozygous parents		Heterozygous no linkage		Heterozygous linked genes	
					V_{GCA}	V_{SCA}	V_{GCA}	V_{SCA}	V_{GCA}	V_{SCA}
100					18	0	9	1	9	1
	100				13	12	6	4	6	4
		100			7	24	3	8	3	8
50	50				15	5	7	2	7	2
	50	50			9	17	4	6	4	6
50		50			12	10	6	4	6	4
33	33	34			13	10	6	4	6	4
100			100	100%	55	2	21	3	26	3
100			100	50%	45	7	19	4	22	4
100			100	0%	40	23	19	8	21	8
	100		100	100%	47	38	18	15	23	15
	100		100	50%	31	44	13	16	15	16
	100		100	0%	27	72	13	24	13	23
		100	100	100%	31	75	12	26	16	26
		100	100	50%	19	26	7	26	8	27
		100	100	0%	14	106	6	34	7	34

4. in such a diallel the more heterozygous the parents, the more inconsistent are the data. Not a great difference is observed whether genes are linked or not.

With H. Feyt we are trying to introduce in this model new quantitative parameters which are closer to biological processes such as: linkat tension: related to aggregative ability of alleles; rate of crossing over as a function of linkat value; sharing of cis and trans epistasy, each being relative to different biological phenomena such as genetic control of crossing over, and balanced heterozygous blocks.

The deep meaning of these distinctions and the efficiency of these parameters are given in Table 1. At this stage we hope to be able to compare other effects and new tests.

Consider now the cellular level of heterosis. I shall insist on nucleocytoplasmic interactions which seem to me the most promising feature.

The total amount of genetic information $\dot{I}$ (nuclear as well as plasmatic) of a plant comes into use following the normal process of a programme. In a given cell, at every moment only a part $d\,\dot{I}$ is working, as if some epigenic context would inhibit most of the genome. We may imagine that this epigenic context results from a set of two kinds of dynamics (Thom): fluid dynamics such as histones, small-proteins, hormones, mineral salts, glucides, etc., and filtering dynamics which are more stable such as membranes, cell walls, epiderms, plastids, hairs, etc. So any effect from environmental and internal conditions will be transferred and translated through successive series of filtering and fluid dynamics to the epigenic context. This will allow the normal sequence $d\,\dot{I}_1$, $d\,\dot{I}_2$, $d\,\dot{I}_3$, ... of a programme (Fig. 5).

The first effect of this programme is to induce a morphological or functional differentiation of the cell. We may interpret this as a "stratification" in cytoplasmic diversity, in other words as putting in order the topologic structure of the plant material. Differentiation thus appears as a decrease in entropy, through "consumption" of information. A young cell is not repressed. A partially stratified cytoplasm induces a specific repressing epigenic context. A whole stratified cytoplasm leads to senescence, total repression and death (Fig. 6). An experimental illustration of these ideas arises from a comparison between normal development of pollen grain until maturity and *in vitro* pollen-culture. The normal process is obviously in its morphology and in its functions. *In vitro,* the change in spindle orientation is connected with a reversal of cytoplasmic arrangement and leads to a return to the first steps of embryogenesis (Fig. 7).

We have already interpreted heterosis (Demarly, 1972) as related to youth. If heterozygous structures are less repressed than homozygous, because of an initial cytoplasmic rejuvenescence, then a new meaning appears for fertilization, male-sterility, maternal effects and heterosis.

Imagine that during normal morphogenesis, the nucleus undergoes differentiation of the structure of cell, organs and individual in a way which leads to coadaptation of a repressing context: both stratification of the epigenic dynamics and of information disposal (Fig. 6). Hybrid vigor may result from an initial superiority of the heterozygous zygote. It can be shown (Demarly, 1972) that a higheı rate of cell activity during the first three mitosis cycles is sufficient to induce the superiority of an individual all its life long.

If we consider that male and female mature gametes are a repressed form, when fertilization occurs two kinds of situations can be generated according to whether self pollination or out-crossing occurs (Fig. 8).

Such a view of heterosis where the hybrid genotype is fully derepressed in the egg as soon as fertilization occurs allows us to say that an egg resulting from inbreeding is cytoplasmatically older than an egg coming from out-crossing.

These hypotheses are not in disagreement with the results of I. V. Sarkissian, Mac Daniel obtained in 1972 and Donley in 1972. Each set of organites in the cell (mitochondria, plastids, etc.) is probably represented by genetic populations. One part of these populations is of maternal origin and the other part was transmitted by the male parent.

There is experimental evidence that when arriving in a foreign cytoplasm, cell organites may have a very high rate of replication. Even in some cases they eliminate the local population of organites. As an example we shall quote the data of Diacumaos (1965) in *Neurospora* (Table 2) when he injected a few "Poky" mitochondria in a normal cytoplasm, after 10 to 15 transfers, all the colonies became of "Poky" type.

Table 2. Transition from Normal to Poky

Number of transfers	Rate of growth in 50 colonies		
	Normal	Medium	Slow and Null "Poky"
10	30	20	
12	28	17	5
13	17	25	8
15	0	3	47

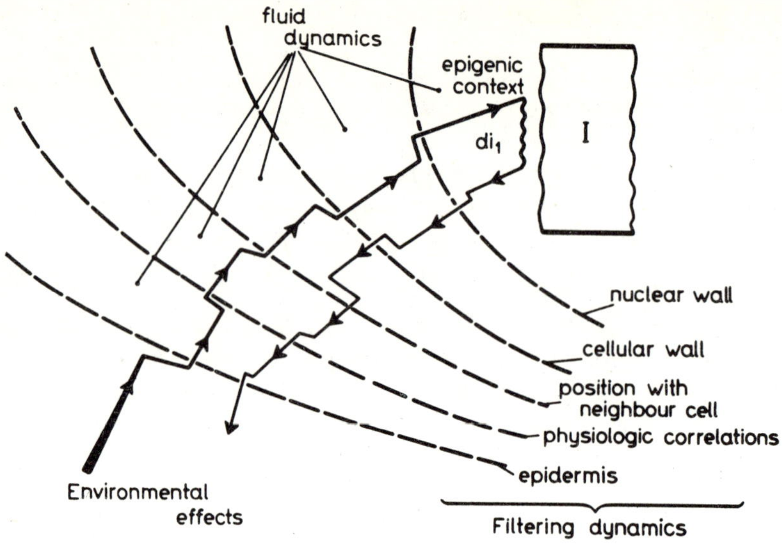
fluid
dynamics
epigenic
context
I
di_1
nuclear wall
cellular wall
position with
neighbour cell
physiologic correlations
epidermis
Environmental
effects
Filtering dynamics

Fig. 5.

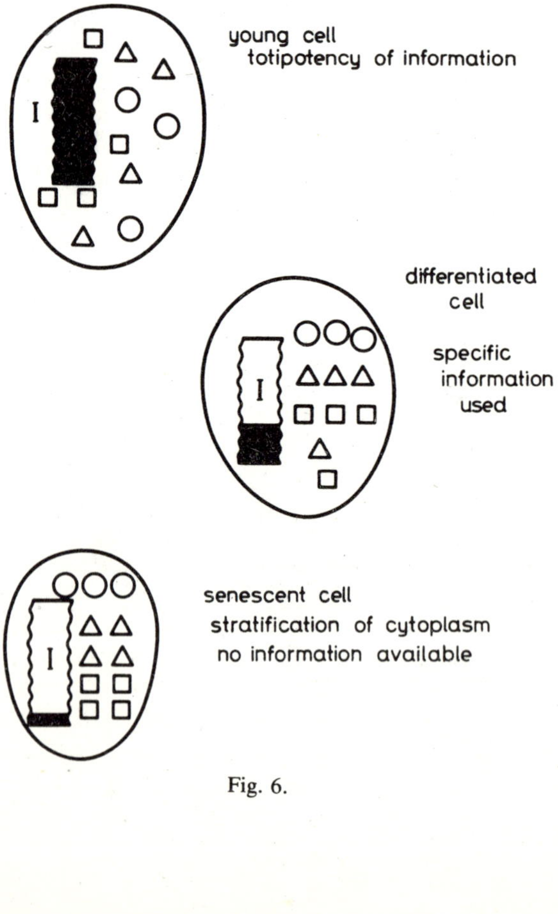
young cell
totipotency of information
I
differentiated
cell
specific
information
used
I
senescent cell
stratification of cytoplasm
no information available
I

Fig. 6.

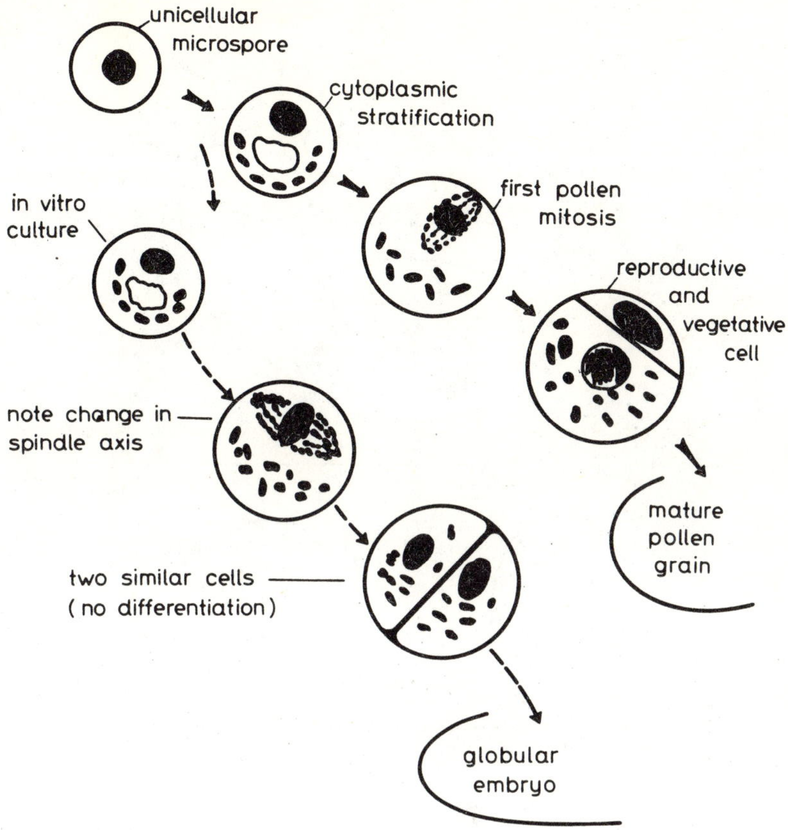

Fig. 7.

In Gymnosperms Camefort (1966) observed the process of fertilization with electron microscope. He showed that a trail of cytoplasm enters around the male gamete and that male mitochondria undergo replication and in some species invade the embryo cytoplasm.

From these examples I propose the hypothesis that when a population of organelles enters a cytoplasm different from its own there is a change in its kinetics (and generally an increase of replication). After this first period an adaptative process is set up and a balance between allo and auto organites is reached.

This view could also explain male sterility. In the first explanation of cytoplasmic male sterility (Berville & Demarly, 1970) we suggested at least two (but probably a wider distribution) of kinds of mitochondria: a highly energetic type and a less energetic type. Each genotype–cytoplasm interaction specifically controls the ratio of first to second.

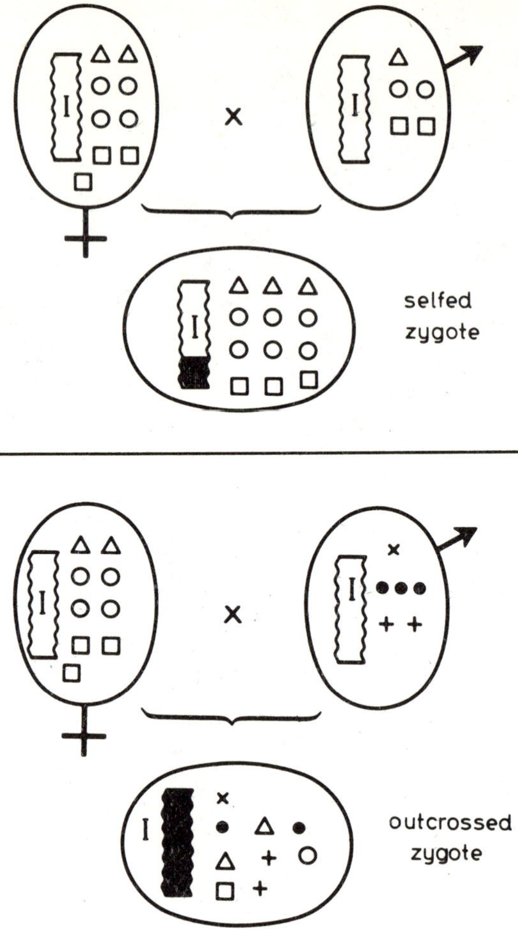

Fig. 8.

A male sterile cytoplasm appears when the proportion of low-energy types becomes too important. Then at the stage of meiosis the turn out of energy is insufficient.

When a normal parent is crossed with a male-sterile female, in most cases the F_1 progeny is sterile. Yet this male-sterile progeny possesses a larger vigour than normal plants at the seedling stage. This high energy level of sterile plants can be due to the rapid replication of male mitochondria when surrounded by alien sterile cytoplasm. This superiority vanishes as the plant grows. Generally at meiosis the cellular energy is lower in male sterile cytoplasms. At this stage of development the equilibrium between cell organites evolves towards maternal equilibrium i.e. sterile cytoplasmic structure. Oligogenic restoration could be viewed as a change in a factor which influences the kinetics of replication. Polygenic restoration would be the effect of modifiers of the quantitative amount of energy needed for gametogenesis. Moreover

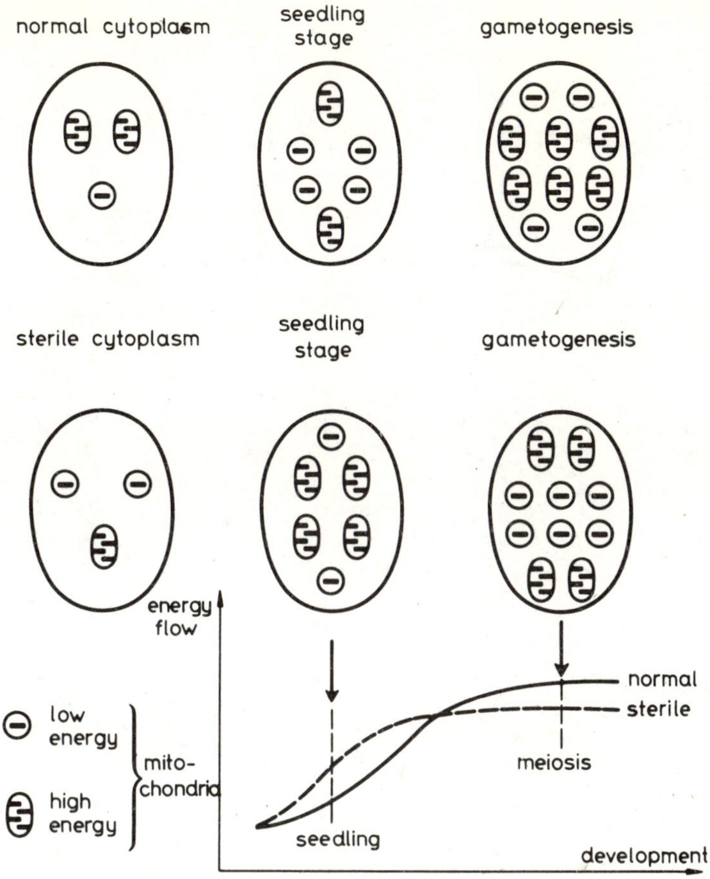

Fig. 9.

this could explain the effects of grafting, temperature and environmental conditions which change the energy flow wanted (Fig. 9).

Then if heterosis is a kind of cytoplasmic rejuvenation how can a plant-breeder increase this level? It has been shown (Fig. 10) that when dynamic tenses are suppressed, the cytoplasmic organization is disturbed. The cellular entropy increases again; this is known as dedifferentiation process. A meristem, an organ, or an embryo culture in artificial medium induces such effects. Beside virus cleaning off, these techniques display an increase of vigour. When plant tissues are grown in a specific medium inducing a high rate of cellular mitosis without any morphogenesis, the plasmatic disturbance can be more efficient and we can get selective drifts of cytoplasmic structures. Sibi (1974) in our laboratory tried to induce such cytoplasmic variants in lettuce (*Lactuca sativa,* cultivar Val d'Orge). The initial plant was homozygous and gave no segregation by selfing. A few plants of this progeny were

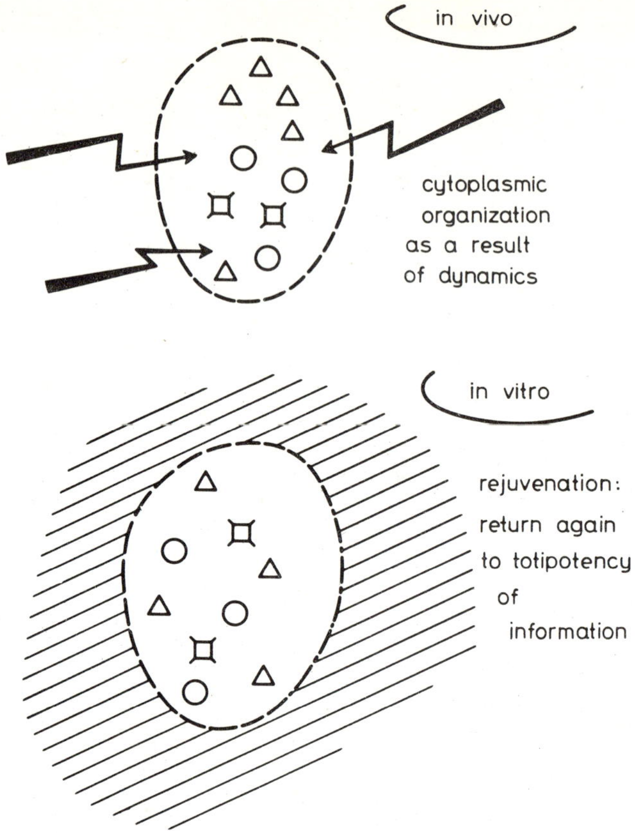

Fig. 10.

planted and selfed to be kept as standard, a part of a cotyledon of the others was used for tissue culture and gave amorphous cellular strains during many transfers in medium I (Fig. 11). In a medium II regenerated plants were obtained from these strains. Most of them showed lethality and sterility, but for a few plants normal or nearly normal phenotypes were obtained. Chromosomal counts showed a normal number but in the selfed progeny some of special kind were seen.

No segregation occurred through three successive generations of selfing: the whole progeny showed the variant phenotype. By reciprocal crosses the cytoplasmic origin of these variants was verified. But the most interesting feature of the experiment was that in five crosses between the variant and the standard a strong heterosis-like effect was found (Fig. 11).

If these results can be generalized to other species, the possibility to get cytoplasmic heterosis in homozygous conditions would be opened. In the scope of these ideas on hybrid vigor, plasmatic structuralization and differentiation, many aspects are under experiments at Orsay (University Paris XI).

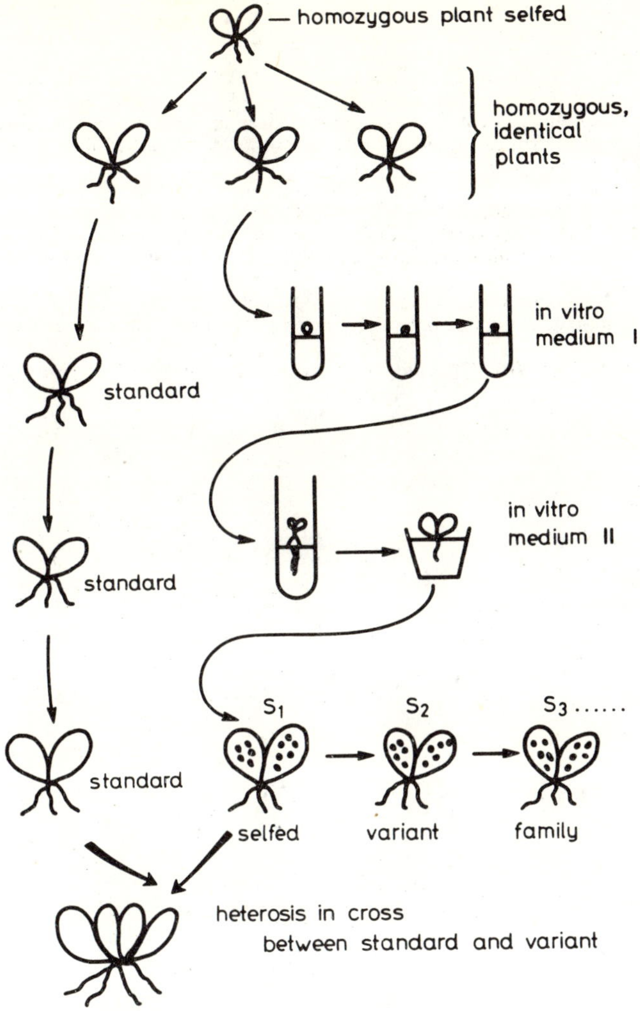
— homozygous plant selfed
homozygous, identical plants
in vitro medium I
standard
in vitro medium II
standard
S1
S2
S3 ......
standard
selfed
variant
family
heterosis in cross between standard and variant

Fig.11.

REFERENCES

BERVILLE, A. & DEMARLY, Y. (1970). La stérilité mâle cytoplasmique Congrès EUCARPIA – Versailles – Juin 1970

CAMEFORT, H. (1966). Origine et évolution structurale d'un cytoplasme propre à l'embryon ou " néocytoplasme" chez les pins. Certificat de Botanique, exposé le 24 février (Reims) 14 pages

DEMARLY, Y. (1972). Régulation et hétérosis. *Annales de l'Amélioration des Plantes,* **22** (2), 143–166.

DIACUMAKOS, E.G. *et al.* (1965). A cytoplasmic character in *Neurospora crassa.* The role of nuclei and mitochondria. *J. Cell. Biol.,* **26,** 427–443.

FISHER, R.A. (1949). The theory of inbreeding. Oliver and Boyd, Edinburgh.

GOODWIN, B.C. (1963). A statistical mechanism of temporal organization in cells. *Towards a theoretical Biology,* **2.**

MAC DANIEL, R.G. & SARKISSIAN, I.V. (1968). Mitochondrial heterosis in Maize. *Genetics,* **109,** 4.

SIBI, M. (1974). Création de variabilité par culture de tissus *"in vitro"* Thèse 3ème Cycle – Université de Paris-Sud, Orsay.

TILNEY-BASSETT, R.A.E. (1970). The control of plastid inheritance in *Pelargonium. Genet. Res.,* **16,** 49–61.

EUCARPIA VII (1976)

F_1-hybrids in some out-pollinating crops

C. DORSMAN

Institute for Horticultural Plant Breeding, Wageningen, The Netherlands

SUMMARY

Modern horticulture not only needs productive varieties but first of all varieties adapted to mechanical harvesting, showing reliability of yield and good quality. In variety trials with onions, carrots, cole crops and witloof F_1-hybrids were compared with open pollinated varieties. It is concluded that the productivity of F_1-hybrids is in some cases clearly superior to the best open pollinating varieties, in other cases about equal. Moreover, F_1-hybrids seem to open up the prospect of breaking down negative correlations which may exist between productivity and earliness. As to uniformity F_1-hybrids of cole crops and witloof are clearly superior to open pollinating varieties. For several quality aspects again F_1-hybrids are superior.

Some possible disadvantages of F_1-hybrids are mentioned: the possibility of genetic erosion by the general use of F_1-hybrids, the possibility of a narrow adaptation to environmental conditions and high seed costs. Several technical breeding problems are discussed: inbreeding depression, combining ability, male sterility (genetics, stability, insect pollination) and sporophytic incompatibility (seed production on inbreds, stability, relations between *S*-alleles, insect pollination).

As to the cultivation of F_1-hybrids of carrots, onions and cole crops in Europe it is concluded that a break-through of the F_1-hybrids seems imminent.

Finally, special attention is called for a number of research objectives in connection with F_1-hybrids.

INTRODUCTION

Since at our institute we are interested in plant breeding primarily in view of the possibilities to improve horticultural products by providing better varieties, I intend to discuss in the first place the pros and cons of F_1-hybrids in horticulture. Next I shall try to put forward several bottlenecks in the breeding of F_1-hybrids, thus formulating those problems which in my opinion need special attention from the plantbreeder. In my talk I shall concentrate on those out-pollinating crops which are most important to horticulture and of which we have the most experience at our institute, namely, onions, cole crops and carrots.

I. WHY F_1-HYBRIDS?

When speculating about the merits of F_1-hybrids I think one of the first essentials is to analyze the modern trends in European horticulture. After this we should look into the question how F_1-hybrids can fulfill the needs of the grower of today and more especially of tomorrow.

To my mind the development of modern horticulture in Europe is influenced basically by three dominant factors:

1. quickly rising costs of labour and energy,
2. increasing international competition for products produced far from the big consuming areas,
3. changing demands by the consumer and the processing-industry.

Taking into account these trends we endeavour to derive the consequences for the grower. From an economic point of view *high productivity* of our crops will remain important in the future just as it has been in the past. However, the interesting point is that other characters may receive equally high or even higher priority. In my opinion the future grower will require from the plantbreeder new varieties which are superior in the following characters.

a) *Adaptation to mechanical harvesting*

This means that all plants must reach a certain stage of maturity at the same time, which implies a large amount of uniformity of the crop. Moreover, the constitution of the parts to be harvested should be such that they survive the rather rough handling by machinery.

b) *Reliability of yield*

This means in particular resistance to diseases and pests. This need received strong emphasis in the last few years by the growing concern about fungicidal and pesticidal residues. Freedom from residues might of course also be considered as a quality aspect.

c) *Better quality*

As both the consumer and processing-industry become more quality conscious, breeding for better quality will become far more important.

Before attempting to answer the question if F_1-hybrids meet these modern aims of horticulture better than open-pollinated varieties I want to mention quite another special reason for the interest in F_1-hybrids shown by the breeder, namely, the *built-in protection* of the variety. Breeding is a fascinating job but an extremely expensive one. For the private breeder it is essential that he should benefit from his efforts. Essential for himself to make the necessary investment to be able to keep on producing new varieties but also essential to the grower and consumer who need a steady flow of better varieties. Whereas legal protection by breeders rights is only possible in a few countries, F_1-hybrids provide the breeder with worldwide protection.

II. ADVANTAGES OF F_1-HYBRIDS

Let us now take a closer look at the possible advantages of F_1-hybrids regarding the characters mentioned above.

1. PRODUCTIVITY

International data based on experiments about the productivity of F_1-hybrids in comparison with open-pollinated varieties are rather scarce. In the USA an example is given by Munger *et al.* (1959) of a 9–30% higher productivity of onion hybrids. However, in the USA there does not seem to be a consensus of opinion that F_1-hybrids of onions, carrots and cole crops are clearly superior in productivity. On the other hand, a rather impressive number of trials with cole crops, onions and witloof in the Netherlands, France, the German Democratic Republic and Poland suggest that with these crops we can obtain F_1-hybrids which are more productive than open-pollinated varieties although some results are more promising than others. Our experimental hybrids with carrots point in the same direction.

We may assume that productivity is polygenically inherited and may be determined partly by additive factors and partly by non-additive factors. Heterosis *sensu stricto,* by which I understand the non-fixable heterosis, depends only on the non-additive acting factors and finds its expression in the specific combining ability (SCA) of the inbred lines. It is important to know what part of the productivity depends on this (SCA) effect because additive effects can also be obtained by other means than F_1-hybrids. Unfortunately, such analyses are rather few. Johnston (1971) experimented with inbred lines of different *Brassica* types. He concluded that the GCA effect was highly significant for all yield components studied in the F_1-hybrids and that the SCA effect although also significant, was of a much lower level. At our institute M. Nieuwhof studied the GCA and SCA effects on the head weight of pointed headed cabbage, and obtained similar results. However, again at our institute Eenink obtained quite the opposite results with several growth components of Brussels sprouts (personal communication).

These data seem to indicate that the main influence of additive or non-additive factors on productivity varies with the population and that F_1-hybrids are therefore not always the one and only possibility to reach a high productivity.

2. EARLINESS

Now the absolute level is not the only interesting aspect of productivity. For several crops earliness is very important too. Several cases of a better earliness in F_1-hybrids are mentioned for kohlrabi (Fabig & Nowak, 1969), white cabbage (Hoser-Krauze, 1972) and pointed headed cabbage (Dutch variety trials, 1972, 1973). At our institute an analysis of the GCA and SCA effects on earliness of pointed headed cabbage proved the SCA effects to be about equal to the GCA effects. Moreover, it is extremely interesting that with pointed headed cabbage the negative correlation between earliness and productivity that exists in open-pollinated varieties could be partly broken down in the hybrids.

3. UNIFORMITY

The extreme importance of the genetical uniformity of F_1-hybrids is illustrated dramatically by the success-story of Brussels sprouts hybrids in the Netherlands. After the introduction by our institute of the hybrids Thor and Olaf in 1963, the first hybrids produced by private seed firms appeared in our Variety List of 1969. Nowadays, five years later, F_1-hybrids have won over 80% of the Dutch market. The main reason for this breakthrough was the uniformity of the F_1-hybrids which made mechanical harvesting possible. Many trials with white cabbage, pointed headed cabbage and witloof also show the F_1-hybrids to be far superior in uniformity in comparison with the open-pollinated varieties.

It is less evident that the genetical uniformity of the F_1-hybrids will express itself phenotypically equally strongly with crops like carrots and onions. The essential parts of these crops develop within the soil and we could expect that at least part of the genetical uniformity will be overshadowed by environmental factors. However, experimental hybrids both of carrots and onions at our institute give us good hope that with these crops also the genetical uniformity of the hybrids will have a real practical value.

4. RESISTANCE

The crops I am talking about are no happy examples to illustrate the advantages of F_1-hybrids in breeding resistant varieties. Many of the most important diseases and pests are extremely difficult to handle and often no source of resistance has been found in the same species up to now (onion-fly, carrot-fly, nematodes, viruses, *Botrytis allii*).

However, Williams *et al.* (1968) have given an elegant example of the possibilities with their cabbage hybrids Hybelle and Sanibel. Another example is the breeding work by W. H. Gabelman with onions, in the USA, who tried to combine within an F_1-hybrid resistances against pinkrot *(Pyrenochaeta terrestris)*, fusarium *(Fusarium oxysporum f. cepae)*, softrot *(Erwinia carotovora)* and tipburn (personal communication).

5. QUALITY

Again the Brussels sprouts hybrids have made it quite clear that F_1-hybrids provide an excellent opportunity to improve quality, provided that we possess a broad collection of parent lines which permits us to select these parents for important quality characters. Equally favourable results were obtained with witloof, pointed headed cabbage and white cabbage. With onions again the hybrids tested for our Variety List are of good quality, one of them having a far better skin quality than the open-pollinated varieties. An interesting point is that onion hybrids seem to open up the possibility to combine high productivity with high dry matter content whereas in the open-pollinated varieties a negative correlation exists between these two characters.

Our experimental hybrids with carrots give us good hope that we will obtain better quality characters such as colour homogeneity and firmness combined with high productivity.

From the many experimental data at our disposal we may conclude that F_1-hybrids in many respects meet with the demands of the modern grower. They offer new possibilities to improve productivity, earliness, uniformity, quality and sometimes the resistance of our crops. The opportunity of breaking down negative correlations between productivity on the one hand and earliness and quality on the other is interesting. Moreover the general aspect that F_1-hybrids open the way for the breeder to modify his variety assortment, to make it more flexible and better adapted to changing demands and special circumstances is extremely important, it is perhaps the most pronounced advantage of F_1-hybrids.

III. DISADVANTAGES OF F_1-HYBRIDS

Of course, all is not gold that glitters. F_1-hybrids have their problems too. Before going into technical breeding problems I want to discuss three general points.

1. GENETIC EROSION

When we see the all-out effort to change over to the breeding and growing of F_1-hybrids, as happens today in the Netherlands with Brussels sprouts, there is a real danger that hundreds of growers' "strains" might get lost without anybody noticing it, resulting in the loss of genes which we might need desperately in the future. I consider it a responsibility of our national research institutes to prevent this possible genetic erosion. It could be done quite easily by collecting old open-pollinated varieties and producing seed by collective flowering. An allocation of task to the European plant breeding institutes could be an elegant solution of this problem.

2. ADAPTABILITY

From a breeder's point of view the genetical uniformity of F_1-hybrids would be a general drawback if this should diminish their adaptability to widely divergent conditions of soil and climate. It might be supposed that this danger is counterbalanced by the heterozygosity of the F_1-hybrids. However, there are indications that this is not always the case. In the USA, the maize single cross hybrids seem to have a narrower adaptability than three-way and double-crosses.

We might derive other indications from vegetatively propagated crops which are genetically absolutely uniform by being clones but which are often extremely heterozygous. When we look at apples we find that of our two most famous varieties the Golden Delicious is grown from Marseille to Hamburg and shows a wide adaptability. However, Cox's Orange Pippin only grows well in those north-western regions which have a cool sea climate and consequently has a far more limited distribution. This shows convincingly that heterozygosity by itself is no guarantee of a wide adaptability. On the other hand, it proves that even absolute genetical uniformity, when the whole crop has the same genotype, does not exclude a remarkable adaption to widely differing environments.

A typical example of differences in adaptability is provided by two witloof hybrids, produced by H. Bannerot (personal communication), one of which can be forced successfully at one forcing period ("Flambor"), while the other performs quite well at several forcing periods ("Zoom"). Taking also into account the successes of Japanese cabbage hybrids in widely different countries, I feel that adaptability is not necessarily a problem inherent in F_1-hybrids but a subject which needs more study.

3. SEED COSTS

A very real drawback of F_1-hybrids might be the costs of the seed. It is inevitable with crops like onions, carrots and *Brassica* that the production of hybrid seed is expensive in consequence of many labour-consuming techniques. We have no exact calculations but in general we estimate the production costs of hybrid seed of these crops to be 3–6 times as high as those of the seed of open-pollinated varieties. This makes the prospects of hybrid seed rather doubtful with crops where a lot of seed is needed per hectare and when the normal open-pollinated varieties already have a high standard. Such is for instance the case with the high quality carrot variety Amsterdam Forcing where we need 17 kg seed per hectare, which amounts to approximately 7% of the total production costs. With bulk crops like onions, cole crops and the "coarser" varieties of carrots the prospects are more favourable but there are strong differences between these crops. With cole crops where the seeds take only 1.5% of the total production costs the seed price is of rather minor importance. However, with onions the seed costs constitute 7% of the production costs and a 3–5 fold price of the F_1-hybrid seed could easily be prohibitive to its use. It is therefore essential to keep the costs of hybrid seed as low as possible. This could be done by changing from single-crosses to three-way and double-crosses. It is a matter of weighing the lower costs against any possible losses in productivity and uniformity. Unfortunately, only with a few of the crops mentioned do we have an impression of this aspect. In a trial with pointed headed cabbage we found a marked decline in uniformity, a loss of about 3 days in earliness and a tendency to lower productivity in the double-cross hybrids as compared with single-crosses. A trial with Brussels sprouts pointed in the same direction. Considering the important economic consequences the need for more research in this direction seems indicated.

IV. TECHNICAL BREEDING PROBLEMS

As has already been mentioned both costs and quality of F_1-hybrids are closely related to the technical problems which confront the breeder. Without going too much into details I shall try to give a survey of some most important ones.

1. INBREEDING DEPRESSION

The degree of inbreeding depression not only influences the amount of seed to be produced and consequently the seed price but also the reproducibility of the hybrid. Moreover, it may restrict the possible number of inbreeding generations, thus raising a barrier against optimal uniformity of the hybrid. The sensitivity to inbreeding differs with the species. Whereas cole crops permit inbreeding to an I_5 or even further, although seed production may then drop to an extremely low level, carrots and onions in our experience hardly stand inbreeding to an I_3. B. Schweissgut at CNRA, Versailles found with leek again severe inbreeding depression of 35% in the I_1 and 50% in the I_2 to be a severe barrier to a hybrid program (personal communication).

To my mind the whole problem of inbreeding depression, which is a crucial point in all hybrid-breeding of cross-pollinating crops, is highly neglected in research. The phenomenon seems to be generally accepted as one of the less agreeable whims of Nature, a fickle woman better left alone, and only interesting to speculate about in genetical theories. I feel that it would be most interesting to tackle this problem from the physiological side, to look into the causes of the unsatisfactory growth of the inbreds and to try to stimulate growth by external treatment with growth regulators. We ourselves did some experiments in this direction. Although they were on too small a scale they gave some indications that positive results might be within our reach. I am convinced that a strong research effort in this field could lead to a break-through in hybrid breeding.

Other methods to by-pass the inbreeding problem could be the production of homozygotes from anther-culture or by inducing macromorphs. However, up to now the efforts to produce plants from anther culture have been negative with onions and carrots, while the Japanese results with *Brassica* could not be reproduced. Furthermore, a thorough investigation by Eenink at our institute with cole crops proved that macromorphs induced by prickle pollination are heterozygous.

The only way open to us is to select our inbreds for vigour with an open eye to the danger that by selecting the strongest plants from our lines we might select the most heterozygous ones.

2. COMBINING ABILITY

I already mentioned that an analysis of the GCA and SCA effects on important characters like productivity, earliness and quality is important to distinguish fixable and non fixable components. For the efficiency of a hybrid program it is moreover essential that we get this information of the breeding value of our potential partners as early as possible. In the case of male sterility this should be done as soon as reliable A lines are available. We can use for instance the North Carolina design No. 2, first in a top-cross between the A lines and a number of commercial varieties, followed by crosses between the A lines with selected lines from the best top-cross varieties.

In the case of self-incompatibility a diallel crossing program allows an analysis to be carried out already after one generation of inbreeding. If differences between reciprocal crosses are not to be expected, a half diallel without reciprocals will do. However, in practice many set-backs may cause gaps in the system. That is why I want to focus attention on the method for analysis of the genetic variation in incomplete diallel crosses as described by Garretsen & Keuls (1973). This is a general method developed according to the theory of the least squares, applicable to any set of diallel crosses even when a number of combinations is missing from the system.

3. MALE STERILITY

Although in practice we gladly use male sterility as a tool to breed hybrids of carrots and onions, this does not mean that this phenomenon entails no problems.

a) *Genetics*

Knowledge of the genetics of male sterility is a first condition for using it efficiently. Although acceptable models were developed by Jones & Clark (1943) for onions and by Banga *et al.* (1964) for carrots, there are indications that other systems also exist.

Berninger (1955) found phenomena with onions which he explained by assuming a cytoplasmic factor *T* which, when present, always causes male sterility even when the chromosomal *Ms* factor is present in its dominant form. This is in contrast with the accepted theory of *S* plasma, manifesting itself only in combination with the homozygous recessive chromosomal sterility factor. If Berninger is right this *T* plasma would greatly facilitate the development of A lines as every male-fertile line could be used as a maintainer line.

Again with carrots M. Nieuwhof (personal communication) obtained strong indications that there are deviations from Banga's *ms* model. Banga stipulates that *ms* occurs in the presence of *S* plasma combined with a recessive chromosomal factor *a* or a dominant chromosomal factor *B* and the simultaneous absence of two other dominant chromosomal factors *C* and *D*. Nieuwhof supposes that other systems, probably polygenically determined, may also be present. These hypotheses have yet to be worked out but may have consequences for the practical hybrid breeding of carrots.

b) *Stability*

Especially with onions and to a lesser degree with carrots the stability of *ms,* an essential condition for its use, is a source of concern.

Experiments in 1963 and 1964 with carrots in the phytotron of our institute with both the petaloid and the brown-anther type of *ms* with a range of temperatures from 10°C to 26°C and daylengths between 8 and 24 hours did not reveal any influence of these circumstances on the stability of either type. All the same in the field we got the impression that the environment could influence the sex-expression. And indeed new experiments in our phytotron last year with temperatures of 10°C and 23°C proved to contain some A-lines which became fertile at 23°C whereas others stayed sterile. Dr. Strelnikova (Tiraspol, USSR) told me that in her experience the petaloid *ms* type was more stable and less sensitive to environment than the brown-anther type.

Quite another point with carrots is the fact that rather numerous maintainer (B) lines may produce a smaller or larger number of male sterile plants after some generations. Perhaps by careful selection this problem could be solved.

With onions we encountered many difficulties, again particularly with B-lines. Q.P. Van der Meer (personal communication) proved that with these lines a thermo-labile sex-expression is more the rule than the exception. With the variety Wolska he hardly found any pure male fertile lines at 14°C. With the variety Rijnsburger lower temperatures in general caused more *ms* than high temperatures. In view of these experiences we should watch the stability of *ms* in the A-lines and of male fertility in the B-lines both at low and at high temperatures with onions as well as with carrots. This is an essential precaution to prevent annoying set-backs at a later stage of our program.

It is quite possible also that other environmental factors such as the relative air humidity are of importance. This is a field still to be explored.

As a last point I wish to draw attention to the interesting possibilities of using gibberellic acid as a gametocide in onions. Van der Meer obtained 100% male sterility by spraying only once (4th June) 1% GA 4/7 at a dose of 2.5 cc per plant without any harmful influence on seed production. Although as yet this costs $ 17 per kilogram of seed, it is possible that more research on the use of spreaders and concentrations will bring down these costs quite appreciably. This could open up new prospects of creating male sterile partners in a very simple way.

c) *Insect pollination and seed set*

Our experience has taught us that many insects, particularly bees and bumble-bees, prefer male-fertile plants to male-sterile ones; Q.P. Van der Meer (personal communication) observed with onions that bees showed a 2:1 preference and bumble-bees even a 3:1 preference. Differences in quantity and quality of nectar seem to be involved, moreover, pollen collecting insects are obviously not attracted by male-sterile plants.

This of course seriously hampers seed production as the proportion of B-lines in maintaining A-lines and the ratio of C-lines in the production of hybrid seed have to be relatively high. Experiments with carrots by M. Nieuwhof gave the following results (personal communication): putting the seed production at a 3:1 plant ratio of *ms* and *mf* plants at 100%, a plant ratio of 6:1 gave 80% and a 9:1 ratio only 40% seed. There is an urgent need for more research into the causes of different behaviour of insects on *ms* plants.

However, we must keep an open eye for other sources of differences in seed setting. Quite recently we found quite surprising differences in seed production between the brown-anther type and the petaloid type of *ms* in carrots. This point, also mentioned by A. Bonnet (INRA, Montfavet, personal communication) certainly needs further research as well.

4. *SPOROPHYTIC INCOMPATIBILITY*

Although successfully used with cole crops and having the advantage over *ms* that it allows us to harvest hybrid seed from both parents, it is quite clear to all, who are "nourri dans le sérail" of sporophytic incompatibility that this system too has its problems varying from sib-production of plants to sib-production of seeds.

a) *Seed production on inbreds*

The techniques of bud-pollination are laborious, the seed production of the inbreds is often disappointing. Perhaps these difficulties could be overcome if we had a better insight into the real nature of the incompatibility reaction. Roggen (1972) studied the stigma-pollen interaction of Brussels sprouts by scanning electron microscopy. He put forward convincing arguments that the essential barrier in this incompatibility system is formed by the wax-layer covering the stigma and that the "recognition reaction" between this wax layer and the enzyme of the pollen is an extremely quick one. This explains why self-fertilization is possible when the wax layer has not yet been completed, as in the flower-bud stage, or when it is not quite unimpaired anymore as at the end of flowering. It also explains why several pollination methods by which we damage the wax layer, such as the electric aided pollination described by Roggen, are more or less successful. However, we still lack an absolutely full-proof method which permits us to obtain a satisfactory seed set after selfing in a simple way. To my mind a concentrated research effort may solve at least this problem. Another method to reduce the hand pollination of the lines by a factor 50–100 could be the system of using isogenic lines which only differ in their *S* factors. However, with this system the dominance relationships between the *S* factors may easily give complications. Weak *S* factors are mostly sufficiently stable when homozygous, but when present in a heterozygous form they often show a mutual weakening with the result that the self-incompatibility of such parents is insufficient. Probably with strong *S* factors this mutual weakening is far less.

b*) Stability*

The breeder must have every confidence in the stability of the self-incompatibility of the parent lines. Practical experience has made it highly probable that at least in certain cases this stability can be impaired by environment. Nasrallah & Wallace (1968) mention one case in which the seed set from selfing was increased from 1 seed per pod at a constant temperature of 27°C to approximately 10 seeds per pod when the plants were kept for 6 days at 27°C and then pollinated at 15°C. At our institute van Marrewijk (personal communication) found that a weak *S* factor was influenced far more by changes in the day/night temperature regime than a strong *S* factor. Hardly any research seems to have been done on the influence of relative air humidity. This is an absolute "must" for future research. For when the wax layer of the stigma is indeed the essential barrier to pollentube growth, it is quite conceivable that abrupt changes in temperature combined with changes in relative air humidity might cause cracks in this wax layer thus breaking down the self-incompatibility system.

c*) Relations between S alleles*

The outstanding research especially by Thompson (Thompson & Taylor, 1966; Thompson, 1968) and I. G. van Hal (personal communication) based on the concept of Bateman (1955) has explained much of the complicated relationships between *S* alleles, such as codominance and dominance in pistil, pollen or both. Nowadays an elegant method is available to the breeder to measure in a quick and easy way by UV microscopy the incompatibility reaction, and to judge the practical breeding value of the *S* alleles present in the parent lines. This knowledge has considerably diminished the percentage of selfs among the hybrid seed. At present the UV method is already applied as a normal procedure by many private breeders.

d*) Insect pollination*

Maximal production of hybrid seed by both parent lines is of course important. It might be expected that in the case of self-incompatibility, where both parents are male fertile, insect pollination will give no problems. However, field observations by Faulkner (1974) indicated that bees do indeed distinguish between lines and may stick to particular lines and consequently do not contribute to the necessary cross-pollination. He observed that bees distinguish shades of yellow flower colour not to be seen by the human eye but which could be detected by ultra violet light reflection. This could mean that in future we shall also have to pay attention to small differences in flower colour or other flower characters in our selection to improve seed set.

V. CULTIVATION OF F_1-HYBRIDS IN EUROPE

Looking at the impressive victories of maize hybrids, bearing in mind that already 52 years ago Jones started his breeding of onion hybrids, considering that in the USA an estimated 40% of the onion acreage and 50% of the white cabbage acreage are sown with hybrid seed, recalling the Japanese successes with hybrid cole crops, we must admit that Europe is lagging behind with the cultivation of hybrid varieties of cross-pollinating horticultural crops. Of course, we can advance several reasons for an explanation of this state of affairs. In the first place it is not quite honest to compare the scientific efforts of an army of research workers busy on a worldwide crop like maize with the efforts of a few lost soldiers toiling in the fields of horticulture. Secondly we must admit that at least with onions and carrots the surplus value of F_1-hybrids up to now has not been proved as convincingly as with some agricultural crops. However, there are quite definite signs that also in Europe F_1-hybrids of cross-pollinating horticultural crops are on the move. As I have only reliable data from the Netherlands, I shall give you these as an example:

CARROTS

This year a seed-firm presented the first commercial F_1-hybrids for our national variety-trials. Of course, nothing can as yet been said about their cultural value. More hybrids are soon to be expected as several seed-firms are active in this field. Moreover, the IVT has made available to the Dutch seed-firms a wide range of A-lines and maintainers. This surely will accelerate breeding activities in this crop.

ONIONS

The first single-cross onion hybrid was included in our Variety List in 1969. Today this number has only grown to 4. However, with onions it is the same as with carrots: several private firms being active and having been stimulated by the recent issue of a number of A- and B-lines from the IVT.

COLE CROPS

I already mentioned that in Holland, the Brussels sprouts hybrids have taken over 80% of the market and that they cover the whole scale from early to late varieties. Also with most other cole crops, such as white and savoy cabbage, we are on the brink of a breakthrough of the hybrids which can be expected at any moment. Considering the enormous acreage of cabbage all over Europe this will be a happy and memorable event.

VII. FINAL REMARKS

Weighing all pros and cons it is my strong belief that the breeding of F_1-hybrids of cross-pollinating horticultural crops is extremely valuable in the face of its economic prospects. This breeding presents many fascinating problems to the scientist with an open eye for the practical needs both for the breeder and the grower of these crops. In this paper I have purposely emphasized these practical sides, being prejudiced myself in favour of scientific research with practical aims such as is the task of an institute like ours. In this respect I called special attention to the following research objectives:

– the danger of genetic erosion by the loss of old open-pollinated varieties,
– adaptability of F_1-hybrids,
– artificial stimulation of the growth of inbreds,
– analysis of GCA and SCA effects,

– the cultural value of three-way and double-cross hybrids,
– the genetics of *ms,*
– the influence of the environment on the stability of *ms* and *si,*
– the artificial break-down of the *si* barrier,
– the behaviour of insects on *ms* and *si* lines.

I hope that I have succeeded in stimulating enthusiasm with regard to these problems and wish all success to those research workers active in the breeding of F_1-hybrids. They have a difficult task but one which benefits all who breed, grow or eat cross-pollinating horticultural crops.

REFERENCES

Banga, O., Petiet, J. & Bennekom, J.L. van (1964). *Euphytica* **13,** 75–93.
Bateman, A.J. (1955). *Heredity* **9,** 53–68.
Berninger, E. (1955). *Annales d'Amélioration des Plantes* **15,** 183–199.
Fabig, F. & Novak, M. (1969). Proceedings Heterosis Symposium, Sofia.
Faulkner, G.J. (1974). *Annals of Applied Biology* **77,** 181–190.
Garretsen, F. & Keuls, M. (1973). *Proceedings 1st Meeting Section Biometrics in Plant Breeding of Eucarpia,* Hannover, pp. 24–35.
Hoser-Krause, J. (1972). *Genetica Polonica* **13,** 109–136.
Johnston, T.D. (1971). *Euphytica* **20**, 81–85.
Jones, H.A. & Clarke, A.E. (1943). *Proceedings American Society of Horticultural Science* **43,** 189–194.
Munger, A.M., Jones, H.A. & Davis, E.W. (1959). *Farm Research* **25,** 14.
Nasrallah, M.E. & Wallace, D.H. (1968). *Euphytica* **17,** 495–503.
Nieuwhof, M. & Garretsen, F. (1975). *Euphytica* (in press).
Roggen, H.J.P.R. (1972). *Euphytica* **21,** 1–10.
Thompson, K.F. (1968). Proceedings Brassica Meeting of Eucarpia, 25–28.
Thompson, K.F. & Taylor, J.P. (1966). *Heredity* **21,** 345–362.
Williams, P.H., Walker, J.L. & Pound, G.S. (1968). *Phytopathology* **58,** 791–796.

Heterosis observed for cold tolerance in opaque-2 and analogous normal crosses of maize at single, three-way and double-cross levels

D. GUPTA AND I. KOVÁCS

Agricultural Research Institute of the Hungarian Academy of Sciences, Martonvásár, Hungary

SUMMARY

All possible single, three-way and double cross hybrids obtained from six opaque-2 (o_2) converted inbred lines have been compared with the analogous normal hybrids obtained from their normal counterparts for their ability to germinate at low temperatures. The normal hybrids, in general, were observed to be nearly 60% more heterotic than the o_2 hybrids. It was further observed that nearly 7% of the o_2 hybrids demonstrated more than 100% positive heterosis which could efficiently be used in the opaque-2 maize breeding programmes, but the corresponding figure for the normal hybrids was 29%.

Cold testing of maize in the laboratory is known to provide a reliable estimate of the ability of maize (*Zea mays* L.) to germinate at low temperatures prevalent at the time of sowing (Clark, 1954; Kovács, 1961; Herczegh, 1970). In a study of cold tolerance of all possible 2, 3 and 4 line hybrids made from six opaque-2 (o_2) converted inbred lines and their normal analogues, we have recently shown that the o_2 hybrids and lines are, in general, nearly 35% inferior to their normal analogues (Gupta & Kovács; in press, *a*). In another paper we have shown that the difference between the o_2 and the normal types increases as the period of incubation at low temperatures increases (Gupta & Kovács; in press, *b*). It was further demonstrated that a treatment of seeds with such fungicides as Quinolate V4X, TMTO-Lindan and Vitavax 75W prior to planting improved the cold tolerance of both the o_2 and normal forms, but the improvement of o_2 forms remained far below the level of normal. We consider here a comparison of the heterosis observed in the single, three-way and double cross hybrids obtained from a number of the o_2 converted inbred lines and their normal analogues for their ability to germinate at low temperatures.

MATERIALS AND METHODS

Six of the o_2 converted inbred lines namely, WF9 o_2, R61 o_2, N6 o_2, HMv850–2 o_2, C103 o_2 and W187 o_2 and their normal analogues were used in making a series of all possible single crosses (15 in number), three-way crosses (60 in number) and double crosses (45 in number) without reciprocals, with and without the o_2 gene in recessive homozygous condition. Details of the materials used have been presented elsewhere (Gupta & Kovács; in press, *a*).

Five replicates of 10 seeds of each genotype were placed 5 cm deep in soil from a maize field in plastic boxes and incubated at 8° C for the first 10 days and thereafter at 13–14° C for a further 22 days. Observations were made on seedling emergence on each day of incubation and the cold tolerance expressed mathematically (Herczegh, 1970) as:

$$\text{CT-value} = \frac{\text{Percentage emergence}}{\text{Days of incubation}} .$$

Further details of the methods employed are given elsewhere (Gupta & Kovács; in press, *a*).

Observations on heterosis presented here are all based on the mid-parent value (average of the two parents involved in the cross).

RESULTS

Observations on the percentage of heterosis for CT-value in the o_2 and the normal single cross hybrids, three-way cross hybrids and double cross hybrids are presented in Fig. 1 along with those for all hybrids together. It will be seen that the o_2 recessive homozygotes, in general, demonstrated a poorer heterosis for CT-value than the normal analogues. This trend is visible in case of the double cross and the three-way cross hybrids, but in case of the single cross hybrids, a reversed situation was observed. The o_2 single cross hybrids normally demonstrated more heterosis than the normal analogues. It will be seen that this happened due to poorer performance of the normal single cross hybrids rather than to better performance of the o_2 recessive homozygotes at single cross level.

Table 1. Mean heterosis observed for CT-value

Type of hybrids	% heterosis		Difference between o_2 and normal
	o_2	normal	
Single crosses	—1·55	—33·02	—31·47
Three-way crosses	—16·70	47·37	64·07
Double crosses	41·42	127·74	86·32
All hybrids	7·19	67·63	60·44

The mean observations on the percentage of heterosis for CT-value (Table 1) show that all of the hybrids together demonstrated nearly 68% heterosis for CT-value at the normal gene level, whereas the corresponding figure for the o_2 converted analogues was about 7%. There is thus a difference of nearly 60% in favour of the non-mutant form of maize, though in the single cross hybrids the difference was nearly 31% in favour of the mutant gene. The three-way cross hybrids demonstrated a difference of 64% in favour of the normal form and the double cross hybrids a difference of nearly 86%.

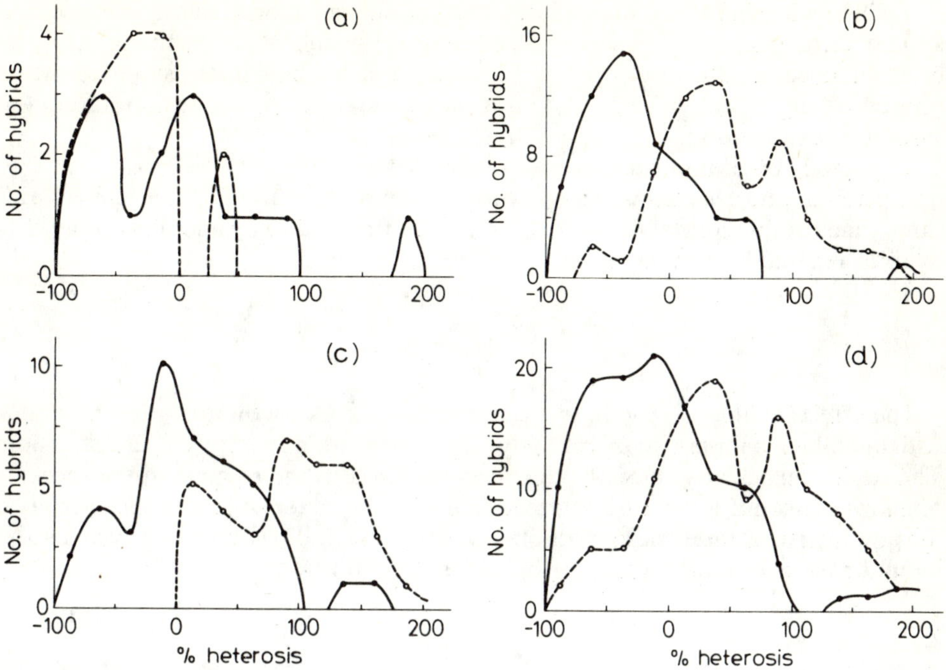

Fig. 1. Frequency distributions of the percentage heterosis observed for CT-values in *(a)* single-cross hybrids, *(b)* three-way cross hybrids, *(c)* double-cross hybrids and *(d)* all hybrids together. ●–● = opaque-2 and ○–○ = normal.

It may be pointed out that there were a number of hybrids both in the o_2 and the normal groups which demonstrate a more than 100% positive heterosis (Fig. 1). The percentage of such hybrids are presented in Table 2. Likewise, a number of hybrids demonstrated a negative heterosis for CT-value (Table 3).

Table 2. Percentage hybrids demonstrating more than 100% positive heterosis for CT-value

Type of hybrids	o_2	Normal	Mean
Single crosses	6.67	0.00	3.33
Three-way crosses	3.39	15.25	9.32
Double crosses	11.11	60.00	35.55
All hybrids	6.72	29.41	18.06

Table 3. Percentage hybrids demonstrating negative heterosis for CT-value

Type of hybrids	o_2	Normal	Mean
Single crosses	53.33	86.67	70.00
Three-way crosses	71.19	16.95	44.07
Double crosses	42.22	0.00	21.11
All hybrids	57.98	19.33	38.65

It will be seen in Table 2 that among all of the normal hybrids nearly 30% demonstrated more than 100% positive heterosis for CT-value whereas among the o_2 hybrids there were only about 7% hybrids to show such a high heterosis. With the exception of single cross hybrids, the three-way and the double cross hybrids demonstrated a similar trend.

In general, 19% of normal hybrids demonstrated a negative heterosis over their mid-parents whereas among the o_2 hybrids there were nearly 58% of such hybrids, three times of the normal (Table 3). As before, the three-way and the double cross hybrids exhibited a similar trend.

DISCUSSION

The effects of the o_2 gene in the recessive homozygous condition are not underestimated in a comparison of crosses of the type used in the present investigations. The result thus clearly show the deleterious effects of the o_2 gene in the recessive homozygous condition on the heterotic ability of maize for cold tolerance. Utility of the few opaque-2 heterotic hybrids observed in these findings may, however, also not be underestimated in breeding of hybrid maize with o_2 gene.

REFERENCES

CLARK, B. E. (1954). Factors affecting the germination of sweet corn in low temperature laboratory tests. *New York (Geneva) Agr. Exp. Sta. Bull.* **769,** 23.

GUPTA, D. & KOVÁCS, J. (In press, *a*). Cold tolerance of parents, single 3-way and double crosses of opaque-2 maize inbreds and their normal analogue.

GUPTA, D. & KOVÁCS, J. (In press, *b*). Dynamics of seedling emergence of opaque-2 and analogous normal maize in cold tests and effect of fungicide seed treatment on their cold tolerance.

HERCZEGH, M. (1970). Some problems of cold tolerance. *In* Some Methodological Achievements of the Hungarian Hybrid Maize Breeding (I. KOVÁCS, ed.). Akadémiai Kiadó, Budapest, 271–281.

KOVÁCS, I. (1961). A "hidegpróba" alkalmazása a kukorica hidegtűrőképességének fokozására irányuló nemesítésben (Application of the cold test in our breeding programme to increase the cold tolerance of maize). *Növénytermelés* **10** (1):27–42.

EUCARPIA VII (1976)

Possibilities of the use of heterosis in lucerne hybrids made with a cytoplasmic male-sterility system

Z. BÖJTÖS

Agricultural Research Institute, Kompolt, Hungary

SUMMARY

Searching for cytoplasmic male-sterile lucerne (*Medicago sativa* L.) breeding material specific crosses were made in more series which have resulted a successful transmission of the male-sterile phenotype to many F_1 lines.

Data are presented on the yield performance of three-way hybrids, and on problems of the identification of male-sterile F_1 lines with a better combining ability required for the production of hybrids. It is assumed that the use of heterosis in true hybrids would be easier for plant breeders if general combining ability is considered the main source of variation.

By using three genetically very divergent open-pollinated varieties as male parents to 15 male-sterile F_1's, three-way hybrids were produced. Seeds of the single crosses were gained by hand-pollination in greenhouse, while the three-way hybrids by leaf-cutter bees (*Megachile rotundata* F.) under cages. Evaluation for yield performance showed that about 50% of the hybrids surpassed highly significantly a well adapted synthetic variety.

Tests for general combining ability showed significant differences between the single crosses, however, the majority of them did not display a fairly stable ranking with each of the tester parents.

In this study only relatively few single crosses and tester parents were tested. The production of superior three-way hybrids requires the use of more single crosses and more tester parents with a wide genetic diversity. Considering that in the presented data the single crosses were produced with one given male-sterile clone only, the maintainer parents may have largely contributed to the results. Field observations on isolated plots and under cages indicated that pollination and seed production of the same male-sterile clone may be very different according to the maintainer parent.

Practical application of a cytoplasmic male sterility system in lucerne (*Medicago sativa* L.) seems to represent an essential progress in the use of heterosis. This report is partly intended to present data on a newer source of cytoplasmic-genic male sterility, of which the evidence was reported by Davis & Greenblatt (1967), Bradner & Childers (1968), and Pedersen & Stucker (1969) in lucerne. On the other hand, it is intended most of all to demonstrate the possibilities of using the heterosis effect in

Table 1. Degrees of sterility in selected F_1 lines from *cms*-19 lucerne clone crossed with 107 male-fertile plants, based on pollen staining and/or germination tests

Male parent	Number of F_1 plants per sterility class					Average % male sterility
	0—50 %	51—75 %	76—85 %	86—95 %	96—100 %	
TK—3	—	—	—	—	25	100
TK—2	—	—	—	2	48	100
CS—33	—	—	—	1	24	99
.						
.						
.						
CS— 9	—	—	—	9	16	96
CS—54	—	—	—	10	15	96
CS—32	—	5	1	4	40	96

three-way hybrids made with a cytoplasmic male sterility system, and to present also some data on problems of the identification of male-sterile single crosses with a better combining ability.

With the object of searching for this type of breeding material specific crosses were made with one fully sterile (*cms*-19) and three partially (about 90%) sterile clones which were found in 1968 (Böjtös, 1973; 1974). In the first cycle the four male-sterile clones were crossed with 107 selected unrelated inbred (mainly S_1 and S_2) male-fertile plants to make an attempt to find nonrestoring maintainers producing male-sterile F_1 progenies. Sterility classifications performed repeatedly by microscope using two staining techniques (acetocarmine and/or Alexander's method) or in doubtful cases by *in vitro* pollen germination, left only 15 F_1 lines of the *cms*-19 female plant in which almost all progenies were fully or partially sterile. In some lines sterility transmission was practically 100% with certain maintainer parents (Table 1).

In a further investigation of male-sterile F_1 lines considered promising for the hybrid combinations 68–281 F_1 plants per line were visually classified (Table 2). This rating in the field was a good supplement to ascertain the environmental effect.

In a second cycle of crossing the *cms*-19 clone was crossed with 94 male-fertile inbred plants in order to find more maintainers for a hybrid breeding programme. After sterility classification 12 maintainers were found (Table 3).

After the male-sterile phenotype had been successfully transmitted to the F_1's, 8–13 random plants of four F_1 lines were back-crossed to a suitable pollen parent. A second backcross combination was made with random plants of two BC_1 lines (Table 4).

Data of Tables 1–4 suggested that – as reported by Davis & Greenblatt (1967) and Pedersen & Stucker (1969) – this source of male sterility can also be considered as a cytoplasmic-genic system. This consideration is based upon 27 male-sterile F_1 lines obtained by crossing the male-sterile *cms*-19 clone first with 107, and later with 94 male-fertile plants, and according to the male sterility data of some tested BC_1 and BC_2 progenies.

Table 2. Degrees of sterility in selected F_1 lines from *cms*-19 lucerne clone crossed with 107 male-fertile plants, based on visual rating

Male parent	Visual rating			
	fertile	partial fertile	partial sterile	sterile
TK—3	—	1	11	163
TK—2	—	—	2	215
CS—33	—	2	7	181
.				
.				
.				
CS—9	1	8	21	139
CS—54	—	1	28	184
CS—32	1	1	12	162

Table 3. Degrees of sterility in some selected F_1 lines from *cms*-19 lucerne clone crossed with 94 male-fertile plants, based on pollen staining

Male parent	Number of F_1 plants per sterility class					Average % male sterility
	0—50 %	51—75 %	76—85 %	86—95 %	96—100 %	
B—29	—	—	—	—	25	100
B—31	—	—	—	1	22	99
B—28	—	—	—	5	19	98
.						
.						
.						
B—21	—	1	1	9	20	95
B—26	—	—	1	8	10	95
B—22	—	—	2	11	15	95

Table 4. Degrees of sterility in backcross populations of selected lines from *cms*-19 × male-fertile crossings, based on pollen staining and/or germination tests or visual rating

Pedigree of males	Population	Plants per sterility class				Total
		0—50 %	51—85 %	86—95 %	96—100 %	
TK—2	BC_1	—	—	—	136	136
CS—44	BC_1	—	—	4	165	169
TK—23	BC_1	—	5	4	172	181
CS—32	BC_1	4	24	14	314	356
.						
.						
.						
TK—2	BC_2	—	—	2	21	23
TK—23	BC_2	—	2	1	24	27

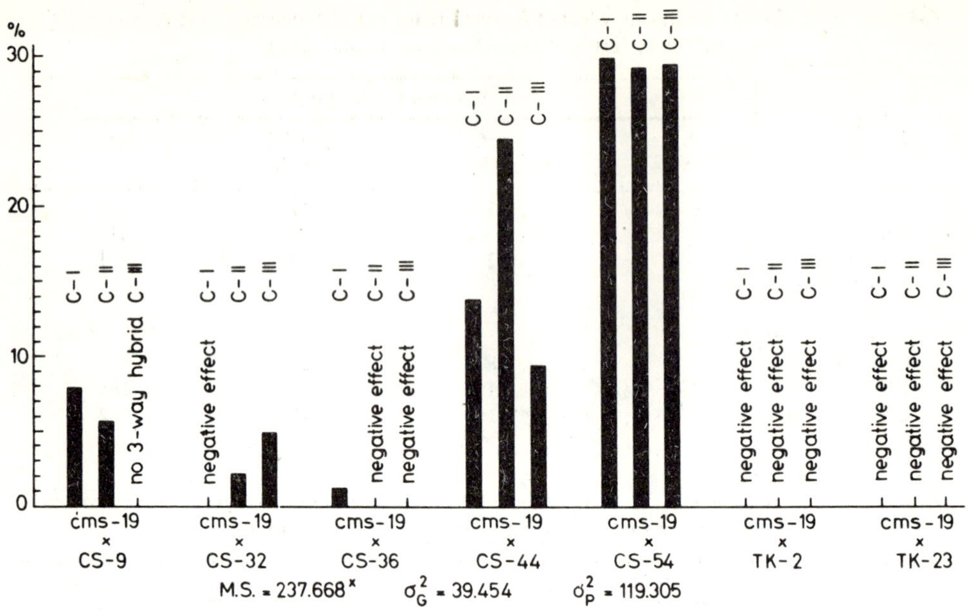

Fig. 1. Mid-parent values in per cent of total green yield increase in three-way hybrids (1971–72).

To study the possibilities of using the heterosis effect in three-way hybrids, male-sterile single crosses were produced by hand-pollination with *cms*-19 clone and with 7 unrelated maintainer clones. Open-pollinated populations of a 15-clones synthetic (C-I), a Hungarian local (C-II) and a Hungarian-Flemish variety-cross (C-III), were used as male-fertile pollen parents also with hand-pollination. Single crosses, three-way hybrids and the open-pollinated varieties were established in a field experiment at one location. The field plot design was a randomized complete block with four replications, each represented by a single row 1 m long with 100 seedlings and spaced 20 cm between rows. Total green yield of 5 harvests in two years was studied. When studying the percentage yield increase in three-way hybrids compared with the two parents we found the heterosis effect of two single crosses especially outstanding in their three-way hybrids (Fig. 1). It seems, however, that we cannot obtain such a 29% heterosis effect if we establish the field experiments with larger plots and at more locations.

By using three genetically more divergent varieties as male-fertile pollen parents with 15 male-sterile F_1's, new three-way hybrids were made. Seeds of the single crosses were produced by hand-pollination in a greenhouse, and of the three-way hybrids by leaf-cutter bees (*Megachile rotundata* F) under cages. Total green yield of 7–9 harvests of 35 three-way hybrids in 1972–73 was studied in the field at 1–4 locations in randomized complete block design trials with five replications and 3 m^2 plot size. About 50% of the hybrids highly significantly outyielded a well adapted Hungarian synthetic variety. Data suggest that such a type of lucerne hybrid may give the

Table 5. Relative mean performance of lucerne three-way hybrids compared to Mv Synalfa cultivar over two or three locations in 1972—73

Three-way hybrids	Relative green yield	
	over 2 locations	over 3 locations
03—1	107**	—
03—2	109***	—
03—3	—	108**
07—1	112***	—
07—2	108**	—
07—3	110***	—
09—1	106**	—
09—2	—	107**
09—3	—	108**
15—1	106**	—
15—2	—	108**
15—3	108**	—

Table 6. Ranking of male-sterile single crosses obtained by *cms*-19×11 different maintainers according to yielding ability, with two tester parents, in 1972—73

Rank	With tester parent		With average of 2 parents	Average yield kg
	C—I	C—II		
1	03	09	03	55·59
2	15	17	09	54·32
3	08	14	15	54·27
4	16	03	07	54·23
5	07	07	16	53·89
6	14	04	08	53·82
7	09	16	14	53·76
8	12	15	17	52·71
9	04	08	04	52·39
10	17	11	12	50·77
11	11	12	11	50·39
M. S.	28·297**	17·452***		
σ^2_G	4·172	2·820		
σ^2_P	11·610	6·173		

possibility of about 8–10% yield increase compared to the improved populations or synthetic varieties (Table 5).

Tests were made at one location in practical sense for general combining ability of male-sterile single crosses, using 2 and 3 genetically very divergent tester parents. It was assumed that the use of heterosis in true hybrids would be easier for plant breeders if general combining ability is considered as the main source of variation (Gallais & Guy, 1970). In this regard there were significant differences among the single crosses, however, the majority of those did not show a fairly stable ranking

Table 7. Ranking of male-sterile single crosses obtained by *cms*-19 × 7 different maintainers according to yielding ability with three tester parents in 1972—73

Rank	With tester parent C—I	C—II	C—III	With average of 3 parents	Average yield kg
1	03	09	15	15	54·82
2	07	17	09	03	54·51
3	15	14	07	09	54·42
4	14	03	17	07	54 09
5	09	07	14	14	53·44
6	17	15	03	17	52·99
7	11	11	11	11	49·85
M. S.	29·149**	11·591*	26·038***		
σ^2_G	4·734	1·651	4·457		
σ^2_P	10·214	4·983	8·211		

with each of the tester parents. Yet, indeed, we could classify them into fairly good and fairly poor groups according to GCA (Tables 6–7). These data were justified by the field performance of three-way hybrids at different locations.

In this study only relatively few single crosses and tester parents were tested. The production of superior three-way hybrids requires the use of a greater number of both the single crosses and the tester parents, with a wide genetic diversity and of course with disease resistance. The male-sterile single crosses showing a good and stable ranking may give good chance for the production of superior three-way hybrids. Considering that in the presented data the single crosses were produced with one given male-sterile clone only, in the results an important role was played by the maintainer parents. It seems that in the case of an adequate cytoplasmic male-sterile parent, the selection of maintainers is one of the most important factors. On the other hand, in the production of male-sterile single crosses, field observations on isolated plots and under cages indicated that pollination and seed production of the same male-sterile clone may be very different according to the bee preference of the maintainer parents.

REFERENCES

Bójtös, Z. (1973). Some problems of male sterility transmission and combining ability in lucerne hybrids made with a cytoplasmic male sterility system. *Meeting of Eucarpia Fodder Crops Section,* Wageningen.

Bójtös, Z. (1974). Some problems of male sterility transmission and combining ability in lucerne hybrids made with a cytoplasmic male sterility system. *Proceedings of the Conference of Eucarpia Medicago sativa Group*, Kompolt, 17–26.

Bradner, N.R. & Childers, W.R. (1968). Cytoplasmic male sterility in alfalfa. *Can. J. Plant. Sci.* **48,** 111–112.

Davis, W.H. & Greenblatt, I.M. (1967). Cytoplasmic male sterility in alfalfa. *J. Hered.* **58,** 301–305.

Gallais, A. & Guy, P. (1970). Use of heterosis in breeding autotetraploids. *Meeting of Eurcarpia Fodder Crops Section,* Lusignan, 105–118.

Pedersen, M.W. & Stucker, R.E. (1969). Evidence of cytoplasmic male sterility in alfalfa. *Crop Sci.* **9,** 767–770.

Evaluation of heterosis in *Cucurbita pepo* L.

B. BORGHI*

Istituto Sperimentale per l'Orticoltora – Montanaso L. (MI), Italy

and

W. PIRONI

Istituto Ricerce Orticole – Minoprio (CO), Italy

SUMMARY

The practical and theoretical aspects of inbreeding and heterosis in *C. pepo* have been studied.

The practical aspects concern: 1. selection of new lines; 2. evaluation of general and specific combining ability of the lines; 3. production of F_1 hybrid seeds on a commercial scale by means of Ethephon (2-chloroethylphosphonic acid).

The more promising lines are dwarf, with erect habit, small leaves, spineless and very lobed. Several hybrids, in replicated trials, carried out in different seasons and locations, appeared to be more productive and earlier than the traditional varieties. Our findings suggest that the traditional pedigree selection could yield positive results but a breeding program devoted to the production of hybrids appears more practical. This program appears to be more realistic considering the relative facility for the commercial production of F_1 seeds since the discovery of Ethephon.

Our results suggest that seed parents and pollinators should be grown with a ratio of 10:1 rows in isolated fields. The best Ethephon treatment schedule is: 400 p.p.m. on active principle, sprays at 4–5 leaf stage and 7–10 days later. Following this technique the cost of squash hybrid seed is slightly more than that of a traditional variety.

The genetical structure of the marrow populations were studied by the diallel cross method. The lines appeared generally less vigorous than their hybrids, except for a limited number of lines which appeared as productive as their best hybrids. These lines carried dominant genes acting mainly in favour of the best expression of the characters. The positive correlation between dominant genes and favourable phenotypical expression of the characters accounts, at least in part, for the superiority of the hybrids to the parental lines.

The results obtained suggest that *C. pepo* suffers from inbreeding depression like other allogames, and so, generally speaking, the heterozygosity appears more favourable. However, there are also inbred populations with favourable phenotypical expression of the characters related to fitness. A possible explanation is that in *C. pepo* populations characterized by completely different genetical structure could coexist, in some cases similar to the allogames and in others more related to the autogames.

* Present address: Istituto Sperimentale per la Cerealicoltura S. Angelo Lodigiano (MI), Italy.

The Cucurbit family includes a conspicuous group of plants with about 90 genera and more than 500 species. Some species of the *Cucurbita* genus are among the first examples of domesticated plants and their evolution has been greatly influenced by man (Whitaker, 1956; Whitaker & Bemis, 1965).

From the genetical point of view the Cucurbits have been little studied and not in the recent past. The experiments of Bushnell (1922), Cumming & Jenkins (1928); Haber (1928); Scott (1934); Hutchins & Croston (1941); Curtis (1939, 1941) have revealed that, at least in the related species *C. pepo* and *C. maxima* degeneration and reduction of vigour do not necessarily arise from prolonged inbreeding. In consequence, the conviction has grown among plant breeders that the Cucurbits react to inbreeding in a different way from maize, onions or other cross-pollinated species. Allard (1960) in his classical book on plant breeding wrote that Cucurbits, although monoecious, do not suffer inbreeding depression and in many ways their population structure is more similar to that of inbreeders than outbreeders, and he gave several explanations to justify this behaviour.

Several cases of heterosis for characters related to yield and earliness have been reported in the Cucurbit family, especially in the genus *Cucurbita* (Whitaker & Davis, 1962; Borghi, 1974). Cucurbits are among the first examples of hybrid seed production in vegetables. Right back in 1939 Curtis suggested the production of F_1 squash by growing pollinator and female parents in isolation and removing male flowers of the seed parent by hand prior to anthesis.

Consequently, breeding activity in the last 20 years has been mainly devoted to the selection of lines for hybrid production. At present F_1 hybrids are available for all the most important Cucurbits such as cucumber, watermelon, melon, squash and marrow. In *C. pepo* hybrid seed production on a commercial scale has been greatly facilitated by the discovery of Ethephon (2-chloroethylphosphonic acid), a new gametocide able temporarily to suppress male flowers on seed parent plants.

The purpose of this communication is the evaluation of heterosis in *C. pepo*. The practical and theoretical aspects in relation to the choice of the breeding method and the genetical structure of the populations will be discussed.

MATERIALS AND METHODS

The material came from a breeding program commenced in 1966 (Borghi, 1970) with a view to obtaining improved marrow varieties, suitable for the Italian market. The initial material came chiefly from the old populations cultivated in Italy together with some improved foreign varieties selected in France, Holland and USA. The American material seems to have been derived from Italian populations: this is confirmed by the Italian names of some varieties ("Zucchini", "Cozzelle").

The morphological characters taken into consideration during selection were: dwarf and erect habit, and small, spineless and very lobate leaves. During selection the most promising lines were crossed with testers (2 or more lines). The hybrids thus obtained were grown in replicated trials (3 replications, 10 plants per plot 1 m apart) for several years, in different locations and seasons, in order to evaluate their general (GCA) and specific (SCA) combining ability.

Nine inbred lines were crossed in all possible combinations, reciprocals excluded (diallel crosses), and the 45 derived families were grown during the summer of 1970 in a replicated trial (3 replications, 15 plants per plot).

The characters recorded in all experiments were: number of fruit and total weight of the fruits. The fruits were harvested every 2nd or 3rd day at a weight of about 100–150 g, for a period varying from 1 to 2 months. For convenience the data have been divided into "early production" (about half of the total harvesting period) and "total production".

Data from the first experiments were used for combining ability analysis. A more detailed genetical interpretation was obtained from diallel data on the basis of the Mather & Jinks (1971) analysis.

RESULTS

Table 1 gives the results of production trials in which the parental lines were included. The effects of general and specific combining ability, omitted for the sake of brevity, are highly significant in almost all experiments. The number of hybrids surpassing mid-parent value is very high and varies from a minimum of 57% in the trial at Montanaso L. in the summer of 1968 and a maximum of 100% in the latest experiment. Several hybrids surpass also their best parent with a variation between experiments ranging from 26% to 86%. On the other hand, only a limited number of hybrids appear superior to the best parental line included in each experiment.

The type of gene action involved was studied further by means of a diallel analysis. Figure 1 shows an example of diallel analysis for the character: "early production in fruit per plant." The figure contains the results of the statistical analysis, the graph of the V_r/W_r relations, and the estimated components of phenotypic variance. Moreover, a histogram shows the mean value of each line, its mean array and its best hybrid value. The lines show very different yield capacities, varying from about 4 to 10 fruits per plant; the hybrids appear less variable, with a minimum of 6.5 and a maximum of 12.5 fruits per plant. Additive and dominance effects are presented (a and b highly significant). Dominance is unidirectional in its effects (b_1 highly significant) and acts in favour of productivity. In fact the correlation between the array mean and $(V_r + W_r)$ is significant and negative ($r = -0.736$).

The highly significant term b_3 suggests the presence of non allelic interactions, as confirmed by the low value of the regression line in the V_r/W_r graph ($b = 0.408$). Omitting the arrays 1 and 3 showing non-allelic interactions, the various array points appear distributed close to the regression line whose slope is not significantly different from 1.

Lines 8 and 6 carry mainly dominant genes, line 2 carries the recessives. 6 of the 7 lines are clustered in three groups. Within each group lines derived from varieties grown in the same area are present.

Analysis of variance

item	df	MS
c	8	5304**
b_1	1	12 757**
b_2	8	1708*
b_3	27	1541**
b	36	1890**

Components of phenotypic variance

D	= 1087	H_1	= 874
F	= -589	H_2	= 956
h^2_N	= 50.88	E_2	= 530
h^2_B	= 66.14		

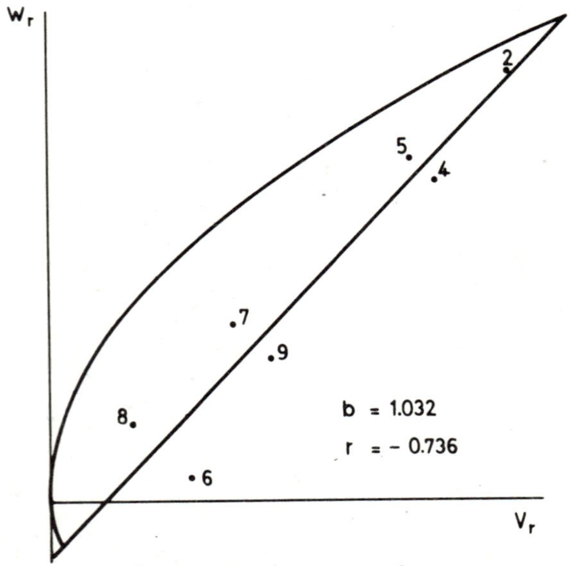

Variance / covariance graph

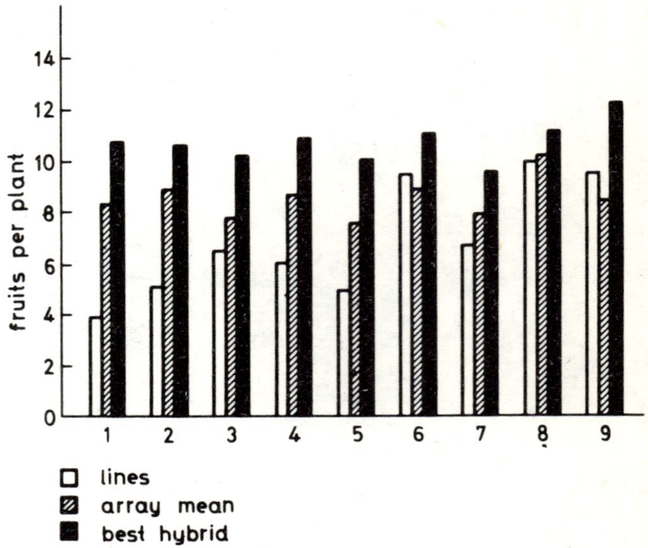

Fig. 1. Diallel analysis of the character: early production in fruit per plant.

Table 1. Percent of nybrids superior to the mid parent value ($\bar{P}$), to the best parent (best P) and to the best parental line (best line).

Location	Hybrids tested	Total production						Early production					
		in weight			in number			in weight			in number		
		$\bar{P}$	best parent	best line	$\bar{P}$	best parent	best line	$\bar{P}$	best parent	best line	$\bar{P}$	best parent	best line
Minoprio (CO), summer 1968	16	69	56	6	81	75	0	63	31	6	86	86	12
Minoprio (CO), summer 1970	36	84	69	25	78	56	25	84	61	8	80	53	22
Montanaso L. (MI), summer 1971	21	57	28	0	71	52	5	—	—	—	—	—	—
Pandola (SA), summer 1971	21	90	71	19	81	62	0	—	—	—	—	—	—
S. Angelo L. (MI), autumn 1973	36	94	83	36	100	86	25	84	83	42	92	81	30

Our experiments reveal that as a whole the hybrids give a high yield, and, on the average, are much more productive than the parental lines. Most hybrids surpass the mid-parent value and in several cases also the best parent. The yield capacity of the lines is extremely variable: from very poor values to values well approximating the best hybrids.

These results suggest two possible breeding programmes: traditional pedigree selection or F_1 hybrid production. Apart from obvious commercial factors (in capitalist countries especially), there are several valid reasons in favour of hybrids. 1.It is possible to select vigorous lines from a limited number of varieties. For instance, inbreeding depression was not found in our breeding program for "Grigia genovese", whereas a more or less pronounced vigour depression was evident in all the other varieties. 2. Hybrids, with few exceptions, reveal high phenotypical values for all the characters considered above. 3. Despite the very similar organoleptic characteristics, the customers of different zones show a marked tendency to prefer fruits of a particular colour or shape. The best way to meet the different demands would be to select a limited number of well differentiated lines and to use them to produce suitable hybrids. 4. Marrows are grown in a wide range of climatic conditions. Furthermore, in the same area marrows are cultivated in different seasons, in open fields and in several types of glasshouses. Borghi (1971) showed the possibility of obtaining hybrids with a high yield and good phenotypical stability.

There are no serious obstacles to seed production on a commercial scale following the discovery of the ability of Ethephon to suppress temporarily male flowering in several species of Cucurbits (McMurray & Miller, 1968; Robinson *et al.* 1969, 1970).

Our experiments carried out in this field (Borghi & Pironi, 1971; Borghi *et al.*, 1973a) allow us to point out the following conditions which should be taken into consideration: 1. using Ethephon it is possible to produce hybrid seeds by growing the pollinator and female parent in isolation without removing by hand male flowers on the seed parent, exploiting the cross pollination performed by bees; 2. the long flowering period of the female flowers facilitates cross pollination and eliminates the need for successional sowing in case female lines differ in earliness; 3. the best ratio between seed parent and pollinator is 10 : 1 in rows, 1 m apart; 4. spray Ethephon at 400 parts per million of active principle at the 4–5 true leaf stage and again 7–10 days later; 5. remove female flowers on the pollen parent during the pollinating period.

Following this technique the cost of squash and marrow hybrid seed is slightly higher than that of a traditional variety.

The genetical interpretation of the present findings can lead to a better knowledge of the genetical structure of the populations in question.

The effects of inbreeding depression are detectable but they are not so evident as in other cross pollinated species like maize or alfalfa. However, most of the lines produced by inbreeding show a very poor yield capacity in comparison with their hybrid crosses. This is an evident indication of inbreeding depression. The low or undetectable inbreeding depression effects recorded by Bushell (1922), Cummings & Jenkins (1928), Haber (1928), and Scott (1934) could be explained by assuming that the breeding material they used, mainly composed of commercial varieties, was still highly inbred. In a more recent experiment Borghi *et al.* (1973b) studied the genetical control of fruit number and total fruit weight, seed number and total seed weight in a diallel cross. For all the characters studied the findings were similar to those given in Table 1; on the average, the hybrids were better than the parental lines but few of these were as good as the best hybrids. From the V_r/W_r graphs it is possible to see that points corresponding to the best arrays are mostly situated near the intersection of the axes; in other words, in the best lines most of the dominant genes, acting mainly in favour of the best expression of the characters, have been fixed. Moreover, the correlation between dominant genes and favourable phenotypical expression of the characters always exists.

These results throw further light on the particular behaviour of the Cucurbits as regards inbreeding. Most of the 50 populations of *C. pepo* studied by us suffer from inbreeding depression and so, generally speaking, the heterozygosity appears more favourable. However, there are some varieties in which inbreeding has induced a high degree of homozygosity which is fully compatible with fitness. In other words, populations characterized by completely different genetical structures could coexist in *C. pepo* in some cases similar to the allogames, in others more related to autogames.

REFERENCES

Allard, R. W. (1960). Principles of plant breeding. New York, Wiley.

Borghi, B. (1970). Miglioramento genetico dello zucchino (*Cucurbita pepo* L.) Parte prima: studio varietale. *Se. Elette,* **16,** (5); 23–32.

Borghi, B. (1971). Miglioramento genetico dello zucchino. 2°. Comportamiento di incroci intervarietali allevati in diverse condizioni ambientali. *Genetica Agraria,* **25,** 17–30.

Borghi, B. (1974). Miglioramento genetico delle Cucurbitacae. Tratt. Genetica Agraria Speciale. EDAGRICOLE (In press).

Borghi, B., Maggiore, T., Boggini, G. & Bonali, F. (1973a). Inbreeding depression and heterosis in *Cucurbita pepo* evaluated by means of diallelical analysis. *Genetica Agraria,* **27,** (In press).

Borghi, B., Maggiore, T., Soressi, G.P. & Poma Treccani, C. (1973b). Tecnica di produzione di seme ibrido nelle cucurbitacee mediante l'impiego dell'Ethrel. *Riv. Ortoflorofrutt. Ital.,* **37,** 750–759.

Borghi, B. & Pironi, W. (1971). Les effects des divers biostimulateurs sur la biologie florale des certaines Cucurbitacées. In "La stérilité male chez les plantes horticoles": 191–203. Ed. Schweisguth. Versailles.

Bushnell, G. W. (1922). Isolation of uniform types of Hubbard squash by inbreeding. *Proc. Amer. Soc. Hort. Sci.,* **19,** 139–144.

Curtis, L. C. (1939). Heterosis in summer squash *(Cucurbita pepo)* and the possibilities of producing F_1 hybrid seed for commercial planting. *Proc. Amer. Soc. Hort. Sci.,* **37,** 827–828.

Curtis, L.C. (1941). Comparative earliness and productiveness of first and second generation summer squash *(Cucurbita pepo)* and the possibilities of using the second generation for commercial planting. *Proc. Amer. Soc. Hort. Sci.,* **38,** 596–598.

Cummings, M.B. & Jenkins, E.W. (1928). Pure line studies with ten generations of Hubbard squash. *Bull. Vt. Agric. Exp. Sta.,* **280,** 39.

Haber, E.S. (1928). Inbreeding in the Table Queen (Des Moines) Squash. *Proc. Amer. Soc. Hort. Sci.,* **25,** 111–114.

Hutchins, A.E. & Croston, F.E. (1941). Productivity of F_1 hybrids in squash *(Cucurbita maxima). Proc. Amer. Soc. Hort. Sci.,* **39,** 332–336.

Mather, K. & Jinks, J.L. (1971). Biometrical genetics. Chapman & Hall, London.

McMurray, A.L. & Miller, C.H. (1968). Cucumber sex expression modified by 2-chloroethylphosphonic acid. *Science* **162,** 1397–1398.
Robinson, R.W., Shanon, S. & Guardia, M.D. de la (1969). Regulation of sex expression in cucumber. *Bio. Sci.,* **19,** 141–142.
Robinson, R.W., Whitaker, T.W. & Bohn, G.W. (1970). Promotion of pistillate flowering in *Cucurbita* by 2-chloroethylphosphonic acid. *Euphytica,* **19,** 180–183.
Scott, D.H. (1934). Observations on some inbred lines of bush types of *C. pepo. Proc. Amer. Soc. Hort. Sci.,* **32,** 480.
Whitaker, T.W. (1956). Origin of the cultivated *Cucurbita. Amer. Natl.,* **90,** 171–176.
Whitaker, T.W. & Bemis, W.P. (1965). Evolution in the genus *Cucurbita. Evolution,* **18**(4), 553–559.
Whitaker, T.W. & Davis, G.N. (1962). The cucurbits-botany cultivation and utilization. World Crops Books. London, Leonard Hill.

EUCARPIA VII (1976)

Remarks on heterosis breeding with grasses

J. ROD

Vyzkumni Ustav Základni Agrotechniky, Hrusovany u Brna, ČSSR

SUMMARY

A method of breeding wind-pollinating grasses was developed for obtaining high performance population from components, giving a heterosis effect on their mutual crossing.

The components were families genetically stabilized by means of crossing in close relationship and spatial isolation. A polycross-test was employed to determine combining ability. Its special feature consisted in the use of so-called "basic" and "inserted" rows. This layout permits an estimation of general and specific combining ability and even of inbreeding depression.

It was shown, that by this method full use can be made of the genetical reserves of the material. The validity of the method is proved by the fact that the variety used as the initial breeding material was successfully improved. The newly-registered synthetic variety surpassed the original one, which still exists in maintenance breeding by means of individual selection.

Parallel to this breeding programme, a study was made of the combinations of components with various genetical scope, i.e. material bred to different degrees, and some foreign varieties. Cyclical and diallel reciprocal crossing by open pollination of spatially isolated partners was used. Crossing of inbred families with good combining ability gave higher performance than intervarietal crossing. Decisive were direction of crossing, and genetical extent of the components. Best results were obtained when crossing two varieties only, but the performance of intervarietal hybrids was not higher than that of hybrids of inbred families.

This paper describes some long-term breeding and research work, successfully ended last year by the registration of a new variety. This programme was motivated especially by the possibility of checking and looking for new ways of applying procedures known from other cross-pollinated plants to the breeding of fodder grasses. The gaining of a new variety of *Lolium multiflorum,* used here as a model plant, confirmed the correctness of the hypothesis and brought some interesting theoretical and methodical information.

The aim of this study can be specified as follows: Designing and testing a method of breeding a synthetic variety, more effective than individual plant selection with progeny testing (strain building). The idea was the mastering of a procedure aimed to build and by degrees genetically stabilize components, to determine their combining ability and to synthesize them into a high performing synthetic population better than the initial material. In order to prove the effectiveness of the devised procedure, only limited initial material was used, derived from our best registered improved variety, still maintained and produced as a variety. In this way the possibility of making full use of genetic reserves in a material maintained by means of simple maternal-line selection was proved.

This main programme was linked up with further investigations, aimed at checking the hypothesis that yield can be increased to above the level of the breeding material itself, if components possibly not related to the basic breeding material are crossed and genetically stabilized on different levels.

As to the given aims, the following methodical approach has been devised: Selection, testing and gradual genetical stabilization of families as starting components. Selection of components with the best combining ability, intercrossing them for synthesis and additional seed increase. A material and chronological plan was fixed, making it possible to master the given task (Fig. 1).

The breeding programme was based on independent cycles of selection (I, II, III, ...) though they were mutually linked; every cycle of selection opens with a space planted nursery of strains (families) in spatial isolation (SF). These strains, amounting to 100 individuals, represent sexual progenies of one parent plant, selected as the best in the strain nursery in the previous cycle of selection (Fig. 2 – seed No. 1). Spatial isolation is secured by means of double framing, i.e. an inner mantle, representing the progeny of individuals closely related in performance and type to the parent plant (Fig. 2 – seed No. 2) and a second outer mantle, representing progeny of non-eliminated but bulked individuals (Fig. 2 – seed No. 3). The effectiveness of spatial isolation, in addition to this double framing by means of related material is emphasized by a dense double row of rye forming a basic lattice, in which the individual strain nurseries are placed.

The strain nursery is followed by a polycross nursery (Fig. 1; *P*) with progenies of all families contained in the strain nursery of the same cycle of selection. The individual progenies originate from bulked seed of selected individuals from the strain nursery, i.e. the same seed is used in the next cycle of selection for the inner frame (Fig. 2, seed No. 2). The seed of individual families gained under uniform conditions of pollination in a polycross nursery is seeded in performance trials in order to establish their combining ability (Fig. 1; *T*).

Families ranked according to the combining ability are intercrossed and synthesized in separate spatially isolated nurseries (Fig. 1; *I*) to gain seed for further increase of breeder's seed (Fig. 1; *B*). For this purpose the original seed of selected families is used, i.e. the seed bulked from selected individuals and sown in a polycross, or as inner framing of strain nursery in the next cycle of selection, i.e. seed which has been stored over four years (Fig. 2, seed No. 2). The breeder's seed plot is sown from seed produced in an intercrossing nursery of selected families and is used for further seed increase (Fig. 1; *B*).

An auxiliary selection nursery is spatially planted simultaneously with performance trials from seed gained in the polycross nursery, the aim being to find new suitable types, resulting from the crossing of families tested (Fig. 1; *SA*). Parent plants, selected in this reciprocal recurrent selection form the basis of new families as replacement for the inconvenient and therefore gradually eliminated ones. They are then placed in the strain nursery, founded always after two further cycles of selection.

All nurseries were planted in regular two year selection periods as follows: the selection nursery of strains (Fig. 1; *SF, SA*) by pregrown seedlings. The polycross

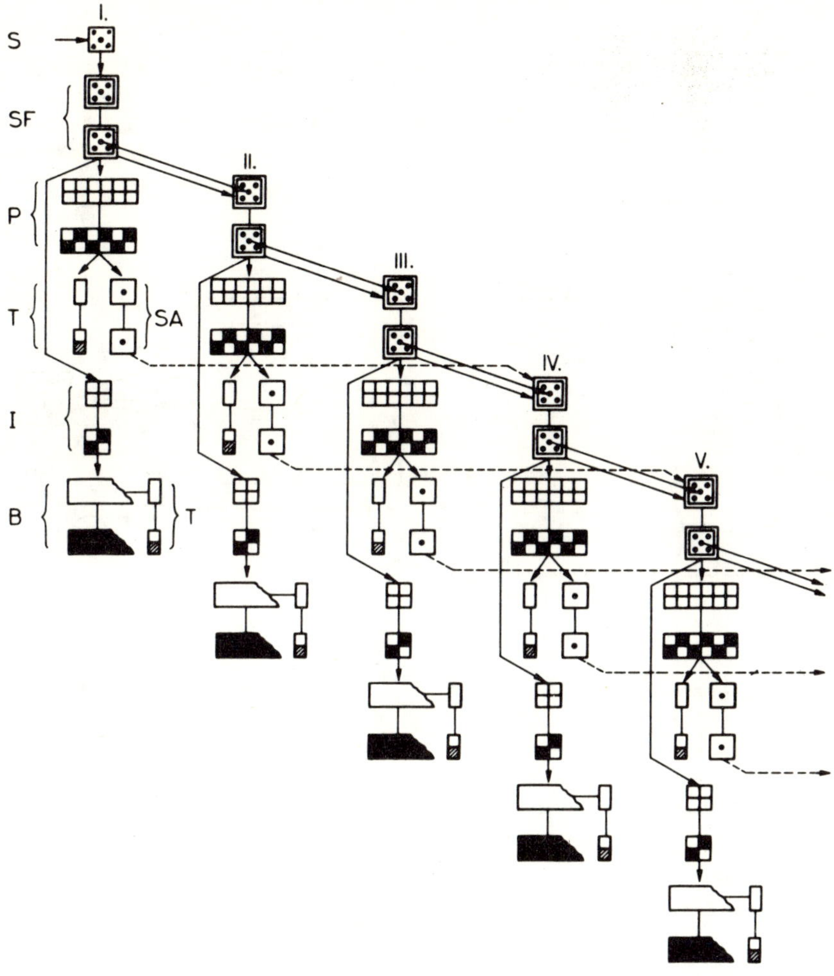

Fig. 1. Material and chronological working scheme for breeding a synthetic variety combining genetically stabilized families.
Key to the symbols: ⚄ S = strain nursery, individual spatial planting; ⊡ SA = auxiliary strain nursery. ⚄ SF = strain nursery, individual spatial planting of families with frame isolation. ▦ P = polycross-test nursery. □ T = performance trial. ⊞ ⊠ I = intercrossing nursery (intravarietal, intervarietal). ◣ ◺ B = breeder's seed plot. ■ = seed harvest. ⌶ = the same nursery from year to year. ⌶ = new nursery set up.

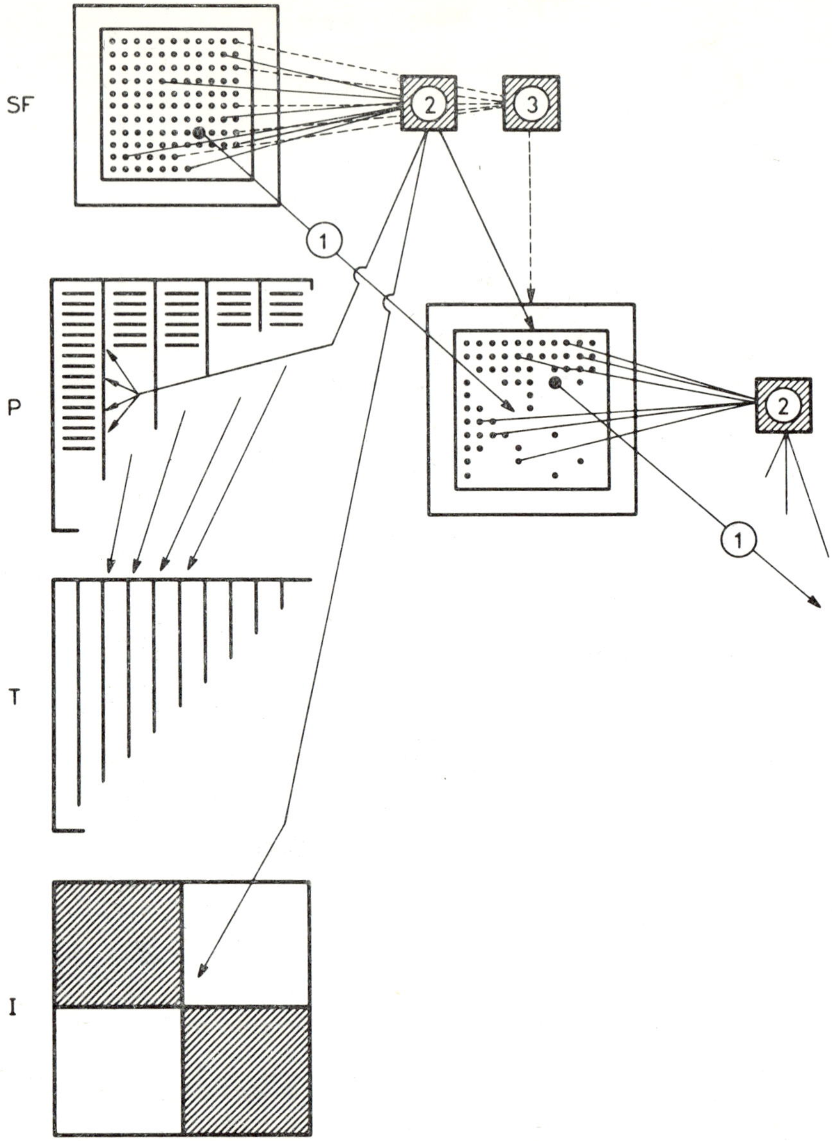

Fig. 2. Seed origin for individual selection steps in one cycle.

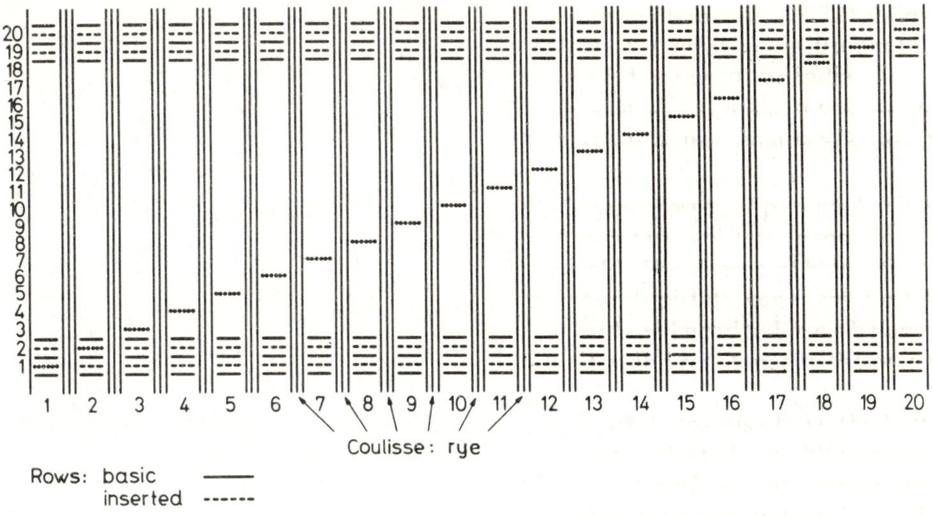

Fig. 3. Lay-out of the polycross-test nursery. Method of "basic and inserted rows".

nursery was layed down by the method of "basic and inserted" rows (Fig. 3). Components tested for combining ability were seed progenies of genetically stabilized families. Thus stabilization and to a certain extent even purity of type were achieved by the above mentioned arrangement of strain nurseries, leading to a prevalent sib-breeding. The proper component-progenies were sown in alternate rows (Fig. 4). One column consisted of rows of a given component alternating once with rows of other tested components. This arrangement guarantees a uniform number of mutual adjacencies of all components and consequently their uniform pollination.

Taking into account the scheme of "basic and inserted" rows, it can be supposed that the different kinds of seed of the same component have different genetical values (Fig. 4; *a, b, c, d,* ....). A representative sample for estimation of unbiased general combining ability is only obtained if equal weights of seed from basic and inserted rows are mixed. Confirmation can be found in the yield differences of the seed types mentioned. The seed of "inserted" rows (Fig. 4; *c, d,* ..., *k*), corresponding according to its location between two rows of another component to a diallel crossing, outyielded other seeds. Progenies of the "basic" rows, i.e. with the other components inserted (Fig. 4; *a*) and thus corresponding to the cyclic crossing, occupied the second place. Last but not least should be mentioned the "self-pollinated" rows (Fig. 4; *b*), the progenies of which were always lower in yield.

The polycross is followed in all cycles of selection by comparative trials laid out as randomized blocks or split plots in which the origin of the seed from "basic" and "inserted" rows is distinguished. The intercrossing of selected families follows chronologically the performance trials and serves for further seed increase (foundation seed). Parallel to breeder's seed plots and foundation seed plots the comparative trials are established with the purpose of comparing different breeding stages in individual cycles of selection, eventually enabling a comparison with the starting material and standard variety.

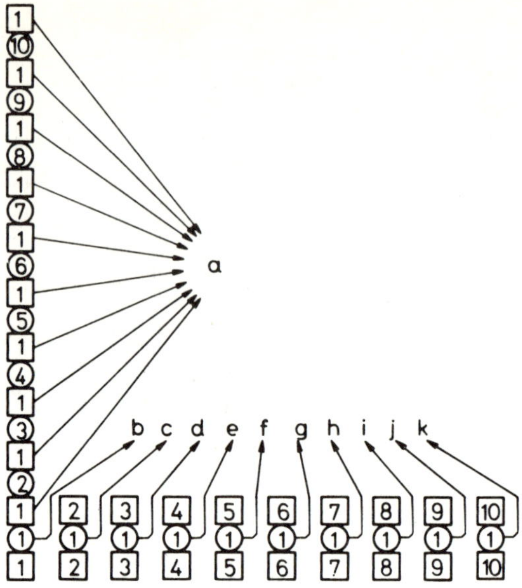

1	2	3	4	5	6	7	8	9	10
10	10	10	10	10	10	10	10	10	10
1	2	3	4	5	6	7	8	9	10
9	9	9	9	9	9	9	9	9	9
1	2	3	4	5	6	7	8	9	10
8	8	8	8	8	8	8	8	8	8
1	2	3	4	5	6	7	8	9	10
7	7	7	7	7	7	7	7	7	7
1	2	3	4	5	6	7	8	9	10
6	6	6	6	6	6	6	6	6	6
1	2	3	4	5	6	7	8	9	10
5	5	5	5	5	5	5	5	5	5
1	2	3	4	5	6	7	8	9	10
4	4	4	4	4	4	4	4	4	4
1	2	3	4	5	6	7	8	9	10
3	3	3	3	3	3	3	3	3	3
1	2	3	4	5	6	7	8	9	10
2	2	2	2	2	2	2	2	2	2
1	2	3	4	5	6	7	8	9	10
1	1	1	1	1	1	1	1	1	1
1	2	3	4	5	6	7	8	9	10

Fig. 4. Scheme and function of "basic and inserted" rows. *a* = seed of the "basic" rows (estimation of general combining ability); *b* = seed of the "inserted" rows (estimation of inbreeding depression); *c, ... k* = seed of the "inserted" rows (estimation of specific combining ability).

The effectiveness of the proposed way of selection and of maintaining of its effects was verified on the presumption, that the starting material had a restricted genetical basis. This had been formed in successive years on the one hand as the maintenance breeding of this improved variety by individual plant selection, and on the other hand as the breeding of a new synthetic variety in the above described way.

The extent of results and studies, including variability and performance of strains, polycross progenies and breeders seed plots cannot be mentioned fully here. But it can be said in summary that the superiority of the newly improved variety manifested itself even in the third cycle of selection, and that by means of consistent sibbing and spatial isolation, partially genetically defined families were obtained, the combinations of which resulted in the synthesis of a population outyielding the starting material, after their combining ability had been tested.

Besides these methodical problems, the practical results also have their relevance. Proof of the usefulness of the chosen procedure has been furnished and the material obtained can be used as a new improved variety. There is evidence of a rising trend in the yielding capacity of the breeding material itself, though within individual selection stages, there is a tendency to a certain yield decrease from the first strain synthesis by way of breeders seed to the seed increase plot. Although the yield differences were not significant, we must presume that definite yield stabilization is reached only after several seed reproductions.

Parallel to this breeding programme, the aim of which was the building of a high-performing synthetic variety, a study was made of the combinations of components with varied genetical scope, i. e. material bred to different degrees, of groups of such components, and some foreign varieties, not related to this basic material. Ways have been sought, leading to a performance rise, exceeding our own unfinished breeding by means of components, stabilized to different degrees.

The intervarietal hybridization was started at the time when the breeding programme was proceeding in five and seven parallel cycles (Fig. 5). In the first cycle, a mixture of 11 families, chosen for a polycross-test in the IVth cycle *(–a–)* was crossed with 39 foreign varieties *(–b–)*. In the next cycle, which had already started on the basis of results on the divergent performance of single combinations, 26 more complicated combinations were arranged, using five components (*h* – breeder's seed from the IInd cycle, *i* – starting variety, *j* – three foreign varieties). Both cyclical crossings were followed by appropriate performance trials with controls included *(A, B, C, ...)*.

The second trial was devised as a polycross-test in which the basic rows were 8 foreign varieties (1, ..., 8), the inserted rows were 20 genetically confined families from the IVth selection cycle (1, ..., 20), and three variously stabilized degrees of the same family, i. e. from the IVth, IIIrd and IInd selection cycle. Individual seed harvesting of the components made it possible to compare diallel combinations in a performance trial and consider their relation to the starting material and to the material bred to various degrees.

The intervarietal hybridization by means of the cyclical method of crossing in free mating of spacially isolated components, and evaluated according to the yield of green matter in the first filial generation, led to direct yield changes in different directions and to different extents. In our own material the components being at a higher level of improvement and with proven general combining ability showed a higher performance. The crossing of genetically stabilized components (self-polli-

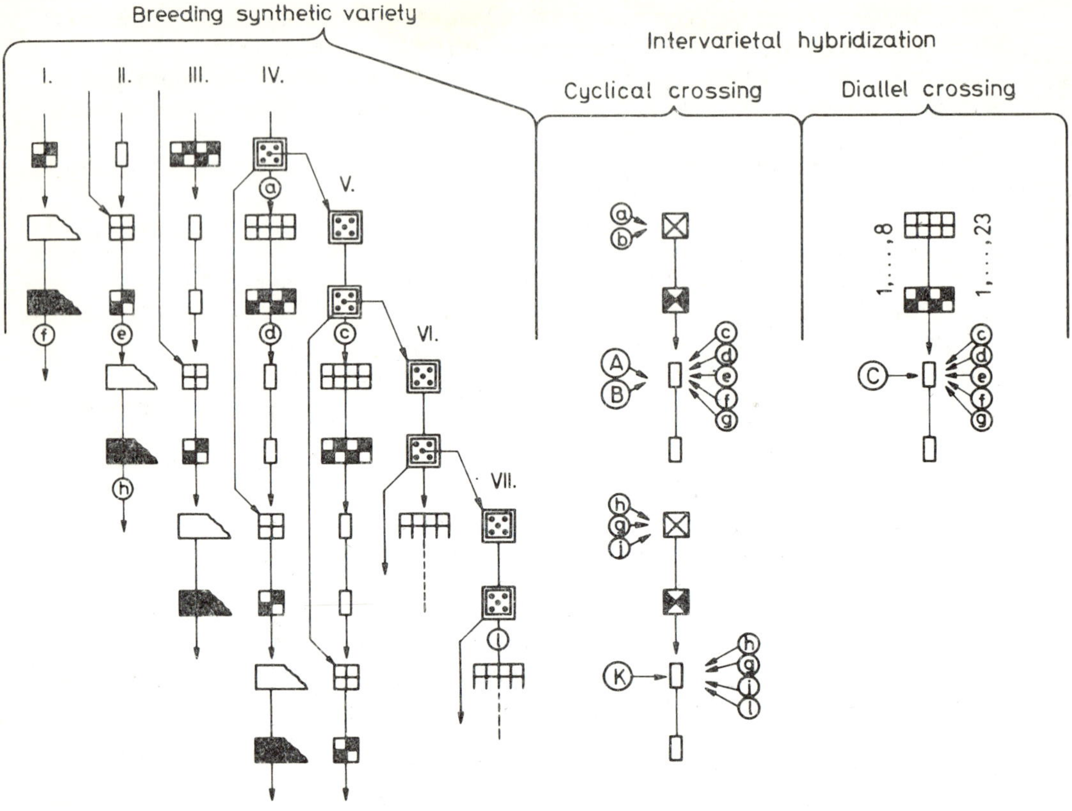

Fig. 5. Material and chronological working scheme for testing the effect of crossing components with different genetic scopes.
Key to the symbols:
Material used for the first hybridization: a = mixture of 11 families from the IVth selection cycle; b = 39 foreign varieties.
Material in the first performance trial: A = progeny of "a" after crossing; B = progeny of "b" after crossing; c = mixture of families chosen for polycross-test in the Vth selection cycle; d = mixture of families after polycross-test in the IVth selection cycle; e = seed gained with intercrossing in the IInd selection cycle; f = breeder's seed in the Ist selection cycle; g = initial variety.
Material used for the second hybridization: h = breeder's seed from the IInd selection cycle; g = initial variety; j = 3 foreign varieties.
Material in the second performance trial: K = components crossed in different combinations; h, i, j = original uncrossed components, l = high performing family from the VIIth selection cycle.
Material for the third hybridization: 1, ..., 8 = foreign varieties; 1, ..., 20 = families from the IVth selection cycle; 21, 22, 23 = three steps of genetic stabilization of the same family from the IVth, IIIrd and IInd selection cycle.
Material in the third performance trial: C = progeny after diallel crossing of 1, ..., 23; c, d, e, f, g = as above.

nated families) with good combining ability makes possible the achievement of a higher performance than does intervarietal crossing.

Crossing only two components with a broader genetical base – i.e. varieties, gave more satisfactory results then crossing more components. To achieve maximum effect specific combining ability should first be tested.

No direct effect of component performance on hybrid performance was proved even though it seems to be helpful to combine components of higher performance.

The performance of the intervarietal hybrids did not surpass that of the high-yielding varieties, nor of genetically restricted families from the selection process in hand.

Our breeding material under development and some foreign varieties were combined by means of diallel reciprocal crossing in free mating of spatially isolated components. The resulting material was evaluated again according to the yield of green matter in the first filial generation and showed that not only the particular combination, but also the partnership of components crossed, was decisive in performance. Corresponding results were not observed with reciprocal combinations. The genetical differentiation of the components crossed, i.e. the degree of selection of the material, showed its influence on the results.

REFERENCES

ROD, J. (1958). Možnosti technického řešení zkoušky hromadným křížením při šlechtění trav. (The possibilities of the technical solution of a polycross test in grass breeding) *Sborník VSZL v Brne, R. A.* 55–69.

ROD, J. & NAŠINEC, J. (1961). Příspěvek k otázce metodik ve šlechtění vytrvalých trav. (Beitrag zu den methodischen Fragen der Züchtung bei Futtergräsern.) *Rostlinná výroba,* 1561–1568.

ROD, J. (1965). Beitrag zu den methodischen Fragen der Heterosiszüchtung bei Futtergräsern. *Bericht über die Arbeitstagung 1965 der Arbeitsgemeinschaft der Saatzuchtleiter im Gumpenstein am 1–3. Dez.:* 235–252.

ROD, J. (1969). Some new aspects in breeding synthetic variety of Italian ryegrass. *Scientia Agriculturae Bohemoslovaca* **1** (XVIII), Nr. 3, 221–232.

ROD, J. (1974). Perspektivy meziodrůdové hybridizace jílku mnohokvětého. (Perspectives in intervarietal hybridization of Lolium multiflorum.) *Genetika a slechtení,* **10,** (In print).

ROD, J. (1974). Význam genetické vyhraněnosti komponent při křížení jílku mnohokvětého. (Significance of the genetical scope of components in the crossing of Lolium multiflorum.) *Genetika a šlechtění* (In print).

Session V

HETEROSIS IN SELF-POLLINATING SPECIES

Chairmen

PROFESSOR H. HÄNSEL AND DR. S. RAJKI

Heterosis in self-pollinating crops, with particular reference to barley

J. D. HAYES AND C. A. FOSTER

Welsh Plant Breeding Station, Aberystwyth, UK

SUMMARY

The genetic basis of heterosis in self-pollinating cereals and an assessment of the possibilities of isolating high yielding pure lines from heterotic F_1 hybrids will be briefly discussed. Evidence for heterosis for yield in barley obtained from trials conducted during 1967–1973 will be reviewed. The limitations of results from trials of F_1 hybrids will be discussed in relation to the expression of genotype × environment interactions.

Progress in developing methods of harnessing F_1 heterosis of barley, in particular, will be considered along with the economic constraints likely to influence the wide scale adoption of hybrid seed production of the small seeded self-pollinating cereals.

The case for continued effort into investigations of the F_1 heterosis phenomenon in the self-pollinating cereals will be presented.

INTRODUCTION

The phenomenon of heterosis has been generally associated with the increased yield and vigour obtained by crossing highly selected inbred lines from heterozygous cross-pollinating crops. With the recent realization of the possibility of producing F_1 hybrids on a large scale increasing attention has been given to heterosis in self-pollinating crops. As far as the breeders of agriculturally important self-pollinating crops are concerned the major considerations are, first, whether or not it is possible to obtain sufficient heterosis for characters of economic importance under conditions which also give high yields per unit area of land and, secondly, whether or not it is possible to fix such heterosis in pure breeding lines.

As MacKey and Demarly in previous papers in this conference discussed in detail the genetic principles governing the expression of heterosis, suffice it to say that heterotic F_1 hybrids (i.e. superior to the better parent), may result from one or more of the following genetic situations.

1. The accumulated action of favourable dominant or semi dominant genes dispersed amongst the two parents, i.e. dominance.
2. Complementary interaction of additive, dominant or recessive genes at different loci, i.e. non-allelic interaction or epistasis.
3. Favourable interaction between two alleles at the same locus, i.e. intra locus, or inter-allelic interaction, referred to as over-dominance.

It will be possible to recover homozygous lines as good as the heterotic F_1 hybrids if either or both of the first two situations are causing heterosis, although the ease with which such lines can be recovered will depend on the linkage relationships of the genes involved and the ability to identify the recombinants when they arise. This will be particularly difficult with close linkage and when heterosis is expressed by a slight improvement in each of the main yield components. If heterosis is due to inter-allelic interactions it is impossible to fix such heterosis in the homozygous condition in subsequent generations. Only by the production of F_1 hybrids, or the duplication of the relevant chromosome segments can heterozygosity be maintained and this type of gene action exploited in self-pollinating diploid species.

In this paper we shall concentrate our attention on work which has been conducted since 1968, because results available previous to this date, for the self-pollinating cereals in particular, were discussed fairly comprehensively at Wageningen in that year (Euphytica, Suppl. 1, 1968).

We pose three main questions:

1. Do self-pollinating crops produce F_1 hybrids that exhibit superior yield and vigour compared with the better parent?
2. What is the genetic basis for such heterosis in the self-pollinating crops?
3. What progress has been achieved in developing systems which will enable F_1 hybrid seed of self-pollinating crops to be produced on a field scale?

1. MAGNITUDE OF HETEROTIC EFFECTS

The breeder is concerned with the magnitude of heterosis obtained rather than with its frequency of occurrence. A review of the recent literature shows that heterosis has been obtained by workers in many different parts of the world with various self-pollinating crops grown under a wide range of test conditions. However, before considering in detail the evidence for heterosis, it is necessary to consider factors which have to be borne in mind when interpreting the results obtained so far.

a) Factors influencing the expression of heterosis

1. The difficulty of producing large quantities of F_1 hybrid seed of a self-pollinating crop, and the need to sample large numbers of different F_1 hybrids, has often resulted in research workers conducting tests for heterosis under conditions which are not representative of normal agricultural practice. Most of the reported yield trials have been carried out at a low seed density and under fertility conditions which do not maximize the yield of the highest yielding genotypes. This is especially important where head, ear or panicle production per unit is the most important component of yield. Similarly, relative susceptibility to lodging and disease infection introduce confounding influences which cannot be easily overcome.

As examples of the problems raised by such genotype × environment interactions in barley we refer first to the effect of differences in fertility level. Garkavyj & Baselhanova (1967) found that F_1 hybrids exhibited 25% heterosis under high fertility conditions, but only 4% under normal conditions. In contrast Conti & Ferraresi (1972) when assessing heterosis at two levels of nitrogen, found that the parents of

20 six-rowed barley F_1 hybrids gave higher yields of grain following the application of additional nitrogen, while the F_1 hybrids gave less, with the result that the apparent heterosis was reduced from 24% at the lower level of nitrogen to 3% at the higher level. Secondly, with regard to seeding density, Severson & Rasmusson (1968) and Dhindsa & Anand (1973) found that the level of heterosis for grain yield varied according to the density of sowing, with the highest expression at the widest spacing. On the other hand, Yap & Harvey (1971) found that heterosis for grain yield was relatively higher in densely sown plots than in plants grown at a wider spacing. Zeven (1972) in reviewing a number of seed rate experiments with wheat, concluded that the level of heterosis only varied slightly with changes in seed rate.

2. Many of the reported results have been obtained from experiments conducted in a single season and at one location. Therefore, the highest values for heterosis are those from the extreme end of a normal distribution curve which, in addition to being subject to the usual experimental errors, are also the product of specific genotype × environment interactions. Consequently, the greatest weight should be given to data obtained from F_1 hybrids tested in trials conducted in different locations over several seasons. However, it has not been possible in many cases to do this, owing to the difficulties of producing consistent quantities of F_1 hybrid seed from one season to another.

3. Differences in seed size are also likely to have a profound influence on yield performance. For example, in barley in the absence of suitable stocks of parents having male sterile genes, F_1 hybrids have to be produced by hand and this necessitates the removal of the awn and the upper portion of the palea and lemma, with the result that the F_1 hybrid seed is smaller than normal. Where F_1 hybrid seed is produced by wind pollination of male sterile plants, seed set is variable and often low, and as a result the size and quality of the F_1 hybrid seed may differ from that of the parents.

An example of the effect of seed size is given by Sage (1973). When large, medium and small seed of three F_1 hybrid wheats produced by wind pollination were sown alongside the appropriate parents, he obtained apparent yield heterosis of +9%, +3% and −3% respectively. A further example of the effect of differences in seed size on the expression of heterosis is provided by data from experiments conducted at Aberystwyth. Unselected seed of nine barley F_1 hybrids obtained from crosses between diploid male sterile plants from Ramage's 63-J-18-17 BTT stock and European varieties was on average 23% lighter than the seed of the male parent. They gave 6% less grain yield at constant plant density with individual F_1 hybrids ranging from −20% to +7% compared with their male parents. However, selected seed having the same weight as the parents gave 8% more grain, with individual hybrids exhibiting from −9% to +20% heterosis.

Even where the spurious effects of grain size on expression of yield are known it is not always possible to obtain exactly similar seed weights of the two parents and F_1 hybrids. Furthermore, the size of seed that should be used for true estimates of heterosis is debatable when it is known that seed size itself is genetically determined. Clearly the comparison which is of fundamental interest to a physiologist is not the same as that which is of interest to the commercial seed producer.

4. The effect of pests and pathogens on F_1 yield trials are seldom reported, yet the combination in F_1 hybrids of two single dominant genes effective against two prevalent diseases would result in the expression of apparent heterosis for increased yield.

b) Estimates of heterosis

Significant levels of heterosis for increased grain yield have been reported for most of the important self pollinated cereals, and a representative sample of results are presented here, with particular emphasis on barley. Barabas in the following paper will deal more specifically with results concerning wheat and sorghum.

Barley: Hayes (1968) reviewed the available data on F_1 hybrids and concluded that heterosis had been obtained from specific combinations of barley, but that most of the trials had been carried out at relatively low densities. The increases in yield were generally associated with an increase in tillering, a component likely to have less effect on yield in drilled crops than under spaced plant conditions. Since that review several workers have reported heterosis in F_1 hybrids of barley.

Garkavyj & Baselhanova (1967) in the USSR reported that F_1 hybrids representing 20 different combinations of high yielding barleys varying in time of ripening, exhibited, on average, 25% heterosis under high fertility conditions, but only 4% under normal conditions. Upadhyaya & Rasmusson (1967) with space planting (30.5 cm × 30.5 m) in the USA, reported that 28 F_1 hybrids had a mean superiority of 9.1% over the better parent. Seven of the F_1 hybrids significantly exceeded their higher yielding parent over 2 environments. Severson & Rasmusson (1968) demonstrated 3.2%, 16.7%, 17.3% and 22.5% superiority of five F_1 hybrids compared to their mid parent values at spacings of 2.5, 7.5, 15.0 and 22.5 cm apart in rows 30.5 cm apart. Wienhues (1968) obtained between 8% to 24% heterosis over 4 seasons in a series of four barley F_1 hybrids grown at fairly close spacing (5 cm × 20 cm). Nettevic & Sergeev (1969) in summarizing their results from 56 different combinations of spring barley found that yields of individual combinations ranged from −48.1% to +41.8% of the better parent.

Gebrekidian & Rasmusson (1970) evaluated 27 F_1 hybrids at normal commercial sowing rates in four environments and obtained an average F_1 superiority of 11%, with a range of −26% to +47% over the better parents. Conti & Ferraresi (1972) evaluated 20 six-row F_1 hybrids obtained from crossing five European cultivars with four Japanese cultivars and found that the best F_1 hybrids exceeded the best parent by 38% when tested at a low level of nitrogen, but, as mentioned previously, the hybrid advantage was reduced when tested at a higher level of nitrogen.

Since 1968 we have also produced numerous F_1 hybrids by various methods and tested them under different conditions of spacing. In all years specific cross combinations have yielded 20% or more heterosis. The results from two series of experiments are of particular interest. In one series F_1 hybrids were produced by pollinating male sterile plants in cultivars segregating for male sterility. Sufficient seed was produced to carry out trials in replicated small plots, but under sowing conditions more closely simulating agricultural practice than was possible in the past (seed rate = 56 kg/ha). The best F_1 hybrids gave up to 28% heterosis in 1972 and up to 36% in 1973. When these F_1 hybrids, derived from crosses between relatively old cultivars

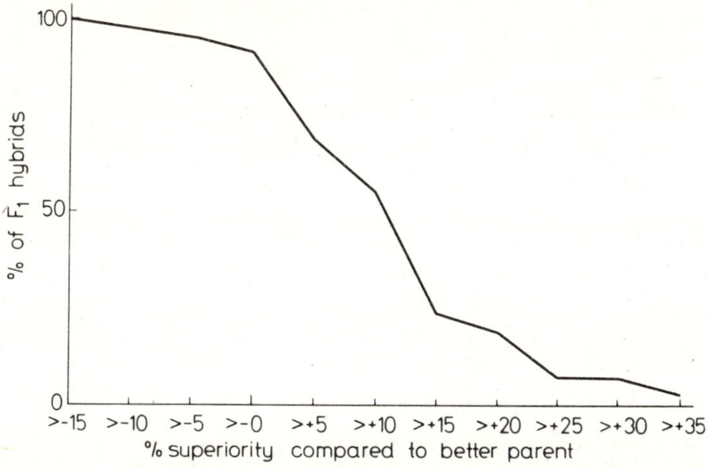

Fig. 1. Distribution of percentage superiority in grain yield over the better parent of 48 F_1 spring barley hybrids.

such as Freja, Herta, Carlsberg II, Maris Concord and Zephyr, were tested against the highest yielding cultivars commercially available in Britain the F_1 hybrids gave up to 19% more grain (Foster & Jones, 1974).

A frequency distribution curve of the number of hybrids exhibiting given levels of heterosis tested in this series shows that there was an approximate normal distribution with an average superiority of around 11% (Fig. 1).

In the other series of experiments Khalifa (1973) studied the F_1 and F_2 progeny from a diallel cross of eight morphologically diverse barley genotypes consisting of Spratt, Zephyr, Midas, Clermont, Akka, Nackta, Jaybel and a dwarf selection – Paladin. The seed for this study was produced by hand and space planted 15 cm apart within rows 60 cm apart. Khalifa obtained an average level of heterosis of 15.4% in this material, with a range from −12.7% to +56.0% of that of the better parent.

However, considerations of heterosis of grain yield alone do not present the full picture as far as economic production is concerned, because the increases in grain yield may be accompanied by undesirable correlated responses. For example, in the 8-parent diallel described above, the average of +15.4% better parent heterosis for grain yield was equivalent to an average F_1 superiority of 26.7% over the mid parent value. The average superiority of the F_1 hybrids over the mid parents was proportionally greater for the vegetative portion, i.e. straw production, viz. +37.9%, and for total plant weight +32.6%.

The F_1 superiority in straw production over the mid parent value was accounted for by an average increase in both straw length (+17.2%) and number of ear bearing culms (+12.0%). As a consequence of the increases in length of straw and ear bearing culms – the later formed of which had less grain – there was a deterioration in the grain:straw ratio from 1.20:1 in the parents to 1.03:1 in the F_1 hybrids. As one of the major objectives in cereal breeding is to reduce the length of straw and improve the grain:straw ratio, clearly the increased length and quantity of straw produced by heterotic F_1 hybrids may be counter productive if this leads to higher yield losses due to lodging. Similar considerations could also apply to certain features of grain quality.

Other self-pollinated crops: For wheat, Livers & Heyne (1968) in Kansas, USA, found the yield of the best F_1 hybrid to be 31% higher than the best variety over a period of four years, ranging from 27% to 37% for individual years. In Russia, Nettevic (1965) tested 40 F_1 hybrids, all of which showed heterosis for yield compared with their parents, and had 26–60% higher yields than the recommended variety. More recently Wells & Lay (1970) reported yields of spring wheat F_1 hybrids in S. Dakota ranging from 14% less than, to 82% more than, the high parent. In Alberta, Cramb & McKenzie (1970) found significant levels of heterosis in spring × winter crosses of up to 40% compared with the higher yielding spring parents. Zeven (1972) in the Netherlands found up to 35% heterosis for grain yield among six F_1 hybrids tested over a range of densities.

Heterosis for grain yield and its components have been reported by several rice workers, but due to difficulties of producing hybrid seed of rice most of the trials have been conducted with spaced plants. Purohit (1972) in India found evidence of significant heterosis for grain yield of up to 93% over the better parent and 38% over the highest yielding parent in the experiment. Chang *et al.* (1971) in Taiwan reported that out of 265 F_1 hybrids 145 gave higher yields than their better parent, with the average heterosis of these hybrids being 18%. In crosses among japonica type rice, Carnahan *et al.* (1972) in California, USA, reported yield increases over the better parent in 17 of the 19 hybrids tested, with maximum values of up to +110%.

The data on heterosis in oats is limited, but the general conclusion is that significant levels of heterosis can be obtained from various crosses. For example, Petr & Frey (1967) studied 15 F_1 hybrids and found that the grain yield of all hybrids exceeded the higher parent by 1% to 30%, with an average of +12%. Similarly Murphy (1966) found that an average of 13% heterosis for increased yield in 72 oat crosses was expressed in increased panicle production. Jenkins (1969) in a study of 36 crosses, many of which were between *A. sativa* and *A. byzantina,* found that F_1 yields ranged from −32% to +39% of the better parent, with an average heterosis of 9%. Hathcock & McDaniel (1973) were the first to produce sufficient hybrid seed to test for heterosis in oats grown at relatively close spacing (4.7 cm apart in rows 30 cm apart), and in their trials the best F_1 hybrid surpassed the better parent by 75% or more at two locations, and the best cultivar by over 35%.

In the main, the heterotic responses of soya beans have also been evaluated in spaced plant trials. The values obtained by Weber *et al.* (1970) are typical of the results reported by other workers. Averaged over tests in four years, 85 F_1 hybrids exhibited mid parent heterosis for grain yield of 25.1% and high parent heterosis of 13.4%, with the range of the high parent heterosis from −39.1% to +90%. Heterosis has also been reported for lint yield in cotton and is generally much greater in interspecific crosses of *G. barbadense* × *G. hirsutum* than in intraspecific crosses (Marani, 1963, 1967).

As would be expected, some workers have reported that none of the crosses which they have investigated have exhibited any heterosis, e.g. for barley, Carleton & Foote (1968) and Yap & Harvey (1970). Some workers have found the maximum expression of heterosis amongst crosses of the lowest yielding parents (e.g. Gebrekidian & Rasmusson, 1970), while others have found no such relationship. Our experience at Aberystwyth suggests that there is a tendency for higher levels to be obtained from low yielding parents, but the relationship was not significant in the 8-parent diallel mentioned previously as $r = -0.25$ for % heterosis vs the mid parent value and $r = -0.27$ for the % heterosis and grain yield of the better parent.

On the basis of the results obtained so far we conclude that crosses between presently available and commercially acceptable cultivars of barley can yield increases of at least 20% of grain under conditions simulating agricultural practice. The results quoted for the other self-pollinating crops present a similar picture, although, with the exception of sorghum, it has not been possible so far to produce sufficiently large quantities of any of the self-pollinating cereals to strictly test the repeatability of the results under normally accepted agricultural conditions.

2. THE GENETIC BASIS OF HETEROSIS IN SELF-POLLINATING CROPS

Most of the detailed genetic information regarding the expression of heterosis in self-pollinating crops has come from diallel crosses of selected parents and F_1 hybrids, occasionally from F_2 generations, and more rarely from subsequent generations. In most cases, the material was space-planted in order to obtain carefully recorded measurements, but such data must be limited in its applicability to breeding for agricultural conditions when there are interactions between the experimental spacings and those used in practice.

In most studies the inheritance of grain yield presents a complex picture, with strong evidence of dominance and epistasis and occasionally apparent over-dominance. However, the main components of yield generally exhibit less complicated patterns of inheritance as they are largely controlled by additive gene action with some dominance and occasionally, epistasis, e.g. Whitehouse *et al.* (1958), Upadhyaya & Rasmusson (1967), Gebrekidian & Rasmusson (1970) in barley, Paroda & Joshi (1970) in wheat, and Murphy (1966) in oats.

A summary of the results obtained by Khalifa with the 8-parent spring barley diallel provides a representative illustration of the type of results obtained by several other workers who have investigated the genetic basis of heterosis of yield and its components in self-pollinating crops. Total plant weight, and straw and grain yield exhibited a high level of heterosis while the components of grain yield displayed little or no heterosis (Table 1). This is reflected in the analysis of variance of the F_1 generation where significant values were obtained for the 'a' additive and 'b' non-additive/dominance effects (Hayman, 1954). The ratio of a:b was far larger for the components of yield than for the complex characters – yield of grain, yield of straw and whole plant weight. These differences in ratios are also reflected in the F_2 generation, although the specific combining ability effects are expected to decline in magnitude in the F_2 generation with the expected 50% reduction in heterozygosity.

These results support the views of Williams (1959) and Grafius (1959) who envisaged yield as the result of the multiplicative interaction at the phenotypic level of components which themselves exhibit relatively simple gene action. These views are

Table 1. Average heterosis and components of variance for yield, height and components of yield of 28 F_1 and F_2 populations from an 8-parent sping barley diallel grown at Aberystwyth, 1972

	$\frac{(F_1—MP)}{(MP)} \times 100$	$\frac{\Sigma \frac{(F_1—BP)}{(BP)} \times 100}{28}$	F_1 Additive:non-additive (a/b)	F_2 General:specific combining ability (GCA/SCA)
Grain yield	+26·7*	+15·4*	1·1	9·5
Straw yield	+37·9*	+21·6*	1·5	14·2
Total plant weight	+32·6*	+18·1*	1·2	9·9
Length of straw	+17·2*	+7·2*	8·9	37·2
No. of culms/plant	+11·4*	—5·1*	14·0	65·6
Wt of grain on main culm	+11·3*	—0·31ns	14·3	46·1
No. of grains/main culm	—2·34	—12·3*	20·5	140·7
1000 grain we/main culm	+13·4*	+4·7*	2·8	13·0

*P=0·05

well illustrated by the data of Whitehouse *et al.* (1958) who found that the components of grain yield were controlled primarily by additive gene action, but that yield exhibited a type of interaction which could be removed by log transformation of the data. The genetic interpretation of yield is often more complicated than this because the interacting components compensate each other at various stages of growth.

Gustaffson *et al.* (1973) recently presented data which suggest that barley F_1 hybrids heterozygous for one, and for two, major gene mutants exhibit over-dominance for yield and its components under various controlled photoperiod and temperature conditions. However, many previous examples of over-dominance can be explained on the basis of dominance and epistasis (Jinks, 1955).

Relatively little work has yet been reported of attempts to recover homozygous pure lines from heterotic F_1 hybrids. Williams (1959) with tomatoes, Aastveit (1964) with spring barley, and Busch *et al.* (1971) with spring wheat, have isolated pure breeding lines as good as or better than the heterotic F_1 hybrids. Clearly, if over-dominance or non-fixable types of epistasis had operated in these instances it would not have been possible to recover such lines. However, because there is strong evidence for a preponderance of additive genetic variance controlling the components of yield in the self-pollinating crops, it should be possible to extract homozygous lines from the majority of heterotic F_1 hybrids, although it could be difficult if many genes are involved and there are tight repulsion linkages.

There is still a great deal of controversy regarding the relative stability of performance of F_1 hybrids and highly selected pure lines of the self-pollinating crops, and until sufficiently large quantities of F_1 hybrid seed are produced for field trials the question will remain unresolved.

Results of experiments conducted in different environments by Yap & Harvey (1970) with barley suggest that non-additive effects are more susceptible than additive effects to changes in the environment, while the results of Liang (1967) with sorghum, and Paroda & Joshi (1970) with wheat suggest that additive effects are most vulnerable to changes in the environment. Swaminathan *et al.* (1972) claimed the F_1 hybrids of maize, sorghum and pearl millet have not only been instrumental in improving the yield, but also the stability of yield of these crops in India.

Based on this genetic information and practical experience, is it possible to predict the parental combinations most likely to produce agriculturally valuable heterotic F_1 hybrids?

Genetically the best F_1 hybrids will be obtained from crosses of parents having a high proportion of additive, dominant or complementary epistatic genes for the favourable expression of the main components of yield. The degree of heterosis will depend on the dispersion of the favourable dominant and epistatic genes amongst the parents. Agronomically and physiologically, the hybrids will need to have a balanced combination of yield components, a high harvest index and resistance to the prevalent diseases.

Carleton & Foote (1968) did not find any significant heterosis above the better parent in crosses between widely contrasting parents selected for differences in components, while Bogomolov & Grib (1971) found heterosis was most frequent in hybrids from morphologically similar parents, and that it was rarely found in highly contrasting parents. On the other hand, Nettevic & Sergeev (1969) stated that crosses between different ecotypes gave the highest levels of heterosis, and Barsukov (1969) obtained the greatest heterosis in hybrids whose parents differed in productivity and came from different geographical or ecological areas. Based on their experience that high yielding cultivars produced a larger proportion of high yielding hybrids than did crosses between low yielding cultivars, Gebrekidian & Rasmusson (1970) recommended that cultivar *per se* evaluation was as good a basis as any for initial screening of potential parents. In contrast neither Wells & Lay (1970) nor Hathcock & McDaniel (1973) in spring wheat and oats respectively, found evidence that high yielding pure lines would necessarily produce high yielding F_1 hybrids.

On a more fundamental basis several workers, including Sarkissian & Srivastava (1967) with maize and McDaniel (1971, 1973) with maize and barley in the United States, and Sage & Hobson (1973) with wheat in the United Kingdom, have measured the oxidative and phosphorylative activity of mitochondria isolated from F_1 hybrids and their parents. They obtained correlations between the activity of the isolated mitochondria and the potential dry matter and/or grain production of the genotypes concerned. Furthermore, McDaniel claims to have found positive interaction *in vitro* between mitochondria of parents which produce heterotic F_1 hybrids. Neither Griffiths (pers. comm.) nor Ellis *et al.* (1973), all of whom have devoted considerable attention to the detailed investigation of this phenomenon, have been able to obtain any evidence of complementation. Consequently, it would appear that these techniques require a great deal of development before they can be used by breeders as routine practice in selecting potential parents for producing heterotic F_1 hybrids.

3. PROGRESS IN DEVELOPING MECHANISMS FOR PRODUCING F_1 HYBRIDS OF SELF-POLLINATED CROPS

A system for producing large quantities of male sterile plants to act as a female parent is an absolute necessity for successful production of F_1 hybrids of self-pollinating crops on a field scale. Several schemes have been proposed and described in some detail by various workers. A system based on cytoplasmically controlled male sterility and fertility restoring genes is being used commercially for producing F_1 hybrids in sorghum. Similarly, the cytoplasmically inherited male sterility resulting from the incorporation of the nucleus of *T. vulgare* in *T. timopheevi* cytoplasm and use of fertility restoring genes first described by Wilson & Ross (1962) appears to be the most promising for wheat and is receiving a great deal of attention from practical breeders (Barabás in the following paper). Cytoplasmically controlled male sterility and fertility restoring genes have also been found in rice and cotton. On the other hand, for barley, major gene controlled male sterility appears to offer the most promising avenue of approach.

There are two main ways that a recessive gene for male sterility can be utilized in self-pollinating crops.

a) Use of Balanced Tertiary Trisomic (BTT) plants

This elegant method was originally devised by Ramage (1965) and has been used successfully in Arizona to produce a series of barley F_1 hybrids on a field scale, the first of which was Hembar. The extra chromosome of the BTT is a reciprocal translocation; it carries the dominant fertile allele of the male sterile gene, and the male sterile recessive alleles are carried on a pair of normal chromosomes. Ideally, the extra chromosome should only be transmitted through the ovules and not through the pollen, with the consequence that all the effective pollen produced by such a BTT plant carries the *ms* gene. Selfing such a BTT effectively produces two types of progeny, namely:

- male fertile BTT plants identical to the parent, and
- diploid plants homozygous for male sterility.

Occasional male sterile primary trisomics are also produced. The proportion of male fertile BTT: male sterile diploid plants obtained in each generation will depend on the translocation used, the length of the extra chromosome and the genetic background of the parent. Generally the trisomic seed is smaller and produces less vigorous plants than the normal diploid seed. The BTT seed can either be used to produce pollen parents for male sterile diploid plants, or it can be multiplied in isolation to produce both male sterile diploid and BTT seed.

Hembar and all the other commercial F_1 hybrid barley cultivars released to date have been produced by the BTT method using Ramage's female stock 63-J-18-17, which contains the T2-7d translocation balanced with ms_2 on chromosome 2.

In practice seed from the BTT is screened and the larger seed is used for sowing the male sterile diploid which acts as the female parent, and the smaller seed is used to produce the BTT stock. The seed for the male sterile female block is sown thickly so that the less vigorous BTT plants are eliminated by competition. The BTT strips are sown thinly to enable the morphologically distinct male sterile diploid plants to be removed by hand – clearly a time consuming and expensive process which has to be eliminated if the full scale production of F_1 hybrids by this method is to be successful. It is difficult to obtain absolute homogeneity of F_1 seed stocks because the separation of the diploid from trisomic seed by weight is not complete, and differences in competitive ability are not sufficient to eliminate all the BTT plants from the male sterile diploid used as the female parent. Furthermore, there may be a small percentage of pollen transmission of the extra chromosome carrying the male fertile allele resulting in the appearance of male sterile plants in the F_1 hybrid crop.

Table 2. Per cent trisomics in large and small seed fractions of semi-adapted Balanced Tertiary Trisomic plants

BC^2 lines	Seed Size				Weighted mean % trisomic
	large		small		
	% of sample	% trisomic	% of sample	% trisomic	
Zephyr (22 lines)	78·6	18·2	21·4	75·8	30·5
Isaria (24 lines)	66·5	31·1	33·5	41·9	34·7

In common with many other breeders and research workers we obtained various BTT stocks from Dr. R. T. Ramage of the University of Arizona. The 63-J-18-17 stock containing the T2-7d translocation balanced for ms_2 has proved to be the most satisfactory, but other translocations/*ms* gene combinations could be used. The 63-J-18-17 derived BTT stocks produce about 70% diploids: 30% BTT on selfing, thus confirming Ramage's results. But in our experience it is difficult to separate the diploid from trisomic seed by weight owing to the considerable overlap in seed from the two categories (Table 2). However, this might be more satisfactory when the extra chromosome is incorporated in the fully adapted 2-row genetic background, rather than the non-adapted 6-row type which tends to produce smaller grains on the laterals than on the medium spikelets of the ear.

We are currently backcrossing T2-7d and ms_2 into adapted European backgrounds in order to improve vigour, agronomic type and disease resistance, with a view to producing female lines which can be used on a large scale as testers for a range of pollen parents. We have encountered some difficulty in getting satisfactory seed set on trisomic plants when artificially hybridized and with a low rate of recovery of trisomics after such hybridization. Despite improvements in genetic background we are also experiencing difficulty in improving the vigour of the plants carrying the extra chromosome.

However, even after recovering a series of balanced tertiary trisomics in an adapted background, the same problems of separating the diploids from the trisomics will remain. Wiebe & Ramage (1971) suggested that the incorporation of a seedling lethal gene closely linked to the *ms* gene would eliminate the need for removing homozygous male sterile diploid plants by hand in the BTT multiplication field and enable more vigorous BTTs to be developed. Dr. R. T. Ramage has already produced one stock containing a seedling lethal fairly closely linked to ms_2 on chromosome 2 (pers. comm.). This is being backcrossed into adapted BTT lines to produce interim stocks for producing male sterile lines.

Certain stocks of tertiary trisomics have changed in character due to loss of parts of the translocated chromosome. This has resulted in much higher transmission rates in the pollen, and as a consequence the efficiency of the system is impaired. As far as the authors are aware there is no information available concerning the factors which control such chromosome fragmentation.

The incorporation of a dominant pollen lethal gene on the translocated chromosome would be of great advantage in that it could prevent completely the transmission of the extra chromosome through the pollen. Stocks heterozygous for *ms* have been treated with mutagens and a search is being made for mutants giving seedling lethals and/or micro-gametic lethals, so that a system based on the BTT scheme can be perfected. However, if suitable mutants could be found closely linked to an *ms* gene on a normal chromosome, systems based wholly on diploid plants could be developed.

b) Systems based on diploid plants

Wiebe (1960) proposed a scheme for producing F_1 hybrid barley using diploid plants only. In this system a genetically controlled susceptibility/resistance reaction to an agricultural chemical can be used to eliminate male fertile plants in a population segregating for male sterility. This requires very close linkage between the recessive genes for resistance to the chemical and male sterility. No such scheme has yet been put into practice owing to the absence of a suitable reaction sufficiently close linked to a male sterile gene.

An alternative is to find a genetic means of preventing the pollen transmission of the dominant male fertility allele of a plant heterozygous for male sterility. A micro-gametic lethal mutation closely linked to the dominant male fertile allele at an *ms* locus and a seedling lethal linked to the *ms* would provide a means of obtaining homozygous male sterile stocks. This would act in much the same way as a BTT, but without the problems associated with the presence of an extra chromosome (Ramage *et al.*, 1972). However, this scheme remains speculative until proof can be obtained of the existence of such haplo-viable mutations. Experiments are in progress at several centres to obtain such mutations by the use of mutagens.

During the course of our investigations on F_1 hybrid barley, a large number of sterile plants have been found in barley stocks, some of which have proved to be male sterile. Many of these have proved to be under single gene control while others appear to be cytoplasmically controlled. However, previous attempts to utilize cytoplasmic sterility derived from interspecific hybridization in barley have not been successful owing to its variable and incomplete expression and its association with undesirable agronomic characteristics.

There have been several attempts to induce male sterility and restore fertility to male sterile plants by the use of chemicals. Sodium 2,3 dichloro-iso-butyrate (F.W. 450), maleic hydrazide, and various gibberellins have been used in the past to induce male sterility, with varying degrees of success, in different species. In most cases the timing and concentration have been critical and it has not been possible to obtain commercially useful levels of male sterility. More recently Ethrel (2-chloroethyl phosphonic acid) has been used in wheat by Rowell and Miller in 1971 and by Hughes *et al.* (1974) but without complete success. Experience at Aberystwyth (Foster & Jones, 1973) and elsewhere has shown that Ethrel is not a satisfactory male gametocide for barley. There is need of more effort in this direction because the availability of such a selective male gametocide would greatly facilitate the production of F_1 hybrids in male fertile parents.

So far attempts to restore fertility to male sterile plants have not been particularly successful, although it is to be expected that the discovery of chemical compounds able to restore fertility to one of the 23 different loci of barley or to the many different types of male sterile cytoplasms found in the various self-pollinating crops is a distinct possibility. The discovery of environmentally sensitive male sterile genes provides the possibility of utilizing this type of male sterility for multiplying up the seed of the female parent without recourse to balanced tertiary trisomics or linked male sterile factors.

4. POSSIBLE LIMITATIONS TO EXPLOITATION OF F_1 HYBRIDS IN SELF-POLLINATED CROPS

a) Cross-pollination of male sterile plants

The basic difficulty in developing hybrids of self-pollinating crops is that the basic floral morphology is not adapted to cross-pollination. In most of the crops pollen will have to be transferred from male fertile to male sterile plants by the wind, although other agencies, such as insects, may be used for specific crops.

The reported values for seed set on wind pollinated male sterile heads of cereals have varied widely, with up to 90% being reported by Wilson (1972) in large crossing blocks of wheat. Clearly, in view of the need to saturate the area around the male sterile plants with pollen, many of the results obtained from the small scale experiments reported so far are unreliable in estimating the maximum level of seed setting that can be achieved in large multiplication fields.

Seed setting following wind pollination is influenced by environmental and genetic factors. De Vries (1971) has reviewed the literature on the floral biology of wheat in relation to F_1 hybrid seed production. In general, moderate temperatures and humidity, as found in Western Europe, seem favourable for high seed setting. High temperatures and dry conditions tend to inhibit flower opening, while low temperatures may restrict anther and pollen development, but do not appear to limit stigma development or flower opening. Humid conditions tend to reduce flower opening, but when the conditions become drier and warmer there is a rapid extension of the anther filaments and simultaneous release of a massive pollen cloud.

Previous work (Hayes, 1968) has shown that seed setting following wind pollination is affected by genetically controlled factors and this is confirmed by results with four male sterile stocks of barley grown in small isolation islands at Aberystwyth during 1971–73 (Table 3). Clipping of the upper parts of the palea and lemma of male sterile florets increased seed set from 50% to 78% on Freja and 27% to 74% on Herta in 1971, indicating that mechanical resistance in the female rather than a shortage of pollen was responsible for low seed set with wind pollination. It has been possible to select heads of male sterile plants from within composite populations which have given full seed set at Aberystwyth, and we forsee no real problem in increasing the efficiency of wind pollination when it becomes necessary. This is confirmed by the results of Thompson (1971) in the United States who has been able to increase the mean seed set to 83% by heavy selection of male sterile plants from composite populations grown alternately in two different environments; 28% of the male sterile plants gave full seed set. Clearly selection for plants with large lodicules and strongly branched protruding stigmas which have a long period of receptivity will increase the seed setting capability of the female parent, while increases in the production of effective pollen for hybrid production can be achieved by increasing anther extrusion (Joppa *et al.*, 1968) and pollen production per anther (D'Souza, 1970). However, there is a suggestion that certain lines having large extruded anthers tend to suffer more ear shattering than other lines.

Table 3. Variation between seasons and parent male sterile stocks in % seed set in small wind-pollinated blocks 1971–73

Male sterile stock and pollinator	1971	1972	1973	Mean
Cb 1000 Freja (*ms* 4)	50±5	57	36	48
Cb 1002 Herta (*ms* 1)	27±4	32	25	28
Cb 1005 Svalof 50/109 (*ms* 11)	36±3	14	29	26
Cb 1003 Carlsberg II (*ms* 5)	25±3	27	9	20
Mean	34	32	25	30

It is also possible that closely related cross-pollinating species could be used as donors of specific characteristics to increase the effectiveness of wind pollination of the cultivated species, for example, the outbreeding and prolific pollen producing characteristic of *H. bulbosum* could be transferred to the cultivated barley. We feel that this aspect of F_1 hybrid production will present no real plant breeding problem once the selected lines which exhibit the required level of heterosis have been identified.

b) Disease susceptibility

The open flowering male sterile lines of normally self-pollinating crops are usually extremely susceptible to floral infecting fungi. In the cereals, loose smut (*Ustilago* sp.) and ergot *(Claviceps purpurea)* are likely to present very great problems when F_1 hybrid production is carried out on a large scale. It is possible that systemic fungicides such as the 1,4-oxathiin derivative, Vitavax, will continue to control loose smut, but the danger of a new virulent form able to overcome the action of this chemical is ever present. The highly noxious ergot presents an even greater threat because one of the available genetic sources of resistance to this pathogen in wheat, viz. that of Kenya Farmer, has already been overcome (Svensson, 1971).

However, if disease infection of the flowering organs can be overcome, F_1 hybrids present an efficient way of using dominant genes for resistance to diseases and insect pests. Different dominant genes could be introduced through the two parent lines and combined in the F_1 hybrid, thus providing one of the only practicable methods of bringing together allelic genes such as those at the *mla* locus in barley. By developing iso-genic male lines having different dominant genes for resistance and female lines having a high level of race non-specific resistance, it is possible to continually change the spectrum of resistance without altering the potential yielding ability of the cross combination. Done (1973) assessed the disease reaction of F_1 hybrids of wheat and barley grown on a large scale in field plots and showed that the expression of the dominant genes for resistance could vary according to the genetic background of the F_1 hybrids.

c) Economics

With regard to the commercial exploitation of heterosis in self-pollinating species, it is possible that for certain crops relatively low levels of heterosis may be sufficient to make F_1 hybrid production profitable. This is especially true for crops such as the tomato where uniformity for mechanical harvesting and/or earliness are important attributes, and these are greatly improved in F_1 hybrids. Relatively low levels of heterosis may also be sufficient for crops producing a large number of seeds per pollination, or which have a high multiplication rate or where the cost of seed constitutes a relatively small fraction of the total cost of cultivation of the crop. However, in self-pollinating crops having a relatively low rate of multiplication, such as the cereals, the amount of heterosis will have to be of considerable magnitude to cover the increased cost of the seed. Several workers including Chase (1971) and Svensson (1971) have attempted to estimate the amount of heterosis required in the small grained cereals to make F_1 hybrids economically competitive with conventional varieties. Our impression is that there is a concensus of opinion that a consistent level of heterosis of the order of 20–25% in increased grain yield will be necessary. This value coincides with the levels of heterosis obtained in experimental work. However, even if the required levels of heterosis can be achieved consistently, it is not likely that a large acreage of F_1 hybrids of self-pollinating cereals will be grown in Europe in the near future owing to the risks inherent in their production.

In view of its magnitude and the hope it offers for raising the yield ceiling of many of our intensely bred crops, the possibilities of exploiting heterotic vigour must continue to be explored. Suitable systems must be developed which will enable fairly large quantities of hybrid seed to be produced, so that the nature and magnitude of various types of gene action controlling characters of economic importance are investigated in early generation trials under agriculturally meaningful conditions. Equipped with such information, breeders of conventional pure line varieties and prospective hybrid varieties will be better able than at present to make rational judgements as to the correct strategy to adopt for improving the yield potential of various self-pollinating crops.

AASTVEIT, K. (1964). Heterosis and selection in barley. *Genetics,* **49,** 159–164.

BARSUKOV, P. N. (1969). Field and grain quality in barley hybrids. (Russian) Selekc Semenovad Mezved nauc Sborn No. II, 64–68 (cited in *P. B. A.* **41,** 2862).

BOGMOLOV, A. J. & GRIB, S. I. (1971). The manifestation of heterosis in F_1 hybrids of spring barley (Russian). Cited in *P. B. A.* **43,** 2695.

BUSCH, R. H., LUCKEN, K. A. & FROHBERG, R. C. (1971). F_1 hybrids versus random F_5 line performance and estimates of genetic effects in spring wheat. *Crop Sci.* **11,** 357–361.

CARLETON, A. E. & FOOTE, W. H. (1968). Heterosis for grain yield and leaf area and their components in two-x six rowed barley crosses. *Crop Sci.* **8,** 554–557.

CARNAHAN, H. L., ERICKSON, J. R., TSENG, S. T. & RUTGER, J. N. (1972). Outlook for hybrid rice in USA. In *Rice Breeding.* The International Rice Research Institute, Los Banos, Philippines. 603–607.

CHANG, W. L., LIN, E. H. & YANG, C. N. (1971): Manifestation of hybrid vigour in rice. *J. of Taiwan Agric. Res.* **20,** (4) 8–23.

CHASE, S. S. (1971). Estimates of commercial profitability in hybrid seed production of small grains. *Econ. Bot.* **25,** 330–334.

CONTI, S. & FERRARESI, A. (1972). "Evaluation of heterosis and components of genetic variance in barley." (Italian) *Genet. Agr.* XXVI, Fasc, I–II.

CRAMB, M. N. & MCKENZIE, H. (1970). Heterosis in F_1 hybrids between spring and winter wheats. *Can. J. Pl. Sci.* **50,** 137–140.

DE VRIES, A. P. (1971). Flowering biology of wheat, particularly in view of hybrid seed production – a review. *Euphytica* **20,** 152–170.

DHINDSA, G. S. & ANAND, S. C. (1973). Effect of spacing on heterosis in wheat. *Ind. J. Agric. Sci.* **43,** 177–180.

DONE, A. C. (1973). Implications of the introduction of hybrid cereals on disease patterns. *Ann. app. Biol.* **75,** 123–149.

D'SOUZA, L. (1970). Untersuchungen über die Eignung des Weizens als Pollenspender bei der Fremdbefruchtung verglichen mit Roggen, Triticale und Secalotricum. *Z. PflZücht.* **63,** 246–269.

ELLIS, J. R. S., BRUNTON, C. J. & PALMER, J. M. (1973). Can mitochondrial complementation be used as a tool in breeding hybrid cereals? *Nature* **241,** 45–47.

FOSTER, C. A. & JONES, J. E. (1973). Chemical manipulation of male fertility for hybrid barley production. *Rep. Welsh Pl. Breed. Stn. for 1972.* 89–90.

FOSTER, C. A. & JONES, J. E. (1974). Heterosis for grain yield in F_1 hybrid barley. *Rep. Welsh Pl. Breed. Stn. for 1973.* 58–59

GARKAVYJ, P. F. & BASELHANOVA, V. B. (1967). Manifestation of heterosis in spring barley hybrids in accordance with the growing conditions. *Proc. All-Un. Lenin Acad. agric. Sci.* No. **5,** 2–4. Cited in *P. B. A.* Vol. 38, No. 649.

GEBREKIDIAN, B. & RASMUSSON, D. C. (1970). Evaluating parental cultivars for use in hybrids and heterosis in barley. *Crop. Sci.* **10,** 500–502.

GRAFIUS, J. E. (1959). Heterosis in barley. *Agron. J.* **51,** 551–554.

GRIFFING, B. & LANGRIDGE, J. L. (1963). Phenotypic stability of growth in the self-fertilized species *Arabidopsis thaliana.* In Statistical Genetics and Plant Breeding, ed. by W. D. Hanson and H. F. Robinson. NAS – NRC 982: 363–394.

GUSTAFFSON, A., DORMLING, I. & EKMAN, G. (1973). Phytotron ecology of mutant genes. II. Dynamics of heterosis in an intralocus mutant hybrid of barley. *Hereditas* **74,** 247–258.

HATHCOCK, B. R. & MCDANIEL, M. E. (1973). Yield and yield components. Heterosis in *Avena* hybrids. *Crop Sci.* **13,** 8–10.

HAYES, J. D. (1968). The genetic basis of hybrid barley production and its application in Western Europe. *Euphytica* **17,** (1968 Suppl. 1) 87–102.

HAYMAN, B. I. (1954). The analysis of variance of diallel tables. *Biometrics* **10,** 235–244.

HUGHES, W. G., BENNETT, M. D., BODDEN, J. J. & GALANOPOULOU, S. (1974). Effects of time of application of Ethrel on male sterility and ear emergence in wheat *Triticum aestivum. Ann. appl. Biol.* **76,** 243–252.

JENKINS, G. (1969). Heterosis and combining ability in hybrids of *Avena sativa* L. and *A. byzantina* C. Koch. *J. agric. Sci. Camb.* **72,** 85–92.

JINKS, J.L. (1955). A survey of genetical basis of heterosis in a variety of diallel crosses. *Heredity,* **9,** 223–238

JOPPA, L.R., MCNEAL, F.H. & BERG, M.A. (1968). Pollen production and pollen shedding of hard red spring (*Triticum aestivum* L. em Thell) and Durum (*T. durum* Desf.) wheats. *Crop Sci.* **8,** 487–490.

KHALIFA, M.A. (1973). A genetic study of yield and its components in spring barley. Unpub. Ph. D. Thesis. Univ. of Wales.

LIANG, G.H. (1967). Diallel analysis of agronomic characters in grain sorghum. *Sorghum vulgare* Pers. *Can. J. Genet. Cytol.* **9,** 269–276.

LIVERS, R.W. & HEYNE, E.G. (1968). Hybrid vigour in hard red winter wheat. In Proc. 3rd Int. Wheat Genetics Symp. 1968. Canberra. Ed. by K.W. Finlay and K.W. Shepherd, 431–436.

MCDANIEL, R.G. (1971). Mitochondrial heterosis in barley. In Barley Genetics II (ed. R.A. Nilan) Washington State Univ. Press, 1971.

MCDANIEL, R.G. (1973). Genetic factors influencing seed vigor: biochemistry of heterosis. *Seed Sci. Technol.* **1,** 25–50.

MARANI, A. (1963). Heterosis and combining ability for yield and components of yield in a diallel cross of two species of cotton. *Crop Sci.* **3,** 552–555.

MARANI, A. (1967). Heterosis and combining ability in intraspecific and interspecific crosses of cotton. *Crop Sci.* **7,** 519–523.

MARANI, A. & AIVELI, E.(1973). Heterosis in cotton. *Crop Sci.* **13,** 15–18.

MURPHY, C.F. (1966). Heterotic responses in oats. *Crop Sci.* **6,** 84–85.

NETTEVIC, E.D. (1965). A study on heterosis of spring wheat hybrids in the F_1 (cited in *P. B. A.* **35:** 6080).

NETTEVIC, E.D. & SERGEEV, A.V. (1969). Heterosis in spring barley hybrids. *News agric Sci.,* **1969, No. 10, 108–112. (cited in *P. B. A.,* 40,2977).**

PARODA, R.S. & JOSHI, A.B. (1970). Genetic architecture of yield and components of yield in wheat. *Ind. J. Genet. Pl. Breed.* **30,** 298–314.

PETR, F.C. & FREY, K.J. (1967). Heterosis in oats. *Crop Sci.* **7,** 35–36.

PUROHIT, D.C. (1972). Heterosis in rice. *Madras Agric. J.* **59,** 335–338.

RAMAGE, R.T. (1965). Balanced tertiary trisomics for use in hybrid seed production. *Crop Sci.* **5,** 177–178.

RAMAGE, R.T. & WIEBE, G.A. (1969). Use of chromosome aberrations in producing female parents of hybrids. In ***Induced Mutations in Plants.*** Proc. of IAEA/FAO Symposium, Pullman Wash. IAEA Vienna, 655–659.

RAMAGE, R.T., WIEBE, G.A., ESLICK, R.F. & THOMPSON, R.K. (1972). Use of mutagenic agents in hybrid barley breeding. 1971. ***Barley Genetics* 15,** 73–78. (Quoted with permission of Senior Author)

SAGE, G.C.M. (1973). The expression of heterosis for yield in restored F_1 hybrid wheats and its interactions with seed rate and seed size. *J. agric. Sci. Camts.* **81,** 125–129.

SAGE, G.C.M. & HOBSON, G.E. (1973). The possible use of mitochondrial complementation as an indicator of yield heterosis in breeding hybrid wheat. *Euphytica* **22,** 61–69.

SARKISSIAN, I.V. & SRIVASTAVA, H.K. (1967). Mitochondrial polymorphism in maize. Further evidence of correlation of mitochondrial complementation and heterosis. *Genetics* **57,** 843–850.

SEVERSON, D.A. & RASMUSSON, D.C. (1968). Performance of barley hybrids at four seeding rates. *Crop Sci.* **8,** 339–341.

SVENSSON, G. (1971). Prospects for hybrid wheat and hybrid barley in north western Europe. *The Way Ahead in Plant Breeding. Proc. of VI. Congr. Eucarpia,* 67–76.

SWAMINATHAN, M.S., SIDDIQ, E.A. & SHARMA, S.D. (1972). Outlook for hybrid rice in India. In ***Rice Breeding.*** The International Rice Research Institute, Los Banos, Philippines, 609–613.

THOMPSON, R.K. (1971). Barley as a cross-pollinated crop. In ***Barley Genetics*** II (Ed R.A. Nilan) Washington State Univ. Press 1971, 319–322.

UPADHYAYA, B.R. & RASMUSSON, D.C. (1967). Heterosis and combining ability in barley. *Crop Sci.* **7,** 644–647.

WEBER, C.R., EMPIG, L.T. & THORNE, J.C. (1970). Heterotic performance and combining ability of two-way F_1 soybean hybrids. *Crop Sci.* **10,** 159–160.

WELLS, D.G. & LAY, C.L. (1970). Hybrid vigour in hard red spring wheat crosses. *Crop Sci.* **10,** 220–223.

WHITEHOUSE, R.N.H., THOMPSON, B. & RIBEIRO, M.A.M. (1958). Studies on breeding of self pollinating cereals. II. The use of the diallel cross analysis in yield prediction. *Euphytica* **7,** 147–149.

Wiebe, G. A. (1960). A proposal for hybrid barley. *Agron. J.* **52,** 181–182.

Wiebe, G. A. & Ramage, R. T. (1971). Hybrid barley. In ***Barley Genetics*** II. (ed R. A. Nilan) Washington State Univ. Press 1971, 287–291.

Wienhues, F. (1968). Long term yield analysis of heterosis in wheat and barley: Variability of heterosis, fixation of heterosis. *Euphytica* **17** (1968 Suppl. I), 49–62.

Williams, W. (1959). Heterosis and the genetics of complex characters. *Nature* **184,** 527–530.

Wilson, J. A. (1972). Hybrid wheat breeding. In ***Rice Breeding.*** The International Rice Research Institute, Los Banos, Philippines. 593–602.

Wilson, J. A. & Ross, W. M. (1962). Male sterility interaction of the *Triticum aestivum* nucleus and *Triticum timopheevi* cytoplasm. *Wheat Information Service* **14,** 29–30.

Yap, T. G. & Harvey, B. L. (1971). Heterosis and combining ability of barley hybrids in densely and widely seeded conditions. *Can. J. Plant Sci.* **51,** 115–122.

Zeven, A. C. (1972). Plant density effect on expression of heterosis for yield and its components in wheat and F_1 vs F_3 yields. *Euphytica* **21,** 468–488.

The problems of hybrid wheat and some steps to solve them. The reality of hybrid wheat

Z. BARABÁS

Cereal Research Institute, Szeged, Hungary

SUMMARY

The problems of hybrid wheat were discussed in the Round Table Discussion of recent issues of Cereal Research Communications. We conclude from this that a necessary first step is to transform the self-pollinated species into allogamous ones with large stigmas, big anthers, and lots of well manoeuvring pollen.

Before making such a big investment it is important to test the combining ability very early, in the F_3, F_1 or F_0 generations with micro-plot, tube, and MC-complementation tests.

In testing the parental lines and their hybrids, particular attention must be given to environmental factors and adaptability. Induction and restoration of either genetic or chemical male sterility raises many problems aggravated by the economic problems of hybrid seed production.

The marker gene method may help solve some of these problems, because the hybrid parents can be separated according to their different seed colour (white and purple respectively). The purple colour in special combinations can be removed. Further possibilities for solving the problems may rest, e.g. in the use of female sterility.

It seems to us, that the problems involved with hybrid wheat or the self-pollinated species are among the most fascinating plant improvement problems of this decade. Can hybrid wheat be realized, and if so when?

We asked many colleagues from Mexico to Mongolia. Recent issues of Cereal Research Communications contain a good sample of current ideas.

Among the other self-pollinating plants, hybrid barley and hybrid tomato are most promising. Hybrid sorghum is practically a solved problem. Hybrid triticale, soybean, rice or millet? There will be more words about these till the end of the story of hybrid wheat or hybrid barley. Let us first view the main problems as listed in Table 1.

It would be beneficial for hybrid wheat breeding, if we were able to transform *self-pollinating* wheat into an *out-pollinating* crop. This possibility would have practical value, and we believe it is attainable. The hermaphroditic sorghum was transformed into diclinous (Barabás, 1962) at least morphologically (Table 2).

There are many requirements for a good hybrid partner. Some of them are given in Table 3.

Table 1. The main problems of hybrid wheat production

1. Heterosis effect, combining ability
2. Male sterility (cytoplasmic, genic, chemically induced)
3. Restoration (climatic effects etc.)
4. Pollinating ability (sterility, pollen character etc.)
5. Seed setting (pistil characters, etc.)
6. Adaptability
7. Dwarfness
8. Economy

Table 2. Induced sex divergence in sorghum

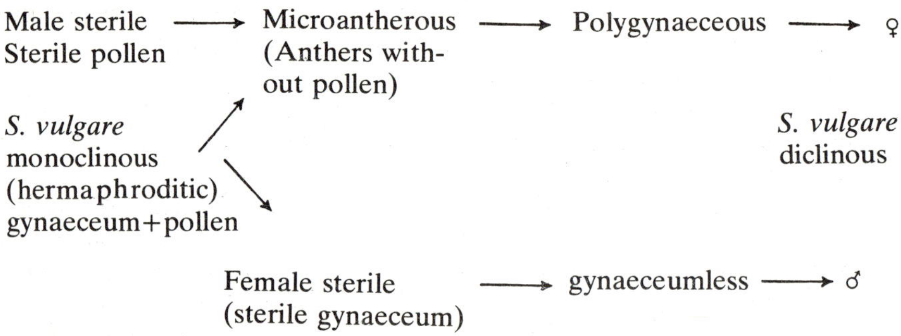

Table 3. Pollinating ability requirements

1. Big anthers
2. Male plant with prolific pollen
3. Good flying ability of pollens
4. Large pistil
5. Receptible female plant
6. Vitality

Probably great variability with respect to all of these characters can be found among and related to the wheat varieties. For example, *anther dimension* of wheat was studied extensively and forms with really big and small anthers have been found. Some very useful correlations were found between the characters. Jost (pers. comm.) claims that the big anthers mean not only large pollen quantity, but also better restoration ability.

With monosomic work it can be proved that chromosome 2A carries a *"big-anther inhibitor"*. It would be a fruitful development to exchange chromosome 2A or the inhibiting segment with the homologous chromosome of another species. Such species with large anthers are some of the *Aegilops, Haynaldia,* and especially *Secale.*

But the pairing between rye and wheat chromosomes is very poor, so that direct translocation is rather difficult. This difficulty may be avoided in two ways (Sears & Kimber, pers. comm.), as shown in Table 4.

Table 4. **Method to translocate a rye chromosome or segment into wheat to solve the inhibitor effects of the wheat chromosome 2A**

1. *Monosomic method*
   T. aestivum monosomic 2A irradiated pollen of *S. cereale*
2. *Aegilops "bridge" method*
   (The *Aegilops* chromosomes pair with both the *Secale* and the wheat chromosomes)
   a) *Secale* × ↓ *Aegilops*
   F_1
   b) T. aestivum ×

HETEROTIC EFFECTS ON HIGHER POLYPLOID LEVEL

Wilson (Kansas, Wichita, pers. comm.) held the opinion in 1970, that the problems of hybrid wheat were generally solved, apart from the suitable level of combining ability.

Mac Key (1973) also doubts whether it is theoretically possible to produce hybrid wheat with practically usable heterotic effects. He says: "Theoretically, the system will allow a perfect diploid F_1 combination to be genotypically fixed at tetraploid level. In a similar way, a hexaploid like our common wheat should be able *to fix the heterozygotic potential* behind a diploid double cross. Such an optimal cooperation and genic adjustment of the different genomes are apparently far from achieved in hexaploid wheat."

If this is always true, then a higher ploidy level can assure *better productivity* by means of *"stable" heterotic effect* in all species. But there are many species in which the situation is other than at the wheat polyploid series. The more productive species can be found among the lower ploidy levels (Table 5).

Table 5. **The polyploid level and productivity of sorghum species**

Species	*No. of chromosomes* 2*n*	*Productivity and type*	
S. versicolor	10	low	wild
S. sudanense	20	very high	cultured
S. halepense	40	medium	semi-wild
S. rigidifolium	60	low	wild

Whether the polyploid series of *Sorghum* is auto- or allopolyploid is under discussion. But it is certain, that there is no strong relation between the ploidy level and the productivity. Kiss (Kecskemét, 1974; pers. comm.) noticed very promising heterotic effect in *Triticale,* which is a really fresh and stabile allohexaploid species.

COMBINING ABILITY OF THE VARIETY VS. LINES

Many wheat breeders think that hybrids with good combining ability come from crosses of productive varieties. The idea was similar to that of the sorghum-sudan-grass breeders, when the species was self-pollinated. Studies have shown otherwise. The good combinations come not from varieties, but from the selected lines with good special combining ability (SCA). For example in sudan-grass (Barabás, 1962). Although not yet well known, it is very promising to select the wheat populations for lines. The varieties are populations unselected for SCA.

EARLY TESTING OF HETEROSIS

Because both parents are co-partners of combining ability, the choice of parents is very complicated, i.e.

a) Either *fertile* combinations must be tested, but to produce a suitable quantity of seeds in this way is slow and expensive.

b) Or only the *male sterile* lines may be crossed with the restorer partners, which also limits our testing opportunity. But to use both of these methods is a necessary condition for this work.

THE MITOCHONDRIAL COMPLEMENTATION TEST

According to our results (Barabás *et al.,* 1973) the MC-test gives us preliminary information about GCA, in spite of its limitation, permitting better planned crosses.

The value of the MC-test in wheat is limited as follows: We were able to gain a practically useful correlation between the MC-test and the F_1 micro-plot test only at low plant density (100 plants/m^2) (Barabás *et al.,* 1973; Sági *et al.,* 1974). But there is a fairly good correlation with respect to GCA between the low plant density and the high (400 plants/m^2) plant density plots (Matuz *et al.,* 1974).

MICRO PLOT

After preselection we are able to obtain a suitable seed quantity for the micro-plot trials. Although low plant density requires less seed per trial, the competition balance is different from that at the high plant densities in use.

We have been studying different micro-plot methods for the last four years (Barabás 1972). Today the plot sizes in our nursery range from 25 cm^2 to 44 m^2. We now think that, at least with enough repetition, 8–10 plants/plot can yield useful information on the productivity of a combination. We need and use a high plant density (400 plants/m^2) also in this very small micro-plot. The data of this trial is more informative and objective than either the low plant density trials or subjective evaluation (Matuz *et al.,* 1974), and is a good base for more extensive work.

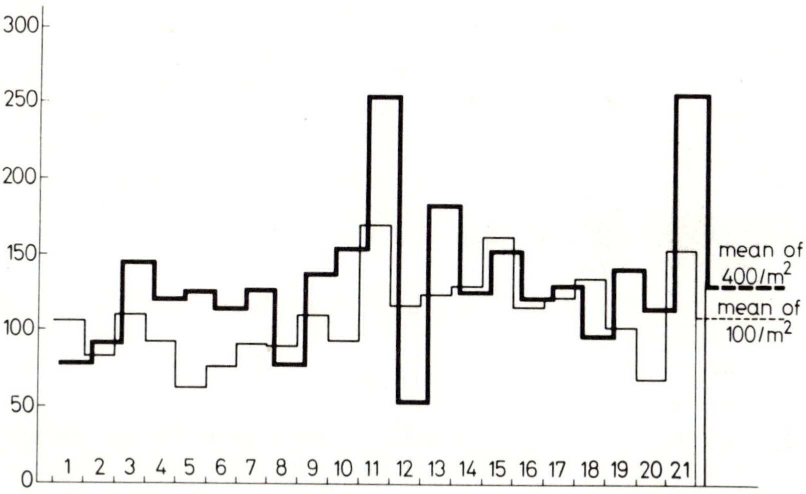

Fig. 1. Hybrid vigour of 21 hybrid combinations at high (400 plants/m^2) and low density (100 plants/m^2) in percentage of the better parent.

ADAPTABILITY OF THE HYBRIDS

It is well known that wheat production conditions in a great part of the world are not optimal. This is particularly true of the hunger-zones, where wheat is produced under stress-conditions because of the influence of climatic conditions and/or agrotechnical level. Under stress-conditions the heterosis effect is relatively higher than with an optimal degree of supply. The productivity of sorghum varieties significantly decreases, for example, with restricted water-supply. The absolute heterosis effect hardly changes under such conditions. This shows that the hybrid vigor can buffer the drought to a higher degree than the variety (Barabás, 1973).

Higher plant density can be a stress-effect for wheat, as the competition ability of the plant is absolutely necessary in this case. A diallel experiment conducted at two different plant densities showed the following heterosis effects (Fig. 1).

a) High plant density is more suitable for selection not only because it approaches more closely to production conditions, but also because it reveals better the differences among the combinations.

b) The average heterosis effect is significantly higher at high plant density.

Also the factors influencing the SCA are significantly different under low and high density. On this problem we have written another paper (Matuz *et al.*, 1974), therefore I mention here only one combination with respect to three factors (Table 6).

Table 6. Special combining ability of the F_1 generation of GT 76150 × Sava at two plant densities.

Factors	100 plants/m^2	400 plants/m^2
Seed weight/spike	+ 0.24	− 0.13
1000 kernel weight	− 4.47	+ 3.30
Plant height	− 1.06	+ 2.72

A hybrid with good adaptability has to be obtained from two varieties or lines with good adaptability. The problem is how to select for this from a great number of hybrid lines. It is appropriate to find methods which simulate the natural stress-effects, affecting the plants in some complex experiments. We started systematic work to discover the important factors, using stress effects that can be artificially limited. Although we have just begun with this work, we note here our basic plan (Table 7).

Table 7. Some induced stress-effects

Studied factors	Methods
1. Diseases	Artificial infection
2. Sowing errors	Extremely deep planting (10–20 cm) Extremely early (late) sowing
3. Chemicals	Superdose at sensitive period of plants
4. Plant damage; hail, frost, insect, etc.	Recutting
5. Nutriment problems	Suboptimal and superoptimal dose of fertilizer
etc.	

ESSENCE OF CYTOPLASMIC MALE STERILITY IN WHEAT

Use of cytoplasmic male sterility entails the use as a complement of a restorer system incorporated into the male parent. The effectiveness of the whole working system could be improved if we knew its genetical, physiological, and morphological basis.

Some people ask if it is a matter of cytoplasmic effects or an operon type of control. The possibility that episomal systems exist between the nuclear DNA and the DNA of the plant organelles in higher plants also merits discussion (Harvey *et al.*, 1972).

In particular, Odenbach *et al.* (1974) suggests that: "In male sterile plants the mitochondrial DNA might produce a special protein which acts as a repressor to the nuclear gene coding this cytochrome oxidase. The nuclear restoration gene might now produce a protein which inactivates the repressor. This might explain fertility and sterility". As the constituent responsible for the inheritance of male sterility has not been found, this is an open possibility.

Atanasoff (1971) and Harvey *et al.* (1972) discuss the possibility of a viral nature of cytoplasmic male sterility in plants. Indirect evidence suggests that it may be true in some of the dicotyledonous plants (*Capsia, Petunia* etc.). In some species it can be transmitted by grafting (Harvey *et al.*, 1972). Heat shock treatment ended the male sterility (Ohta, 1967) in some species. However, so far there are no such results with wheat (Harvey *et al.*, 1972), as heat shock treatments failed to remove the cytoplasmic male sterility in wheat (Barabás, unpublished results). The efforts to repeat the results of Atanasoff (1971) in wheat failed (Barabás *et al.*, 1973). We were unable to transmit cytoplasmic male sterility asexually by rubbing and connecting the leaves of the *cms* and fertile plants. Thus for wheat (and for other monocotyledonous plants) the viral origin of *cms* has so far not been proved. It is necessary to do more intensive basic research to find the essence of *cms* in wheat.

THE PRACTICAL USE OF CMS

Some important circumstances interfere with the practical use of *cms.*

a) The factors concerning the *T*-cytoplasm of corn also influence wheat breeders. Hybrid breeders are working almost only with the *cms* of *Timopheevi* origin. Although many other sources of *cms* have been described (Maan, 1973), they have not come into general use. This is because of certain disadvantages (reduced plant vigor, bad restorer ability), and also because they have not been satisfactorily investigated. Odenbach *et al.* (1974) suggest that we try to induce *cms* within *T. aestivum* itself.

b) The *cms* has some disagreeable *side-effects.* The value of the seed is diminished both physiologically and morphologically (Johnson *et al.,* 1967). The vitality of the plants with *cms* is weaker than that of fertile counterparts. According to Jošt & Milohnić (1974): "After low temperature treatment seedlings of *cms* lines had a higher percentage of dead plants." From the crossing of the fertile C × D a better combination results than if it is produced by the use of male sterility: $C^{ms} \times D^{fr}$. The two combinations differ only in their cytoplasm. Therefore the diminution of the vitality can be attributed only to the side-effect of the male sterile cytoplasm. A lot of ideas and experiments were carried so that male sterility might be used without using *cms.*

GENIC MALE STERILITY

The pioneer work with barley also influenced the production of hybrid wheat. For example Driscoll (1972) elaborated the XYZ method for the use of genic male sterility. The XYZ method raises some further problems. The most important among these in wheat is that although it exists already (Briggle, 1970) there is no convenient genic sterile form. It is possible to induce male sterility with mutagens in monocotyledons (Barabás, 1962), but these are accompanied also by side-effects, for example: female sterility. As far as I know, no induced male sterile line has been used in any species for getting a hybrid (Barabás *et al.,* 1973; Baier, 1974; Keydel, 1974).

RESTORATION

We have more information on the inheritance of the restorers than on male sterility. But in addition to the studied combinations the environment also influences the male-sterility-restoration systems. The effects are different in greenhouse, conditioned chamber and the open air, and are also related to the cultivation place and the yearly weather variables (Johnson *et al.,* 1967; Maan, 1973).

Some of the restorers are particularly interesting. Line T 114 was obtained from the USA as a restorer. In our nursery it produced an unexpected effect in the summer of 1973 and 1974 (Table 8).

Table 8. The behaviour of USA T 114 restorer in Szeged, 1973, 1974

♀		♂	No. of combinations	F_1 No. of plants fertile	*ms*
1. T 114	×	T 114	1	20	0
2*a*. *ms*	×	T 114	3	1	30
2*b*. *ms*	×	T 114	2	12	∅
3. T 114	×	Sterility maintainer	4	∅	37
4. Sterility maintainer	×	T 114	4	39	3
5. T 114	×	restorer	5	57	0

Self-pollinated T 114 is fertile. However, if it served as a pollinator of male sterile varieties, we obtained either fertile or male sterile F_1 generations depending on the mother variety. Cross 2*a* produced male sterile offspring while cross 2*b* produced male fertile offspring. We also obtained different F_1 sterility patterns according to whether T 114 or a sterility maintainer (almost the same as an average variety) served as the male parent. When the sterility maintainer (regular variety) is the male parent, the F_1 is male sterile (cross 3). The F_1 is fertile when T 114 is the male parent (cross 4). Cross 5 shows that T 114 produces male fertile offspring when a restorer serves as the male parent.

In spite of the many problems it seems that restoration is one of the easier problems of hybrid breeding. Using 2–3 genic restorers many workers believe the restoration potential is sufficient (Round table discussion of Cereal Research Communications, 1973, 1974).

DWARFNESS

The general opinion among European breeders is that the wheat of the future will be dwarf. Hybrid wheat would then be the same. The production of the seed, and its use at a super-intensive level needs good stems that don't lodge. But let us see if the requirement of dwarfness is really absolutely necessary?

Some arguments against the dwarf types are:

a) The taller wheats better inhibit weed development, so they need less chemicals.

b) The nutritive value of the dwarf wheats is generally poorer than that of the semidwarfs.

c) Some of the very productive winter wheat varieties have good lodging resistance, eg. Aurora.

In connection with this problem a very fascinating question turns up from Rajháthy in Canada (pers. comm., 1974): "Unfortunately maximum hybrid vigour needs optimum management (fertilizer!) just like dwarf types. We are heading for a period of chronic fertilizer shortages (pollution side effect?) thus we should concentrate on a more efficient chlorophyll and the binding of atmospheric N."

In addition, the best dwarf wheat types are not as good as the best tall and semidwarfs. Although good root systems, effective productivity, suitable restorers, excellent transpiration ability, effective adaptability and good yielding ability exist separately among the dwarfs, no variety contains them all. Of course we have to wait a little for the wheat which will be able to bind the atmospheric N – also among the talls.

Thus, just as *ms* cytoplasm brings unpleasant side effects for the hybrid wheat breeders, so does dwarfness. So we reach the conclusions that:

a) we have to do everything to produce new desirable dwarf types via the convergent way, gene addition, mutational work, etc.

b) The consequence is we do not need to eliminate the outstanding tall and semidwarf types either, where the dominant parent partner is a tall wheat. This may be a promising possibility for the hybrid wheat breeders. May be just this will be the optimal good combiner.

Today, it seems to me that chemically induced dwarfness will not bring a good solution to the problems of dwarfness.

ECONOMICAL PROBLEMS

Hybrid wheat can be produced by any breeder today. Thanks to Kihara, Fukasava, Wilson & Ross (Johnson *et al.*, 1967) the production of hybrids in grand quantities is no longer problematic. Hybrid wheat is commercialized in the USA (Wilson, 1973; pers. comm.).

Most of the specialists have the opinion that the production of hybrid wheat is practically *more an economical than a biological problem.* Regardless of the form of the male sterility, difficulties remain. For example, pollen manoeuvreability in wheat and generally in self-pollinating plants is poor. It means that the pollen is only able to fly over short distances, which diminishes seed production, if the male and female plants are not close enough together.

Barabás (1972) and Barabás *et al.* (1973) proposed a method with *marker colour genes* for help the solution of this problem. The purple seed colour of the male line is significantly different from that of the female line. The two hybrid components can be sown and harvested mixed. The female and the male components can be separated by a photoelectrical method after harvesting.

1. The marker method contributes to the better pollination and higher seed yield.
2. Complicated field manoeuvres, resulting from separate sowing and harvesting can be eliminated.
3. The quantity of male plants per given area can be diminished. This territory can be made available for female plants, contributing to the increase of the yield.
4. The method can be used with any type of male sterility (*cms,* genic, chemical, etc.).

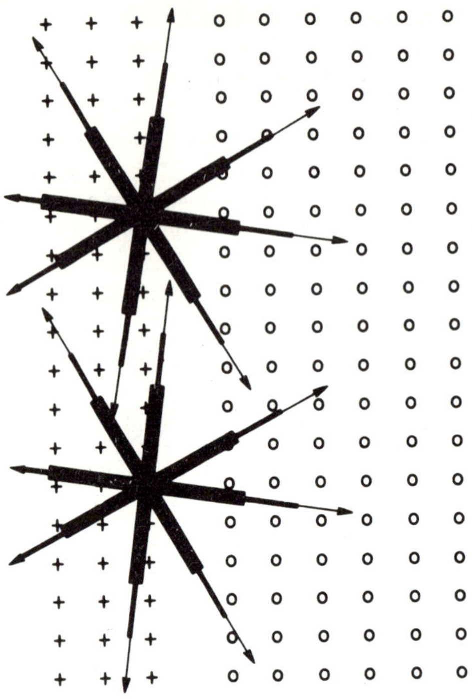

Fig. 2. Pollen dispersal with the strip method.

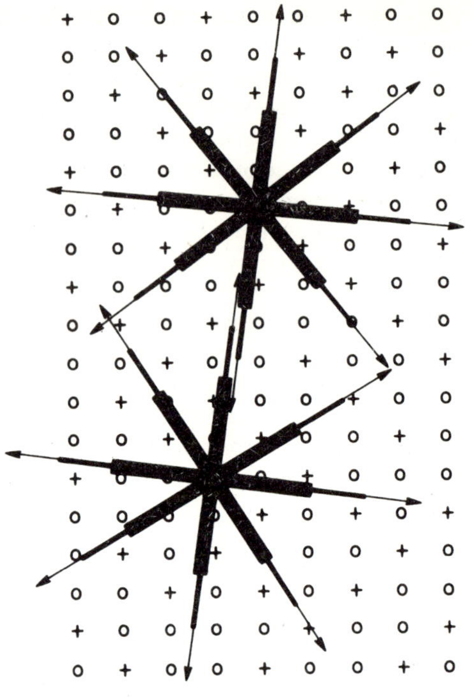

Fig. 3. Pollen dispersal with the marker colour gene method.

5. The use of the method makes hybrid seed production significantly cheaper, and more profitable.

Our experiments have shown that the marker colour can be eliminated with the help of available genes if necessary in, for example, F_2 seeds.

OUTLOOK FOR THE FUTURE

If the problems mentioned can be solved and self-pollinating plants can be transformed into out-pollinating, hybrid production will become more economical in all respects. The marker gene method becomes partially outdated if female sterility becomes available for hybrid wheat production. If the pollen producing plant does not set seed, only the hybrid seed borne on the male sterile mother-plants will ripen in the field. Although this is theoretically possible, it is still far from practical realization. An example is given of how female sterility might be achieved.

Suppose a female sterile inheritance controlled by genes at two loci existed in wheat, analogous to that described by Casady *et al.* (1960), and the following conditions exist:

1. One pair of *fms* does not cause any defect in the single dominant condition.

2. The 2 gene pairs in the heterozygote condition cause female sterility (*Fms*$_1$, *fms*$_2$ *Fms*$_2$ *fms*$_2$).

3. At least one of the *Fms* genes is inherited as a tertiary trisomic, and so only maternally.

Of course we can imagine other ways, too, for example female gametocides patterned after the male gametocides. Such discoveries in the future are absolutely uncertain of course. But it is certain that hybrid wheat breeders are busy people. It takes many irons in the fire to solve the problems. Every day ingenious ideas and methods are born, bringing hybrid wheat closer to reality, hopefully bringing more bread and more food.

REFERENCES

ATANASOFF, D. (1971). The viral nature of cytoplasmic male sterility in plants. *Phytopath. Z.* **70,** 306–322.

BAHL, P.N. & MAAN, S.S. (1973). Chromosomal location of male fertility restoring genes in six lines of common wheat. *Crop Sci.* **13,** 317–318.

BARABÁS, Z. (1957). Male-sterile modifications and forms. (Dissertation) Budapest, typ. 171.

BARABÁS, Z. (1961). A magyarországi hímsteril hibrid takarmánycirok-nemesítés jelenlegi állása. *MTA Mg. Kut. Int. Jubileumi Tudományos Konferenciája* 333–344.

BARABÁS, Z. (1962). Observation of sex differentiation in sorghum by use of induced male-sterile mutants. *Nature* **195,** 257–259.

BARABÁS, Z. (1972). Some ideas on wheat breeding methods. *IWWPN proceedings.* Ankara, Turkey, June 5–9 1972, Univ. of Nebraska,Lincoln, 329–332.

BARABÁS, Z. (1973). Method of producing hybrid wheat by means of marker genes. *Cereal Res. Comm.* **1,** 45–49.

BARABÁS, Z., SÁGI, F. & KERTÉSZ, Z. (1973). Special ways for hybrid wheat breeding. *Proc. 4th Internat. Wheat Genetics Symposium, Columbia, Mo., Missouri, Agr. Exp. Sta.*

BRIGGLE, L.W. (1970). A recessive gene for male sterility in hexaploid wheat. *Crop Sci.* **8,** 693–696.

BROWN, C.M. & EARLEY, E.B. (1973). Response of one winter wheat and two spring oat varieties to foliar applications of 2 – chlorethyl phosphonic acid (Ethrel). *Agronomy Journal* **65,** 829–832.

CASADY, A. J., HEYNE, E.G. & WEIBEL, D.E. (1960). Inheritance of female sterility in sorghum. *Journal of Heredity* **51,** 35–38.

DEPAUW, R.M. & SHEBESKI, L.H. (1973). An evaluation of an early generation yield testing procedure in *Triticum aestivum. Can. J. Plant Sci.* **53,** 465–470.

DRISCOLL, C.J. (1972). XYZ System of producing hybrid wheat. *Crop Sci.* **12,** 516–617.

ELLIS, J.R.S. & BRUNTON, C.J. *et al.* (1973). Can mitochondrial complementation be used at a tool in breeding hybrid cereals? *Nature* **241,** 45–47.

HARVEY, P.H., LEVINGS, C.S. & WERNSMAN, E.A. (1972). The role of extrachromosomal inheritance in plant breeding. *Advances in Agronomy* **24,** 1–27.

INGOLD, M. (1974). The future of hybrid wheat. *Cereal Res. Comm.* **2,** 44–46.

JENSEN, N.F. & ROBINSON, D.F. (1969). Miniature plots for cereal testing. *Crop Sci.* **9,** 288–289.

JOHNSON, V.A., SCHMIDT, J.W. & MATTERN, P.J. (1967). Hybrid wheat in the United States. *Qualitas Plantarum et Material Vegetabiles* **14,** (3). 193–211.

JOŠT, M. & MILOHNIĆ, J. (1974). Influence of *T. timopheevi* cytoplasm on characters of male sterile common wheat. I. Germination and winter hardiness. *Cereal Res. Comm.* **2,** 53–64.

KLATT, A. (1973). Round table discussion on the value of hybrid wheat. I. *Cereal Res. Comm.* **1** (2), 55–56.

KNIGHT, R. (1973). The relation between the hybrid vigour and genotype – environment interactions. *Theor. and Appl. Genetics* **43,** 311–318.

LOWER, R.L. & MILLER, C.H. (1969). Ethrel (2-Chloroethanephosphonic Acid) a tool for plant hybridizers, *Nature* **222,** 1072–1073.

MAAN, S.S. (1973). Cytoplasmic male sterility and male fertility restoration system in wheat. *EUCARPIA Meeting (Cereal Section) Institute of Plant Breeding, Bari, Italy, May, 14–18* 54.

MAC KEY, J. (1973). Possibility and potential of hybrid wheat. *Cereal Research Comm.* **1** (3), 50–52.

MATUZ, J., PARÁDI, L., SCHULTZ, M. & BARABÁS, Z. (1974). General and specific combining ability in winter wheat at two plant densities. *Cereal Res. Comm.* 1974 (in print).
MCDANIEL, R. G. & SARKISSIAN, I. V. (1966). Heterosis: complementation by mitochondria. *Science*, **152,** 1640–1642.
MCDANIEL, R. G. (1972). Mitochondrial heterosis and complementation as biochemical measures of yield. *Nature New Biol,* **236,** 190–191.
ODENBACH, W., KEYDEL, F., BAIER, W. H., MILOHNIĆ, J. & JOŠT, M. (1974). Round table discussion on the value of hybrid wheat. *Cereal Res. Comm.* **2** (1), 37–43.
OHTA, Y. (1968). Effect of virus inoculation and thermoshock treatment on cytoplasmic male sterility in *Capsicum.* Seiken *Ziho* **20,** 63–67.
OHTA, Y. (1967). Effect of thermoshock and virus inoculation on cytoplasmic male sterility in *Capsicum* and *Petunia. Annual Report, National Institute of Genetics, Japan,* No. 18, 26.
RAMAGE, R. T. & NULEEN, N. A. (1964). Balanced tertiary trisomics in barley serve as a pollen source homogeneous for a recessive lethal gene. *Crop Sci.* **4,** 81–82.
SAGE, G. C. M. & HOBSON, G. E. (1973). The possible use of mitochondrial complementation as an indicator of yield heterosis in breeding wheat. *Euphytica* **22,** 61–69.
SAGE, G. C. M. (1972). The inheritance of fertility restoration in male sterile wheat carrying cytoplasm derived from *T. timopheevi. Theoretical and Applied Genetics* **42,** 233–243.
SÁGI, F., BARABÁS, Z. & KERTÉSZ, (1974). Comparative investigations of heterosis effect under laboratory and nursery conditions. *VII. Congress of Eucarpia, Budapest, 24–29 June.*
STOSKOPF, N. C. & LAW, J. (1972). Some observations on ethrel as a method for developing hybrid cereals. *Canadian Journal of Plant Sci.* **52,** 680–683.
STROIKE, J. & MIHALJEV, I. (1973). Round table discussion on the value of hybrid wheat. 2. *Cereal Res. Comm.* **1** (3), 53–55.
THERON, H. & LOUW, J. *et al.* (1968). The effect of spacing on seed production in F_2 families of wheat. *Proc. of the 3rd Congress of the South African Genetic Soc., Pretoria 1966,* 129–131.

Heterosis in wheat following restoration of fertility

E. SÁNCHEZ-MONGE, Mª L. GRANDA, Mª C. ALAMÁN and J.L. PASCUAL

Departamento de Cereales y Leguminosas, I.N.Ì.A., Madrid, Spain

SUMMARY

A total of 61 hybrid wheats (51 with *timopheevi* cytoplasm, 8 with *Aegilops ovata* cytoplasm and 2 with *Ae. caudata* cytoplasm) were obtained in small crossing plots isolated within Triticale seed increase fields. In each crossing plot several male sterile wheats were sown with a common fertility restoring pollinator in alternate rows.

The yields of these hybrids were compared in microtrials in which the checks were the varieties which had provided the male sterile lines used as female parents for the production of the hybrids, but with normal cytoplasm.

Satisfactory fertility restoration was obtained with Primepi wheat and with fertile derivatives of crosses between *T. timopheevi* and hexaploid wheats.

Seed of several hybrid wheats was obtained by using male sterile forms with *Triticum timopheevi* cytoplasm as female parents and a collection of restorers as male parents (Table 1).

Table 1. Restorers

N°	Variety or Pedigree	Origin
179	Lot # 2	Dr. Johnson, Nebraska
195	*T. timopheevi* × Marquis^2	Dr. Willson, Illinois
284	*T. timopheevi* × Marquis^3 × Bison	Dr. Hayes, Kansas
285	*Ae. ventricosa* × *T. timopheevi*^{2} × *T. aestivum*	Dr. Ingold, Lausanne
286	Mir. 805^5	Dr. Rajki, Martonvásár
287	Marchal × Mir 803^3 × Marchal	Dr. Rajki, Martonvásár
301	Primépi	Dr. Rajki, Martonvásár
302	Palmaress	Dr. Rajki, Martonvásár
308	*T. timopheevi* × Minister	Dr. Zeven, Wageningen
320	*T. timopheevi* × Currell	Dr. Zeven, Wageningen
321	*T. timopheevi* × Dominator	Dr. Zeven, Wageningen
331	*T. timopheevi* × Marchal	Dr. Zeven, Wageningen
334	*T. timopheevi* × Estella	Dr. Zeven, Wageningen
339	*T. timopheevi* × Bison^9 × 195	I. N. I. A., Madrid
342	*T. timopheevi* × Bison^3 × 179	I. N. I. A., Madrid
363	Estremo Sur Argelino × *T. timopheevi*	Aula Dei Station, Zaragoza
–	Roma	

Table 2. Yield and heterosis in hybrid wheats

Cross = ms × restorer	Hybrid		♀		Heterosis	Fertility
	kg/ha	% ♀	kg/ha	Order		
Barbilla Sevilla × 285	3744	174	2101	19	++++	++++
Aureo 23 × Primépi	3605	124	2900	6	+++	++
Cabezorro 2 × Primépi	3552	141	2512	14	++++	++
Aragón 03 × Primépi	3530	118	2989	5	++	++
Calatrava × Primépi	3286	104	3157	1	+	+++
Caspino 4 × 331 (*T. tim.* × Marchal)	3200	138	2317	18	++++	++++
Cañivano Málaga × 321 (*T. tim.* × Dominator)	3157	103	3068	4	+	++++
Barbilla Sevilla × Mir 808[5]	3128	149	2101	19	++++	++++
Canaleja × Primépi	3058	107	2853	8	+	++
Calatrava × 363 (E. S. Argel. × *T. tim.*)	2999	95	3157	1	—	+++
Aureo 23 × 339 (*T. tim.* × Bison × Marquis)	2999	101	2900	6	+	+
Traquejos × Palmaress	2997	105	2862	7	+	+
Terminillo × Primépi	2803	100	2809	9	/	+++
Candeal Arévalo × 308 (*T. tim.* × Minister)	2755	117	2353	17	++	+++
Autonomía × 339 (*T. tim.* × Bison × Marquis)	2695	89	3088	2	—	++++
Candeal Teruel × 320 (*T. tim.* × Currell)	2670	97	2747	10	—	+
San Bruno × Roma	2609	100	2622	12	/	—
Albimonte × Mir 808[5]	2475	103	2415	16	+	+
Albimonte × Palmaress	2396	99	2415	16	—	+
Ariana 8 × (Marchal × Mir 808[3] × Marchal)	2316	90	2571	13	—	++
Chamorro Hellín × 321 (*T. tim.* × Dominator)	2213	72	3087	3	—	+++
Cabezorro × Primépi	2150	82	2627	11	—	—
Caspino 4 × Primépi	2071	89	2317	18	—	++
Quaderma × Lot # 2	2068	104	1981	20	+	++++
Hembrilla Jaca × (*T. tim.* × Bison × Lot 2)	2001	81	2483	15	—	+
Ariana 8 × Chamorro Hellín	1959	76	2571	13	—	—
Ariana 8 × (*T. tim.* × Marquis × Bison)	1894	74	2571	13	—	—

Heterosis	Fertility
++++ > 30%	++++ > 2·8 kernels per spikelet
+++ > 20%	+++ > 2·6 kernels per spikelet
++ > 10%	++ > 2·3 kernels per spikelet
+ > 0%	+ > 2·0 kernels per spikelet
— < 0%	— < 2·0 kernels per spikelet

For the production of the seed, small crossing plots were established within Triticale seed increase fields. In each crossing plot several male sterile wheats were sown with a common fertility restoring pollinator in alternate rows.

The yields of these hybrids were compared in microtrials in which the checks were the varieties which had been used in the production of the male sterile lines, but with normal cytoplasm.

The average results from trials over three years are given in Table 2. The hybrids are arranged in descending order of yield. The actual figure quoted is yield expressed as a percentage of the corresponding check, but the yield in kg per ha of the check was adjusted according to the 3 to 5 year average yield obtained with this variety in large fields.

From the results recorded in Table 2 several points emerge which may be useful in the exploitation of heterosis in wheat. These are:

1. Under the dry conditions in the centre of Spain the combinations giving a high degree of heterosis are those between indigenous drought resistant varieties and high yielding ones adapted to more humid conditions (Sánchez-Monge, 1968).

2. Heterosis for yield is usually accompanied by a complete restoration of fertility.

3. The Spanish drought resistant wheats Barbilla de Sevilla, Aureo 23, Cabezorro 2 and Caspino 4 are good combiners which completely restore fertility.

4. Satisfactory fertility restoration was obtained with restorer 285 (from Dr. Ingold), the varieties Primepi and Palmaress, and several fertile derivatives of crosses between *Triticum timopheevi* and high yielding hexaploid wheats which were adapted to more favourable cultural conditions.

5. There seem to be no relationship between yield of the hybrid and the yielding ability of the fertile form of the variety which was used in the derivation of the corresponding male sterile line.

REFERENCE

SÁNCHEZ-MONGE, E. (1968). Approaches to hybrid wheat in Spain. *Euphytica Suppl.* **1,** 35–41.

EUCARPIA VII (1976)

Heterosis and combining ability for some yield components in bread wheat crosses

M. KRALJEVIĆ-BALALIĆ, S. BOROJEVIĆ and T. ĆUPINA

Faculty of Agriculture, Institute for Agricultural Research, Novi Sad, Yugoslavia

SUMMARY

A diallel cross was made between four winter wheat cultivars, short strawed Italian variety Mara and Russian Bezostaya-1, our varieties semi dwarf Sava and dwarf NS-732, in order to estimate the expression of heterosis in parameters which are directly involved in the photosynthetic activity, and how this is related to the formation of final yield components, i.e. the number and weight of kernels per spike.

In the leaf area per plant, no heterotic effect was observed in any of the six F_1 cross combinations when compared with the better parent. The analysis of variance showed that the genes with additive effect play the main role in the inheritance of leaf area.

For chlorophyll content, positive heterotic effect was found in one, negative heterosis in two, while three cross combinations showed intermediate inheritance in the F_1. In the inheritance of this character the main role may be attributed to the non-additive components of variance. For the carotinoid content both additive and non-additive components of variance were involved.

The intensity of respiration exhibited rather high variation and no significant differences were found between the F_1 and parents values.

In number of kernels per spike only one cross combination showed a heterotic effect, while in weight of kernels per spike, five out of six F_1 crosses showed positive heterosis. For the first character genes with both additive and non-additive effect were involved, for the second trait genes with non-additive effect were more involved.

Heterosis expressed in chloroplast pigments or in leaf area, does not show relationship with heterosis expressed in the final yield components. Thus, heterosis expressed in parameters related to the photosynthetic activity also cannot be considered as reliable criteria for the prediction of heterosis in final yield components.

INTRODUCTION

In our earlier investigations we have studied the yield components which develop in the early stages of ontogenesis of several different wheat genotypes (Borojević & Ćupina, 1969).

This study presents the data on the inheritance of leaf area, chlorophyll and carotinoid content, intensity of respiration, and in number and weight of kernels per spike, with the main emphasis on the expression of heterosis and combining ability.

MATERIALS AND METHODS

Four winter wheat cultivars, the short straw Italian variety Mara, Russian Bezostaya-1, our varieties semi dwarf Sava and dwarf NS-732 were crossed in a diallel series excluding reciprocals. Parents, their F_1 and F_2 generations were grown during 1972 in a randomized block design with three replications.

The following characters were analyzed: leaf lamina area, chlorophyll content, carotinoid content, intensity of respiration, number of kernels per spike, and weight of kernels per spike.

The leaf lamina area per plant was determined by the weight method (Alekseenko, 1959). The chlorophyll and carotinoid content were estimated according to Wettstein (1957), and the intensity of respiration by the Boysen & Jensen (1934) method.

Estimates of general combining ability (GCA) and specific combining ability (SCA) were made following the technique of Griffing (1956) for method 2, model I. The heterotic effect was estimated by the test of significance of the mean values of the F_1 in relation to the better parent.

RESULTS

HETEROTIC EFFECT

None of the six F_1 cross combination showed heterosis in leaf area, three combinations showed dominance of the parent with larger leaf area and three intermediate inheritance (Table 1). Positive heterosis for the content of *a* and *b* chlorophyll occurred in one combination, negative in two and intermediate in three combinations. Almost the same picture was observed in the F_2. Only two F_1 combinations demonstrated a positive heterotic effect, others expressing dominant or intermediate inheritance for carotinoid content. The intensity of respiration showed rather high variation, and no significant deviations from the parental values were found. Only one cross combination (Mara × NS-732) showed heterotic effect in the number of kernels per spike. However, positive heterosis in the weight of kernels per spike was found in five out of six cross combinations, not only in the F_1 but also in the F_2. Almost directly opposite results have been obtained in another set of crosses (Borojević, 1965).

ANALYSIS OF COMBINING ABILITY

In order to obtain information on the effect of genes involved in the inheritance of studied quantitative characters the analysis of combining abilities was applied on the performed diallel cross.

Leaf area

Analysis of variance for leaf area revealed highly significant differences for the general combining ability (GCA), whereas differences for specific combining ability (SCA) were not significant (Table 2). It seems therefore that genes with additive ef-

Table 1. Mean values for some yield components in bread wheat crosses

Parents and hybrids		Total leaf area cm^2/plant	Chlorophyll $a + b$ mg/dry mat.	Carotinoid mg/g dry mat.	Intensity of resp. mg CO_2/g d. m./h	Number of kernels per spike	Weight of kernels per spike
Sava		118·8	11·95	0·29	2·52	57·8	1·52
NS—732 × Sava	F_1	120·4*d*	9·97-*h*	0·30-*d*	2·50	53·2*i*	2·10*h*
NS—732 × Sava	F_2	110·2*i*	7·89-*h*	0·27-*d*	2·87	50·9*i*	2·00*h*
Sava × Mara	F_1	136·8*i*	13·08	0·33*h*	2·65	57·6*d*	2·00*h*
Sava × Mara	F_2	134·4*i*	10·09	0·28*i*	2·91	63·5*d*	2·09*h*
Bezost.—1 × Sava	F_1	118·9	14·52*h*	0·39*h*	2·76	49·5*i*	2·02*h*
Bezost.—1 × Sava	F_2	107·1	11·17*i*	0·32*d*	2·27	48·2*l*	1·82
NS—732		105·5	13·13	0·41	2·51	42·7	1·42
Mara × NS—732	F_1	141·8*i*	10·09-*h*	0·30*i*	1·92	65·8*h*	2·20*h*
Mara × NS—732	F_2	139·8*i*	8·41-*h*	0·26*i*	2·62	64·7*h*	2·24*h*
NS—732 × Bezost.—1	F_1	128·4*d*	11·53*i*	0·36*d*	2·41	49·1*d*	2·05*h*
NS—732 × Bezost.—1	F_2	112·2*i*	12·57*i*	0·38*d*	2·48	49·5*d*	2·05*h*
Mara		159·0	11·48	0·28	2·77	43·9	1·21
Bezost.—1 × Mara	F_1	152·0*d*	10·18*i*	0·26*i*	2·57	49·4*d*	2·18*h*
Bezost.—1 × Mara	F_2	130·5*i*	13·42*h*	0·32*h*	2·97	56·1*h*	2·03*h*
Bezostaya—1		116·5	10·53	0·25	2·17	38·0	1·69
L. S. D. 0·05		13·06	1·30	0·038	0·48	7·56	0·29
0·01		17·91	1·78	0·052	0·66	10·37	0·40

Table 2. Analysis of variance for combining ability in leaf area

Source of variation	DF	SS	MS	FE	Ft	
					0·05	0·01
GCA	3	2,281·98	760·66	29·59**	3·16	5·19
SCA	6	354·00	59·00	2·29	2·66	4·01

GCA/SCA = 12·87

General combining ability

Parents	GCA	Rank	L.S.D.	
			0·05	0·01
Sava	— 5·90	3		
NS-732	— 7·91	4	6·15	8·44
Mara	+16·58**	1		
Bezostaya-1	— 2·80	2		

Specific combining ability

Parents	(2) NS-732	(3) Mara	(4) Bezost.-1	L.S.D.	
				0·05	0·01
(1) Sava	+4·39	+12.94*	—32·23		
(2) NS-732		+ 3·31	+ 9·29	12·29	16·85
(3) Mara			+ 8·39		

fect have the main role in the inheritance of the leaf area. The same results have been obtained by Singh & Gupta (1969). Ratio between GCA/SCA shows the additive component of genetic variance to be 12 times higher than non-additive components (dominance and epistasis).

The variety Mara had the highest value of the GCA for the leaf area; this is considered to be an expression of very good combining ability. The highest SCA value was obtained for the cross Sava × Mara. One often gets such results when crossing parents with good and poor general combining ability (Singh & Gupta, 1969).

Chlorophyll and carotinoid content

The variance of GCA for chlorophyll content was not significant, while the variance for SCA was highly significant (Table 3). Thus, it may be concluded that the main role in the inheritance of the chlorophyll content can be attributed to the non-additive components of variance, which include dominance and epistasis. In the crosses which manifested heterosis the non-additive effect of genes for chlorophyll content may result from interallelic interaction (Table 1). Chapman & McNeal (1971) have also explained the heterotic effect in plant height in wheat as a consequence of nonallelic interaction.

Table 3. Analysis of variance for combining ability in chlorophyll content

Source of variation	DF	SS	MS	FE	Ft 0·05	Ft 0·01
GCA	3	2·53	0·65	2·58	3·16	5·19
SCA	6	18·74	3·13	12·49**	2·66	4·01

GCA/SCA = 0·21

General combining ability

Parents	GCA	Rank	L.S.D. 0·05	L.S.D. 0·01
Sava	+0·540	1		
NS-732	—0·063	2	0·607	0·832
Mara	—0·320	4		
Bezostaya-1	—0·157	3		

Specific combining ability

Parents	(2) NS-732	(3) Mara	(4) Bezost.-1	L.S.D. 0·05	L.S.D. 0·01
(1) Sava	+2·15	+1·21	+2·49**		
(2) NS-732		—0·18	+0·10	1·21	1·36
(3) Mara			—0·99		

The F_1 cross Bezostaya-1 × Sava exhibited significantly greater positive heterotic SCA effect than other crosses.

Both additive and non-additive components of variance were involved in the inheritance of the carotinoid content (Table 4). Genotypes NS-732 and Mara had significantly higher GCA values than Bezostaya-1 and Sava; this may be interpreted by their possessing better general combining ability for the carotinoid content. In SCA the F_1 hybrid Bezostaya-1 × Sava proved to be better than other cross combinations.

Intensity of respiration

Analysis of variance of combining ability for the intensity of respiration showed significant differences in SCA, but since that was only on the level of 5%, it may be stated that in addition to the non-additive effect of genes, those with an additive effect have quite an important role (Table 5). Again the cross Bezostaya-1 × Sava showed the highest SCA value. However, the differences were not high enough to be significant.

Table 4. Analysis of variance for combining ability for carotinoid content

Source of variation	DF	SS	MS	FE	Ft 0·05	Ft 0·01
GCA	3	0·0104	0·00346	16·48**	3·16	5·19
SCA	6	0·0161	0·00268	12·76**	2·66	4·01

GCA/SCA = 1·29

General combining ability

Parents	GCA	Rank	L.S.D. 0·05	L.S.D. 0·01
Sava	+0·0017	4		
NS-732	+0·0317**	1	0·018	0·024
Mara	+0·0233*	2		
Bezostaya-1	+0·0133	3		

Specific combining ability

Parents	(2) NS-732	(3) Mara	(4) Bezost.-1	L.S.D. 0·05	L.S.D. 0·01
(1) Sava	—0·05	+0·03	+0·08**		
(2) NS-732		—0·03	+0·02	0·036	0·055
(3) Mara			—0·02		

Number of kernels per spike

Both additive and non-additive components of genetic variance were involved in the inheritance of the number of kernels per spike (Table 6). The same findings have been reported by Bhatt (1971). Variety Sava had the highest GCA and the cross Mara × NS-732 significantly better SCA than the other crosses.

Weight of kernels per spike

Significant differences for SCA and non-significant for GCA indicate that non-additive components of genetic variance are more involved in the inheritance of the weight of kernels per spike (Table 7). Gyawali *et al.* (1968) reported significant differences for this parameter both for GCA and SCA.

The best SCA for weight of kernels per spike was shown by the crosses Mara × NS-732, Bezostaya-1 × Mara and Sava × Mara, but heterosis for weight of kernels was found in all crosses (Table 1).

Table 5. Analysis of variance for combining ability in intensity of respiration

Source of variation	DF	SS	MS	FE	Ft 0·05	Ft 0·01
SCA	3	0·15	0·050	2·08	3·16	5·19
GCA	6	0·54	0·090	3·75*	2·66	4·01

GCA/SCA = 0·55

General combining ability

Parents	GCA	Rank	L.S.D. 0·05	L.S.D. 0·01
Sava	+0·10	1		
NS-732	—0·10	4	0·19	0·27
Mara	+0·06	2		
Bezostaya-1	+0·03	3		

Specific combining ability

Parents	(2) NS-732	(3) Mara	(4) Bezost.-1	L.S.D. 0·05	L.S.D. 0·01
(1) Sava	+0·06	+0·06	+0·26		
(2) NS-732		—0·54	—0·06	0·40	0·55
(3) Mara			+0·10		

DISCUSSION AND CONCLUSION

It was our intention to obtain information on the expression of heterosis in parameters which are involved in the photosynthetic activity and on how this is related to the formation of the final components of yield, i.e. to the number and weight of kernels per spike. Moreover, it was interesting to observe the relationship between the heterotic effect and combining ability.

Analysis of variance for combining ability confirmed earlier investigations (Griffing, 1956; Falconer, 1967) that significant values for the general combining ability (GCA) may be interpreted as the additive effect of genes and the absence of heterosis. In our examples this can be applied to the inheritance of leaf area and intensity of respiration (Tables 1, 2, 5). If genes with non-additive effect are predominantly involved, the heterotic effect is expressed quite strongly. This was demonstrated for the chlorophyll content and weight of kernels per spike (Tables 1, 3, 7).

Table 6. Analysis of variance for combining ability in number of kernels per spike

Source of variation	DF	SS	MS	FE	Ft 0·05	Ft 0·01
GCA	3	221·77	73·92	11·12**	3·16	5·19
SCA	6	405·08	67·51	10·15**	2·66	4·01

GCA/SCA = 1·99

General combining ability

Parents	GCA	Rank	L.S.D. 0·05	L.S.D. 0·01
Sava	+3·76*	1		
NS-732	+0·03	3	3·13	4·29
Mara	+0·91	2		
Bezostaya-1	—4·69	4		

Specific combining ability

Parents	(2) NS-732	(3) Mara	(4) Bezost.-1	L.S.D. 0·05	L.S.D. 0·01
(1) Sava	—2·11	+ 2·50	+0·01		
(2) NS-732		+14·44**	+3·34	6·23	8·55
(3) Mara			+2·76		

When the analysis of variance shows significant differences both for general combining ability (GCA) and specific combining ability (SCA), indicating the presence of genes with additive and non-additive effect, the heterosis is observed in some F_1 cross combinations, while in others dominance or intermediate inheritance are present. Such situations were found for carotinoid content and number of kernels per spike (Tables 1, 4, 6).

Furthermore, a variety possessing good general combining ability may not exhibit good specific combining ability. Data indicate that the best specific combining ability is sometimes obtained in crosses between varieties with good and bad general combining ability. In our case, this was found for leaf area in the cross Sava × Mara (Table 2), and for chlorophyll content in Sava × Bezostaya-1 (Table 3). However, the same was not observed for the number and weight of kernels per spike (Tables 6, 7).

The heterosis is expressed in leaf area and in chloroplast pigments does not show regular relationship with heterosis in the final yield components, i. e. in the number and weight of kernels per spike. Only one cross (Bezostaya-1 × Sava) showed

Table 7. Analysis of variance for combining ability in weight of kernels per spike

Source of variation	DF	SS	MS	FE	Ft	
					0·05	0·01
GCA	3	0·08	0·03	3·00	3·16	5·19
SCA	6	1·06	0·18	18·00**	2·66	4·01

GCA/SCA = 1·66

General combining ability

Parents	GCA	Rank	L.S.D.	
			0·05	0·01
Sava	—0·03	3		
NS-732	—0·01	2	0·10	0·14
Mara	—0·06	4		
Bezostaya-1	+0·09	1		

Specific combining ability

Parents	(2) NS-732	(3) Mara	(4) Bezost.-1	L.S.D.	
				0·05	0·01
(1) Sava	+0·21	+0·26*	+0·11		
(2) NS-732		+0·44**	+0·19	0·23	0·32
(3) Mara			+0·34**		

heterotic effects in three out of the six parameters studied (Table 1). Heterosis in both final yield components number and weight of kernels per spike was found in only one F_1 cross (Mara × NS 732) and this cross showed negative heterosis in chlorophyll content (Table 1).

It seems therefore that the expression of heterosis in the parameters which represent photosynthetic activity cannot be used any more than heterosis in plant height or earliness as a certain indicator of heterosis in the final yield components (Fowler & Heyne, 1955; Lupton, 1961; Borojević, 1966).

REFERENCES

ALEKSEENKO, L. N. (1959). K metodike opredelenija plošadi listjev mnogoletnih trav. *Dokl. VASHIL, Moscow* **9,** 27.

BHATT, G. M. (1971). Heterotic performance and combining ability in a diallel cross among spring wheats (*Tr. aestivum* L.). *Austr. J. agric. Res.* **22,** 359–368.

BOYSEN & JENSEN (1934). *Planta, Arch. of. Wiss.* **21,** 368.

Borojević, S. (1965). Mode of inheritance and heritability of quantitative characters in crosses of different wheat varieties. *Savremena polj. Novi Sad* **7–8,** 587–607.

Borojević, S. (1966). Combining ability in wheat crosses. Proc. 2nd Int. Wheat Genetics Symp., Lund, 1963, *Hereditas, Suppl.* **2**.

Borojević, S. (197). Building a model of high yielding wheat varieties. *Contemp. Agric.* **6,** 33–48.

Borojević, S. & Ćupina, T. (1969). Genetic differences in some physiologico-biochemical parameters and their importance in yield formation in different varieties of vulgare wheat. *Savremena polj. Novi Sad* **11–12,** 31–39.

Chapman, S. R. & McNeal, F. H. (1971). Gene action for yield components and plant height in spring wheat crosses. *Crop Sci.* **11,** 384–386.

Falconer, D.S. (1967). *Introduction to quantitative genetics.* The Ronald Press Company, New York.

Fowler, W. S. & Heyne, F. G. (1955). Evolution of bulk hybrid tests for predicting performance of pure line selections in hard red winter wheat. *Agron. J.* **47,** 430–434.

Griffing, B. (1956*b*). Concept of general and specific combining ability in relation to diallel crossing systems. *Aust. J. Biol. Sci.* **9,** 463–493.

Gyawali, K.K., Qualset, C.O. & Yamazaki, W.T. (1967). Estimates of combining ability in winter wheat. *Crop Sci.* **8,** 322–324.

Lupton, F.G.H. (1961). Studies in the breeding of self-pollinating cereals. 3. Further studies in cross prediction. *Euphytica* **10,** 209–224.

Singh, H. B. & Gupta, V. P. (1969). Combining ability in wheat. *Indian J. Genet.* **29,** 227–232.

Wettstein, D. (1957): *Exp. Cell Research* **12,** 437.

EUCARPIA VII (1976)

Heterosis in soft winter wheat (*T. aestivum* L.) under conditions of the non-chernozem zone of the USSR

E. T. VARENITSA AND T. K. ZIMINA

Agricultural Research Institute, Nemchinovka, USSR

SUMMARY

Ninety-two main hybrid combinations and 430 preliminary combinations were studied in the Winter Wheat Breeding Laboratory during 1967–1973. The hybrid wheat outyielded, on the average, the standard variety by 12.3%, the better parent by 13.3%, and the less productive parent by 39.9%.

Some heterotic combinations outyielded the better parent by 38 and 40% and the average standard (the best commercial variety Mironovskaya 808) by 23 and 25%.

In most cases the hybrids were equal to the better parent in winter survival, and in the extremely severe winter of 1968–1969 they even significantly surpassed the best parental forms. The negative correlation between winterhardiness and productivity is less pronounced in hybrids than in both parents.

While continuing studies involving *CMS*, research on gametocides also was initiated. In 1972–73 promising results were obtained with the application of Ethrel in water solution at a rate of 0.6–0.8 g active substance per sq.m. resulting practically in complete male sterility (99.6–100%). Most of the egg cells remained fertile and 60 to 75% seed set was shown under open wind-pollination.

The prospect of growing heterotic winter wheat hybrids in the non-chernozem zone of the USSR is of key importance, since winter wheat breeding objectives in this zone should couple high productivity with good winterhardiness.

During 1967–73, 74 main and 430 preliminary combinations were studied. In the main trial (on a fertile basis) hybrids yielded on the average 12.3% more than the standard Mironovskaya 808, 13.3% more than the better parent and 38.9% more than the lower yielding parent (Table 1).

The level of heterosis varied by years. Thus, in 1967 the best F_1 heterotic combination yielded 141% of the better parent and 137% of the commercial variety; in 1968 – 129 and 120; in 1969 – 150 and 132; in 1970 – 125 and 110; in 1971 – 141 and 130; in 1972 – 124 and 132; and in 1973 – 136 and 145%, respectively.

During this period, the following hybrid combinations proved to be the best; Mironovskaya Yubileynaya × Nemchinovskaya 154 and Mironovskaya Yubileynaya × Veselopodolyanskaya 485, which in 5 years' trials produced a yield increase of 38 and 40% compared to the better parent and 28 and 25% compared to the average commercial variety (Mironovskaya 808).

One of the factors determining yields in wheat is its winterhardiness. Investigations have shown that the hybrids are generally closer to the better parent in winter survival or even surpass that.

The superpriority of hybrids is especially spectacular in years with severe winter stresses (Table 2). Thus, in 1968–1969 out of 40 hybrids, 14 surpassed the better parent by 10 to 30%. 4 of these hybrids had 30% more overwintered plants than the standard Mironovskaya 808.

It should be noted that no hybrids inferior to the poorer parent in winter survival were revealed in our investigations. The degree of overwintering in hybrids is dependent on the seed parent. The 1968–1969 data indicate that crosses involving as the female component Nemchinovskaya 495, Nemchinovskaya 154 and Kuntsevskaya 45 showed the highest percentage of winter survival. Hybrids involving less hardly varieties (Bezostaya 1 and Mironovskaya Yubileynaya) showed poorer winter survival.

One of the important factors in the winter wheat problem is the relationship between winterhardiness and yield per unit area. Simultaneous display of heterosis for the two characters – winterhardiness and yield – are comparitively rare. Of primary importance in such cases are not only number of plants overwintered and remaining to harvest, but also their performance. Analyses have revealed that hybrids combine higher productivity with higher viability and winterhardiness. A negative correlation between the two characters is less pronounced in hybrids than in both parents or is absent altogether.

Grain quality is important in evaluating F_1 hybrids. Investigations have demonstrated that heterosis for yield is generally accompanied by deterioration in grain quality (in protein content, sedimentation value, etc.) (Table 3). The majority of the F_1 hybrids were inferior to the lower parent or were intermediate in protein content and sedimentation. The highest grain protein content (14.52%) was registered in the hybrids in 1968, the lowest (12.42%) in 1970. In 1970 grain yields reached the record figure (831 g/m^2). That year was also noted for the highest heterotic effect. Thus, 93.8% of the hybrids surpassed the better parent in all characters studied (compared to 52.2% in 1968).

Heterosis for protein content is determined primarily by the choice of parents and is more frequently displayed when high performing lines are crossed. The female component exerts a predominating influence on the incorporation of quality characteristics in F_1 hybrids.

Studies of the interactions between biological and economic traits have shown that increased grain yields are associated with a decrease in grain protein ($r = 0.66 \pm 0.09$). This inverse relation does not imply, however, that breeding for increased grain protein is useless, it only emphasizes the complexity of the problem. Our hybrids Mironovskaya 808 × Salzmünder Bartweizen, Mironovskaya 808 × Bokcharskaya, Nemchinovskaya 495 × Nemchinovskaya 121, Mironovskaya 808 × Carstein VI, Riley × Mironovskaya 808 and others exhibited heterosis both for productivity and protein content (14.25 to 19.61%).

Table 1. Yields of F_1 winter wheat hybrids

Average yield in all combinations, in % of	1966—67	1967—68	1968—69	1969—70	1970—71	1971—72	1972—73	Mean of 7 years
Average standard variety	116·0	101·4	107·6	125·0	100·0	112·0	124·5	112·3
Higher yielding parent	113·6	107·0	112·2	110·0	114·8	115·9	119·5	113·3
Lower yielding parent	137·0	141·4	130·8	154·0	132·0	134·6	143·0	138·9
Average yield of parents	125·3	124·2	122·8	132·0	121·6	124·6	130·2	125·8
Number of combinations studied	23	23	40	16	14	14	17	21

Table 2. Winter survival in F_1 hybrids

Varieties tested	Mean winter survival in all combinations (%)						Mean
	1967—68	1968—69	1969—70	1969—70	1970—71	1971—72	
Hybrids	90	63	85	90	84	96	84·6
Better parent	92	59	68	90	80	95	80·6
Lower parent	87	53	63	84	73	93	75·5
Average of standard variety Mironovskaya 808	94	57	86	94	79	94	84·0

Table 3. Quality of F_1 hybrids

F_1 hybrid groups	Distribution of F_1 hybrids by major quality characteristics, %														
	1968			1969			1970			1971			1973		
	Yield g/m²	Protein content %	Sedim. value ml	Yield g/m²	Protein content %	Sedim. value ml	Yield g/m²	Protein content %	Sedim. value ml	Yield g/m²	Protein content %	Sedim. value ml	Yield g/m²	Protein content %	Sedim. value ml
Surpassing the better parent	52·2	14·8	5·0	65	—	27·2	93·8	—	—	44·7	13·3	13·4	66·0	14·0	46·2
Equal to the better parent	30·4	—	10·0	20·0	10·0	5·4	6·2	7·2	14·3	9·5	8·9	2·5	9·2	1·8	6·2
Intermediate	8·7	25·9	35·0	7·5	42·5	38·2	—	42·8	—	16·1	28·9	29·5	15·1	38·3	23·8
Equal to the poorer parent	8·7	7·4	10·0	2·5	5·0	9·2	—	—	14·3	20·9	22·3	36·4	1·3	4·7	4·4
Inferior to the poorer parent	—	51·8	40·0	5·0	42·5	20·0	—	50·0	71·4	8·7	26·6	19·2	8·4	41·2	19·4
Mean value	573	14·52	7·2	655	12·95	7·6	831	12·42	6·0	449	12·97	9·1	404	14·0	5·9

Table 4. Results of testing hybrid combinations at various seeding rates

Combinations	Seeding rates million seeds/ha	Grain yield g/m² 1970			Yield in % of		Grain yield g/m² 1971			Yield in % of	
		F_1	better parent	standard variety	better parent	standard variety	F_1	better parent	standard variety	better parent	standard variety
Mironovskaya Yubileynaya × Nemchinovskaya 154	1·25	812	590	629	138	129	571	430	391	133	146
”	2·1	916	590	785	163	123	583	441	523	132	111
”	4·0	891	631	561	142	103	535	435	596	123	90
Mironovskaya Yubileynaya × Nemchinovskaya 495	1·25	827	566	553	149	149	578	510	510	113	113
”	2·1	879	699	699	126	126	560	511	511	109	109
”	4·0	827	710	710	117	117	558	538	538	104	104
		L.S.D. ($P < 0{\cdot}05$) 131·4					L.S.D. ($P < 0{\cdot}05$)			79·3	

It is beyond all question that, due to high cost of hybrid seed production, priority will be given to hybrids which are capable of producing high yields with low seeding rates.

Two best combinations, Mironovskaya Yubileynaya × Nemchinovskaya 154 and Mironovskaya 808 × Nemchinovskaya 495 were studied over 2 years at three seeding rates: 1.25, 2.1 and 4.0 million seeds per ha. During the two-year experimental period, 1970 and 1971, data were obtained indicating a marked decrease in the extent of heterosis with increased seeding rates (Table 4). Thus, the potentials of the hybrid plants are possibly realized more completely at lower seeding rates. For the above two combinations the optimal rate is 2.1 million seeds/ha. In this case, although the extent of heterosis slightly decreased compared to the 1.25 million seeds/ha treatment, the actual yield per unit area was higher. The ability of hybrids showing heterosis to produce higher yields with lower seeding rates is of practical importance.

The development of sterile analogues involves the *T.timopheevi* cytoplasm. No difficulties arise here, and 12 varieties promising in breeding for heterosis, incorporate the cytoplasmic male sterile system. With several varieties, the work is coming to an end.

The most difficult aspect of hybrid wheat research is still the development of fertility restorers. As a basis, we use the variety Primepi and Wilson restorer. The Krasnodar scheme of incomplete backcrossing is adopted in the breeding of restorer lines. Highly fertile forms are selected. Following the first backcross in the F_3 and two years' self-pollination, the best forms are selected for repeated backcrossing.

New sources are screened for restorers of fertility. According to our findings, some samples from the VIR collections show fairly high restoring ability. The best samples are involved in hybrid wheat breeding.

The search for gametocides is of primary importance, as their application would allow us to produce hybrid seeds avoiding the CMS system. Promising results were obtained by the American scientists P. L. Rowell and D. T. Miller working with Ethrel (2-chloroethylphosphonic acid). Beginning in 1972, we are studying the efficacy of Ethrel in pollen sterilization. In the course of two-year studies data were obtained indicating that Ethrel sprays at the concentration of 0.8–1.0% at the end of the shooting stage result in practically complete sterility (99.7–100%) of plants. Optimal doses of Ethrel do not affect the fertility of the egg cell. The mean percentage of seed-set over 2 years under wind-pollination of pre-sterilized plants ranged from 53.1 to 73.5 depending on the variety.

We believe that search for effective gametocides should be intensified, for they may show the way of mastering the magnificent natural phenomenon – heterosis in the major bread plant of our planet.

Research on hybrid wheat at Martonvásár, V.

ERNA RAJKI, S. RAJKI AND J. CSÁSZÁR

Agricultural Research Institute of the Hungarian Academy of Sciences, Martonvásár, Hungary

SUMMARY

Heterosis in grain yield and plant height of winter wheat F_1 hybrids obtained on a male sterility and fertility restoration basis, is discussed.

The first F_1 hybrids produced on the basis of male sterile lines and of restorers with a fair restoring ability were put into performance trials in the autumn of 1972.

Percentage heterosis for F_1 grain yield of the best two combinations as related to the male parent and the standard, is demonstrated.

Generally, heterosis of grain yield and plant height appeared simultaneously, and the hybrids were as high as or higher than the higher parent. Only hybrids having a dwarf wheat as one parent at least showed intermediate plant height. This fact repeatedly emphasizes the perspective of our dwarf program which has been carried out for years.

Heterosis in the grain yield and plant height of winter wheat F_1 hybrids obtained on the basis of male sterility and fertility restoration is discussed.

1. As a result of the intensive hybrid wheat research carried out for the past ten years at Martonvásár (Rajki & Rajki 1966, 1967, 1968, 1970) performance trials were set up in the autumn of 1972 on the first F_1 hybrids produced from male sterile lines and restorers with fair restoring ability.

2. The F_1 hybrids studied were obtained in the summer of 1972 by wind pollination on three isolated plots between three restorer lines (Rf *T.timopheevi* × Bison, Rf Primepi, Rf Mironovskaya 808^4S^8) and several male sterile lines.

The combining ability experiments were performed in a rectangular lattice arrangement. The parent varieties, including the standard, were sown together with the combinations in four replications, each replication consisting of four rows, each one metre long, spaced at 10 × 5 cm, with 320 seeds per combination.

3. Table 1 shows the percentage heterosis for the F_1 grain yield of the two best combinations compared with the male parent and the standard Martonvásár 3 (which yielded a few per cent more than either Bezostaya 1 or Kavkaz over an average of twelve replications). This heterosis, which seems promising, was based on three of the yield components: the thousand grain weight, productive tillering and the number of grains per spike.

Generally, heterosis for grain yield and plant height appeared simultaneously, and the hybrids were as high as, or higher than the higher parent. This fact repeatedly

Table 1. Percentage heterosis in the F_1 grain yield compared with the standard and the male parent. Martonvásár, 1973

	Percentage heterosis in the F_1 grain yield compared with		Significance of the grain yield difference between F_1 and	
	Mv 3	Rf Mir^4S^8	Mv 3	Rf Mir^4S^8
Ms_1 Bez^9 × Rf Mir^4S^8	128·0	133·1	**	**
Ms_2 Bez^9 × Rf Mir^4S^8	125·1	130·1	**	**
Mv 3 (standard)	100·0	104·0	—	NS
Rf Mir^4S^8	95·4	100·0	NS	—

NS = not significant
**$P < 0.01$

Table 2. Plant height of F_1 hybrids and their parental lines (spaced at 15 × 15 cm) Martonvásár, 1973

	Plant height (cm)			Significance of the plant height difference between F_1 and	
	female	male	F_1	female	male
Bez. 1 × Rf Mir^4S^8	94·8	120·8	117·0	* * *	NS
Kras. 1 × Rf Mir^4S^8	60·5	120·8	104·0	* * *	□□□
Olson × Rf Mir^4S^8	25·9	120·8	80·8	* * *	□□□
Kras. 1 × Olson	60·5	25·9	52·0	□□□	* * *
Olson × Kras. 1	25·9	60·5	51·8	* * *	□□□

NS = not significant
* * * P> or □□□ $P < 0.001$

emphasizes the prospects of our dwarf programme, which has been in progress for several years. Only hybrids with at least one dwarf parent (Krasnodar 1 – "single" dwarf, Olson – "triple" dwarf) showed intermediate plant height (Table 2).

REFERENCES

RAJKI, ERNA & RAJKI, S. (1966). Research on hybrid wheat at Martonvásár. *Acta Agronomica Acad. Sci. Hung.* **15,** 199–214.

RAJKI, ERNA & RAJKI, S. (1967). Research on hybrid wheat at Martonrásár, II. *Acta Agronomica Acad. Sci. Hung.* **16,** 240–246.

RAJKI, ERNA & RAJKI, S. (1968). Research on hybrid wheat at Martonvásár, III. *Euphytica* **17,** Suppl. 1:43–48.

RAJKI, E. & RAJKI, S. (1970). Research on hybrid wheat at Martonvásár, IV. *Acta Agronomica Acad. Sci. Hung.* **19,** 215–217.

Heterosis in wheat and its hereditary factors

M. A. FEDIN

Laboratory of Heterosis of All-Union Lenin Academy of Agricultural Sciences, Moscow, USSR

SUMMARY

Experiments were conducted in 1970–1971 in three locations for studying different types of gene interaction leading to heterosis, and interactions between nucleus and cytoplasm and between genotype and environment. In the first series of these experiments ten varieties of hard and soft spring wheat and their F_1 and F_2 hybrids have been used in a diallel cross.

The results obtained lead to the following conclusions: dominant gene action is the source of heterosis for the characters "plant height", "number and weight of kernels per ear" and "protein content of the kernels"; dominance and epistasis cause heterosis for the character "number of tillers" and dominance or dominant-epistatic effects for the character "kernel weight per plant". In a number of cases interaction between nucleus and cytoplasm has a stimulating or inhibiting effect on the character. Dominant gene action is more stable to environmental changes than is epistasis.

The main tasks of our research are to define and to distinguish the types of gene interaction leading to heterosis, and to investigate interaction between nucleus and cytoplasm also between genotype and environment.

For this purpose in the first series of experiments the following 10 varieties of spring hard and soft wheat have been used: Krasnozernaya (1), Minskaya (6), Lutescens 758 (10), Saratovskaya 29 (5), Kzil-Bass (9), Justin (2), Selkirk (3), CB 151-a (7), CB 6009 (8), ISWRN-N 225 (4), and their F_1 and F_2 hybrids, obtained from full diallel crosses. In the second series 6 varieties of winter wheat, namely: Mironovskaya 808 (2), Mironovskaya Yubileynaya (1), PPG – 186 (3), Kavkaz (6), Gaines (5), Carstens VIII (4), and 2 varieties of spring wheat, namely: Lutescens 62 (8) and Justin (7) have been used. The experiments of the first series were carried out from 1970 to 1971 and those of the second series in 1973, all of them in three places simultaneously (Moscow, Karabakh and Derbent). The types of gene action have been determined by the method of Hayman (1954a; b; 1957) and by that of Mather & Jinks (1971). The interaction between hybrid nucleus and constant cytoplasm has been determined on the basis of direct and reciprocal crosses. The stability of genetical systems in different environments has been determined by the method of Allard (1956).

We have studied the following characters: height of plant, number of tillers, number of spikelets per main ear, number and weight of kernels per main ear, grain weight per plant, 1000 kernel weight, protein content of grain, and other characters.

There are five types of gene interaction, which determine quantitative and qualitative characters of F_1 of *T. aestivum,* namely: additive, dominant, superdominant, epistatic (non-allelic) and intergenomic. The latter type has not been much investigated for the lack of appropriate mathematical and statistical methods. Still there is some evidence (baking quality and other grain quality indices of common wheat in comparison with durum wheat; frost resistance of a number of varieties of soft wheat in comparison with samples of diploid and tetraploid species of wheat and their relatives, except *Aegilops squarrosa, Ae. cylindrica,* and other species having genome D) which shows the presence of genome interaction.

The analysis of experimental plants demonstrates that there are three types of gene action, controlling productivity of some F_1 hybrids. First we distinguish additive and non-additive gene action which have been determined by evaluation of narrow sense heritability (additive action) and broad sense minus narrow sense heritability (non-additive action) on the basis of variances for general and specific combining ability (Griffing, 1956).

Coefficients of heritability in the broad and narrow senses respectively vary by years and places as follows: height of plants from 0.84 to 0.98 and from 0.57 to 0.93; number of tillers per plant from 0.21 to 0.93 and from 0.04 to 0.16; ear length from 0.20 to 0.95 and from 0.03 to 0.59; grain weight per main ear from 0.59 to 0.96 and from 0.17 to 0.75; grain weight per plant from 0.58 to 0.97 and from 0.10 to 0.79; 1000 kernel weight from 0.62 to 1.00 and from 0.11 to 0.61 etc. Consequently additive and non-additive gene actions which take part in the formation of each character are discrete and independent.

Additive gene action does not ensure heterosis because this type of gene action results in a value of the character intermediate between the parents.

The significance of dominant action is very important for most quantitative characters which determine yield and productivity of the plant.

There are two types of dominant interaction of genes: first dominant alleles may disguise or suppress their recessive alleles or part of them in some loci [the experiment by Keeble & Pellew (1910) is a classical example of this] or produce disguising effect of spring character alleles on winter character alleles; secondly dominance may occur as a result of interallelic action which leads in most cases to a higher character expression if compared with the first type. Non-allelic interaction (epistasis) is also a source of heterosis.

The results of the investigation of some characters are given below.

Height of plants. As has been shown by the analysis, this character of F_1 wheat hybrids is determined by a genetical system, including additive and dominant gene actions (Fig. 1). Height heterosis is a result of crosses of varieties having dominant alleles and differing at least by 1 or 2 dominant oligogenes. This phenomenon can be avoided if crosses are made between genetically conditioned shortstemmed varieties possessing genes controlling dwarfness inherited from the varieties Norin 10, Olsen-Dwarf, Tom Pouce and others, because in this case additive action prevails and hybrids as a whole show incomplete dominance. The analysis of general and rel-

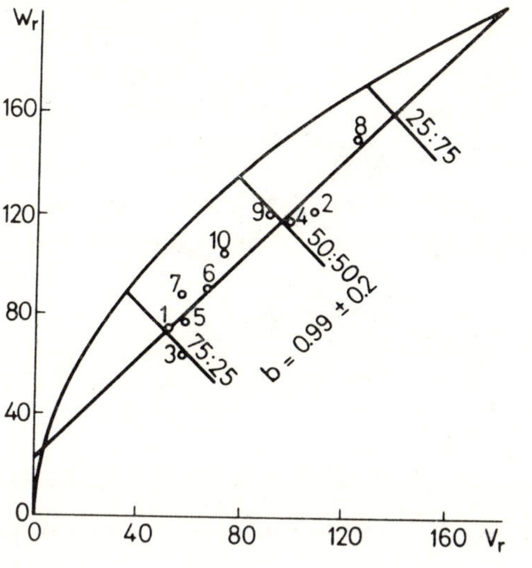

Fig. 1. $(W_r,\ V_r)$ graph for height of spring wheat.

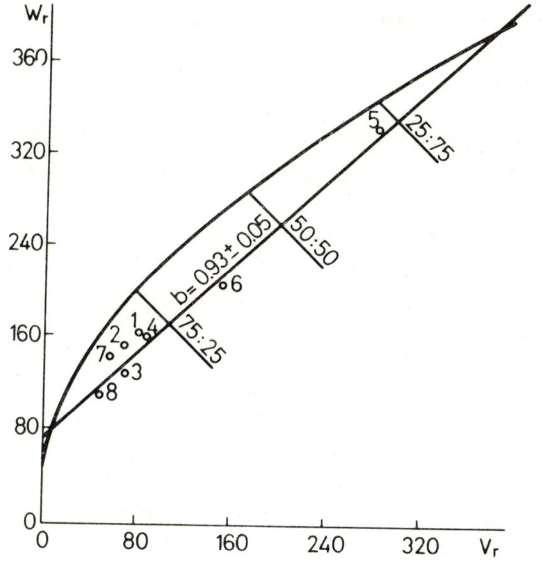

Fig. 2. $(W_r,\ V_r)$ graph for height of winter wheat.

ative dominance has shown that the varieties Krasnozernaya, Selkirk and Saratovskaya 29 possess a relatively complete set of dominant alleles when compared with CB 151a, Minskaya, Lutescens 758 and Kzil-Bass; and all mentioned varieties have a larger number of dominant alleles when compared with ISWRN-N 225 and CB 6009, which is expressed by the coordinates of the varieties on Fig. 1. The same phenomenon has also been observed in the investigation of winter wheat varieties (Fig. 2).

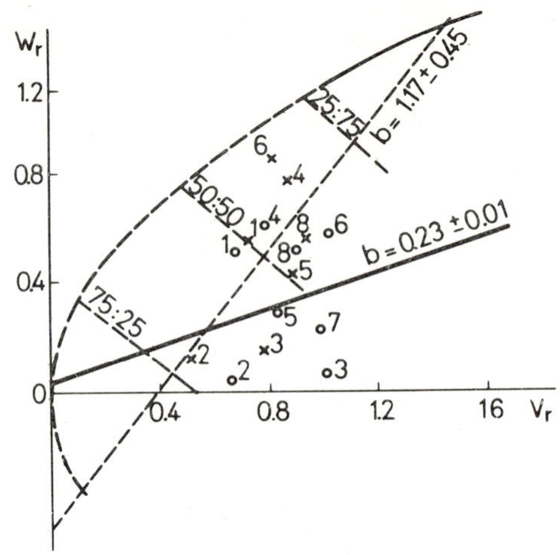

Fig. 3. *(W_r, V_r)* graph for number of tillers/plant of winter wheat.

The number of tillers in F_1 hybrids which showed heterosis for this character is determined by a genetical system including additive, dominant and epistatic gene actions. The epistatic action is proved by the significance of the difference *Wr–Vr* in the arrays when the analysis of variance is made according to Hayman.

Since the epistatic action is proved, it is not possible according to the methods used to interpret general and relative dominance of the varieties on the basis of the regression diagram, because the non-allelic interaction displaces the true position of the regression line (Fig. 3).

Having eliminated from the analysis the data of the variety Justin and F_1 hybrids of this variety, we have established that the non-allelic interaction is caused by the variety Justin. In the hybrid population, obtained on the basis of the other seven varieties, the non-allelic interaction did not take place, and the genetical system controlling the number of tillers, comprises additive and dominant gene actions, and this has allowed us to carry out further analysis.

The part played by additive *(a)* and dominant *(b)* gene actions is significant (Table 1). And in spite of all this the dominant alleles of the varieties studied are not equally directed (b_1 is not significant). The genes, which have positive or negative character expression, are probably not equally distributed between the varieties studied (b_2 is not significant at $P < 0.05$). And, as is seen from the results of the graphical analysis, the varieties Mironovskaya 808 and PPG-186 have a more complete set of dominant alleles than the other five varieties. In the two mentioned varieties the ratio of dominant and recessive alleles is approximately 75:25%, in the varieties Mironovskaya Yubileynaya, Gaines and Lutescens 62 the ratio is 50:50% and in the varieties Kavkaz and Carstens VIII the ratio is 30:70% (Fig. 3). We may suppose that gene expressiveness in some varieties is higher than in the varieties Kavkaz and Carstens VIII.

The mentioned ratio of dominant and recessive alleles, the determination of the character "Number of tillers" is a result of the trend of breeding for this character, followed by the authors of the mentioned varieties.

Table 1. Analysis of variance of diallel crosses

Source of variation	Degrees of freedom	1000 kernel weight		Kernel weight per plant		Degrees of freedom	Number of tillers	
	df	ms	vR	ms	vR	df	ms	vR
a	7	335·41	111·80**	4·46	2·60	6	6·72	19·20**
b	28	21·45	4·28**	5·80	2·94**	21	11·56	4·45**
b { b_1	1	135·67	77·08	69·27	44·40	1	4·29	35·75
b { b_2	7	8·08	4·20**	3·13	4·12*	6	0·88	4·00*
b { b_3	20	20·42	3·27**	3·56	1·48	14	1·65	4·58**
c	7	6·59	1·89	4·58	7·38**	6	0·87	4·35*
d	21	41·80	16·39**	3·71	3·09**	15	0·51	2·40*
II (Replications)	1	0·81		6·71		1	2·55	
II × a	7	3·00		1·72		6	0·72	
II × b	28	5·01		1·97		21	0·35	
II × b_1	1	1·76		1·56		1	0·12	
II × b_2	7	1·92		0·76		6	0·22	
II × b_3	20	6·26		2·41		14	0·36	
II × c	7	3·48		0·62		6	0·20	
II × d	21	2·55		1·20		15		
Reciprocal replication	63	3·79		1·54		48		
Total	127							

* $P < 0{\cdot}05$; ** $P < 0{\cdot}01$.

Table 2. Number of tillers/plant (Derbent, 1973)

Varieties and hybrids	1	2	3	4	5	6	7	8
1. Mironovskaya Yubileynaya	*5·60*	5·85	6·75	5·96	6·51	4·25	5·75	6·44
2. Mironovskaya 808	6·08	*5·67*	4·65	4·90	6·47	5·45	7·75**	5·49
3. PPG 186	6·43	4·01	*5·83*	6·15	6·85	5·45	6·28	5·32
4. Carstens VIII	5·99	5·04	5·25	*4·09*	6·00	4·00	5·18	4·62
5. Gaines	6·18	3·47	6·65	5·90	*5·17*	4·95	7·28**	6·66*
6. Kavkaz	4·61	4·95	5·26	4·00	4·87	*3·19*	6·54	5·07
7. Justin	5·63	4·59	6·28	6·12**	5·56	5·83*	*4·29*	4·59
8. Lutescens 62	7·05*	6·37	5·63	4·50	7·21**	5·75	5·21	*5·15*

L.S.D. $P < 0{\cdot}05$ 1·10
$P < 0{\cdot}01$ 1·56

Table 3. 1000 kernel weight (g) Derbent, 1973

Varieties and hybrids	1	2	3	4	5	6	7	8
1. Mironovskaya Yubileynaya	41·59	44·10	46·80*	42·00	32·60	44·60	43·50	40·80
2. Mironovskaya 808	44·90	41·41	39·90	34·80	41·10	45·70*	42·30	40·70
3. PPG-186	45·20	40·30	36·20	35·90	31·70	45·30*	42·70**	41·50*
4. Carstens VIII	42·35	34·20	34·00	29·07	31·50	38·30	35·90	32·50
5. Gaines	31·50	36·80	31·70	31·50	26·20	39·00	36·10	34·40
6. Kavkaz	45·90*	45·00	45·60	38·10	37·80	41·66	44·60	44·20
7. Justin	43·90	40·06	35·90	39·19	35·80	43·70	35·95	38·10
8. Lutescens 62	43·40	40·40	36·10	32·70	35·50	46·10*	38·60	34·20

L.S.D. $P < 0.05$ 4.00
$P < 0.01$ 5.30

It should be noted that Gaines and Lutescens 62, as well as Mironovskaya Yubileynaya and Lutescens 62 differ in their dominant allele composition, which was the cause of heterosis when these varieties were crossed (Table 2). This makes us suppose that dominant allele of these hybrids have produced their disguising effect on recessive alleles, that is *AABB ccdd* × *aabb CCDD* – *AaBbCcDd*.

The data obtained from the experiment as well as the results of the analysis show that among the studied F_1 hybrids reciprocal differences take place which are caused by the mother effect (c is significant at $P < 0{\cdot}05$), as well as by the interaction of nucleus and cytoplasm (d is significant at $P < 0{\cdot}05$) in crossing Mironovskaya 808 × Gaines, Mironovskaya 808 × Justin, Gaines × Justin (Tables 1, 2).

1000 kernel weight determined in some F_1 hybrids of the studied population included additive and non-additive effect of genes, as shown by the significant deviation of the regression coefficient from 1 ($b = 0{\cdot}69 \pm 0{\cdot}03$, $P = 0{\cdot}001$) (Fig. 4). Statistically significant heterosis was noted in seven combinations: Mironovskaya 808 × Kavkaz, Kavkaz × Mironovskaya Yubileynaya, PPG-186 × Kavkaz, PPG-186 × Justin and others (Table 3). The dominant action of genes controlling a certain character is substantial (b is significant), but the dominant alleles are not unidirectional (b_1 is non-significant). The alleles determining positive or negative value of the character are irregularly distributed in the studied varieties (b_2 is significant), and the differences among the varieties in their specific combining ability are significant (b_3 is significant).

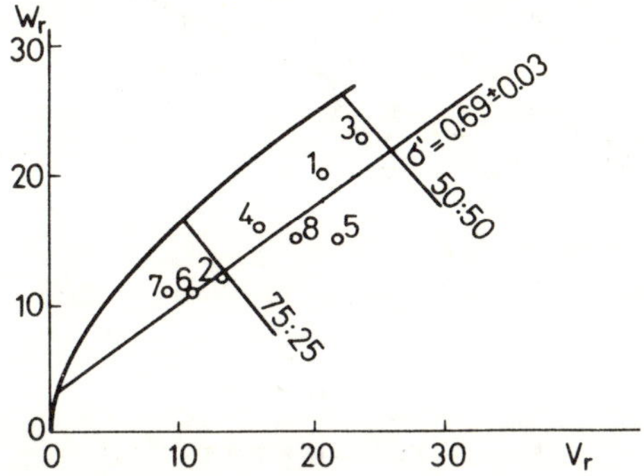

Fig. 4. *(W_r, V_r)* graph for 1000 kernel weight of winter wheat.

The maternal effect on the character "1000 kernel weight" is not observed in F_1 hybrids (c is non-significant). However reciprocal differences in some combinations of crosses, e. g. Mironovskaya 808 × Gaines, PPG-186 × Justin, are quite substantial (d is significant). Genetic differences among the varieties are considerable (a is significant; Table 1). However because of the nonaction of genes, the general and relative dominance of the varieties as shown in Fig. 4 cannot be accepted. The substantial part played by additive gene actions in the studied population makes it possible, in determination of the character "1000 kernel weight", to expect in some combinations of crosses desirable positive transgressions occurring as a result of gene recombination.

Statistically significant heterosis in the character "kernel weight per plant" was observed in the following eleven F_1 hybrids (Table 4): Mironovskaya 808 × Justin, Carstens VIII × Gaines, Justin × Carstens VIII, Gaines × Justin, Lutescens 62 × Gaines, Gaines × Lutescens 62, Kavkaz × Gaines, Justin × Kavkaz, Gaines × Justin, Justin × Carstens VIII, Lutescens 62 × Kavkaz. This heterosis is the result of dominant gene action, i. e. it is controlled by a genetic system comprising additive and dominant gene actions ($W_r - V_r$ does not vary in arrays). It is quite possible that epistatic interaction also takes place as is indicated by the low correlation coefficient ($b = 0.42 \pm 0.01$, and the deviation from 1 is significant at $P < 0{\cdot}001$) (Fig. 5).

The analysis of F_1 and F_2 hybrids and their parental forms from the first series of experiments (spring wheat) for the characters "1000 kernel weight" and "Kernel weight per plant" shows that when F_1 hybrids were grown in Derbent (where spring crops are sown in autumn and irrigation and fertilizers are widely used), the heterosis for the characters mentioned shown by some crosses was caused by dominant and epistatic gene action. The non-allelic interactions may be of three types: interaction between homozygous loci, interaction between homozygous and heterozygous loci and interaction between heterozygous loci. Furthermore the duplicate and complementary genes may serve as a source of epistasis.

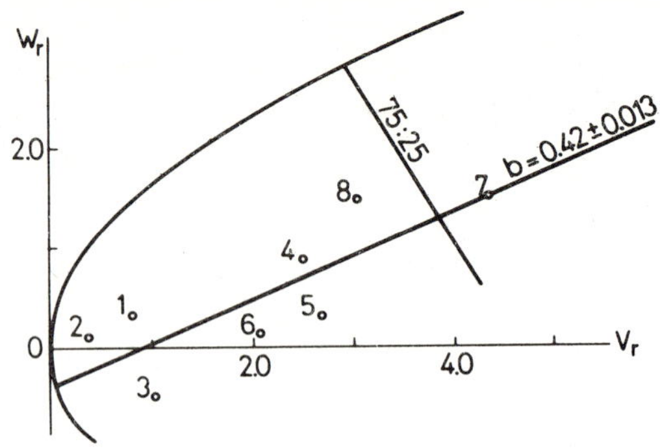

Fig. 5. (W_r, V_r) graph for grain weight/plant of winter wheat.

Table 4. Kernel weight per plant (g)

Varieties and hybrids	1	2	3	4	5	6	7	8
1. Mironovskaya Yubileynaya	6·26	7·56	9·05	7·93	7·35	4·85	7·83	8·34
2. Mironovskaya 808	7·64	7·62	7·21	7·49	9·70	7·44	10·67**	8·35
3. PPG-186	7·56	6·80	6·98	8·80	9·23	8·73	7·95	6·82
4. Carstens VIII	7·26	7·76	6·32	4·34	6·73*	6·62	7·05*	4·78
5. Gaines	6·35	4·19	9·08	6·63	4·27	7·16	10·66**	8·41*
6. Kavkaz	8·05	8·09	7·40	6·42	7·67*	4·89	8·76**	7·13
7. Justin	7·57	4·97	7·88	3·91**	9·32	7·81*	3·16	4·99
8. Lutescens 62	7·41	7·18	7·59	4·70	9·10**	8·89*	5·16	4·48

L.S.D. $P < 0{\cdot}05$ 2·30
$P < 0{\cdot}01$ 3·00

Analysis (Hayman, 1957) of data obtained in Derbent, showed that the character "1000 kernel weight" may in a number of F_1 hybrids be controlled by the additive–dominant–epistatic genetical system. The higher the level of dominant and epistatic gene interaction, the more pronounced is the degree of heterosis. The highest effect of dominance, expressed graphically by perpendicular distance from the horizontal axis (h_{rs}, Fig. 6), was observed in the following combinations: Saratovskaya 92 × CB 6009 (1000 kernel weight – 39·6 g and 117·2% of the best parent), Justin × CB 6009 (40·8 g and 115·9%), Minskaya × CB 6009 (42·3 g and 117·8%) etc. At the same time it has been found that in the four following hybrids, CB–151a × Kzil-Bass (42·8 g and 114·1%) with the probability level of 0·99; ISWRN-N 225 × Saratovskaya 29 (46·2 g and 110·5%); Minskaya × Kzil-Bass (43·3 g and 115·5%), CB 6009 × Kzil-Bass (40·0 g and 107·7%) with the probability level of 0·95, epistatic gene interaction also takes part to a certain degree in the expression of heterosis for this character. As is seen from the results of the analysis, complementary epistasis has been observed in the studied F_1 hybrids (Fig. 6). At the

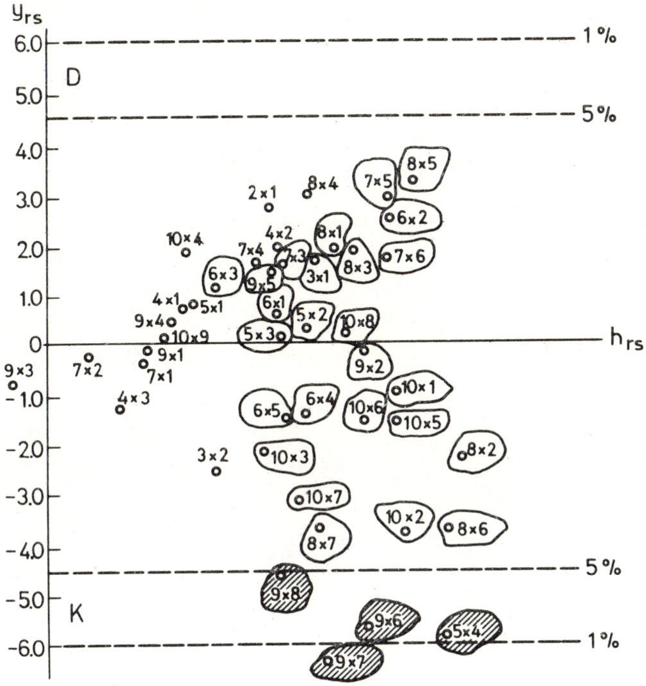

Fig. 6. 1000 kernel weight (Derbent): apparent dominance h_{rs} against epistatic deviation Y_{rs} in each cross of the diallel. The five per cent and one per cent significance levels of Y_{rs} are shown by broken lines. Points lying in the first and third quadrants correspond to crosses exhibiting duplicate *(D)* epistasis and those in the second and fourth quadrants to complementary *(K)* epistasis.

same time epistasis is also possible in homozygous varieties. It occurs as a result of interaction of non-allelic homozygous loci and may be expressed as additive × additive (Hayman, Mather, 1955). Thus in the experiment, carried out in Derbent, this effect is observed in the variety Justin and in samples of ISWRN-N 225 and CB 6009.

The studies of the same collection of varieties and hybrids in Moscow and the analysis of the results obtained have shown that statistically significant epistasis of this character has not taken place. We believe that the epistasis did not take place because the conditions for plant growth were not suitable (Fig. 7).

Now we shall analyse further the types of gene action which determine heterosis in plant productivity. The analysis of the data, obtained from hybrids growing in Derbent, shows that in 12 hybrids, namely: Selkirk × Kzil-Bass (grain yield 138·1% of the better parent), Justin × CB 6009 (132·7%), Selkirk × ISWRN-N 225 (153·5%), CB 151-a × CB 6009 (120·0%), ISWRN-N 225 × CB 151-a (139·4%) etc., heterosis is caused both by dominance and by non-allelic interaction, controlled by complementary genes. In two crosses namely: ISWRN-N 225 × Kzil-Bass

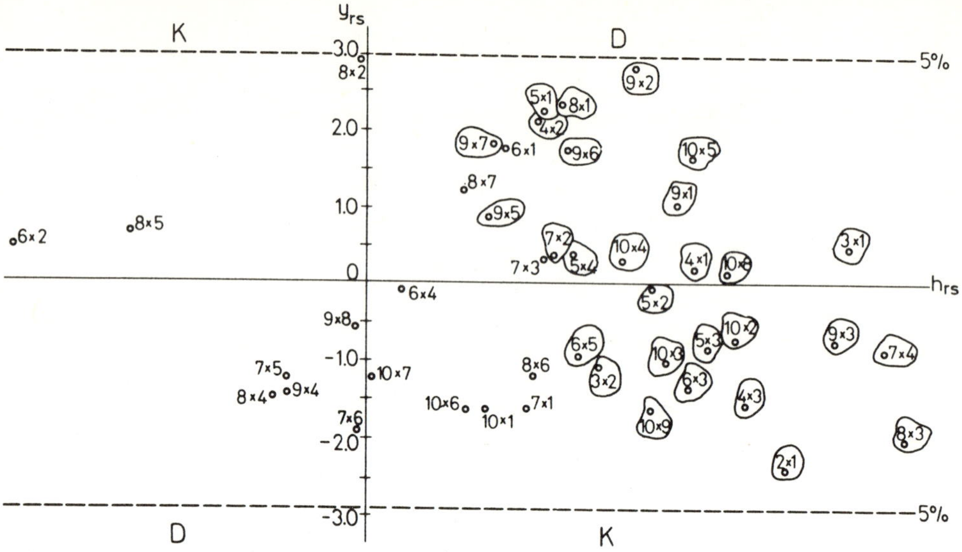

Fig. 7. 1000 kernel weight (Moscow).

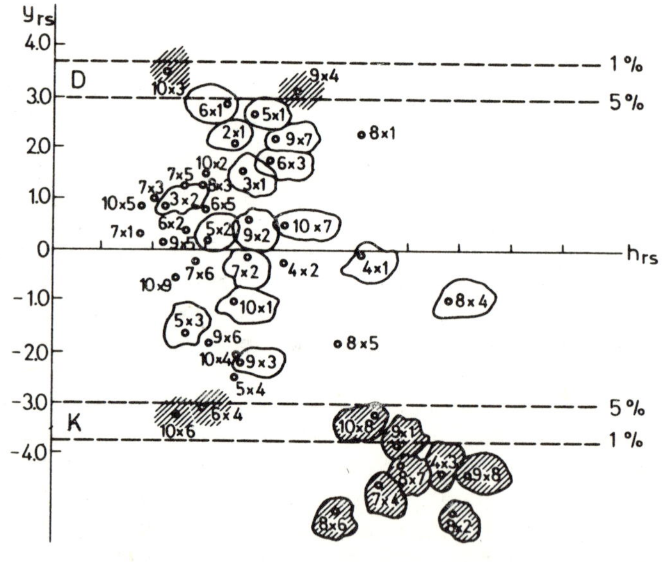

Fig. 8. Grain weight/plant (Derbent).

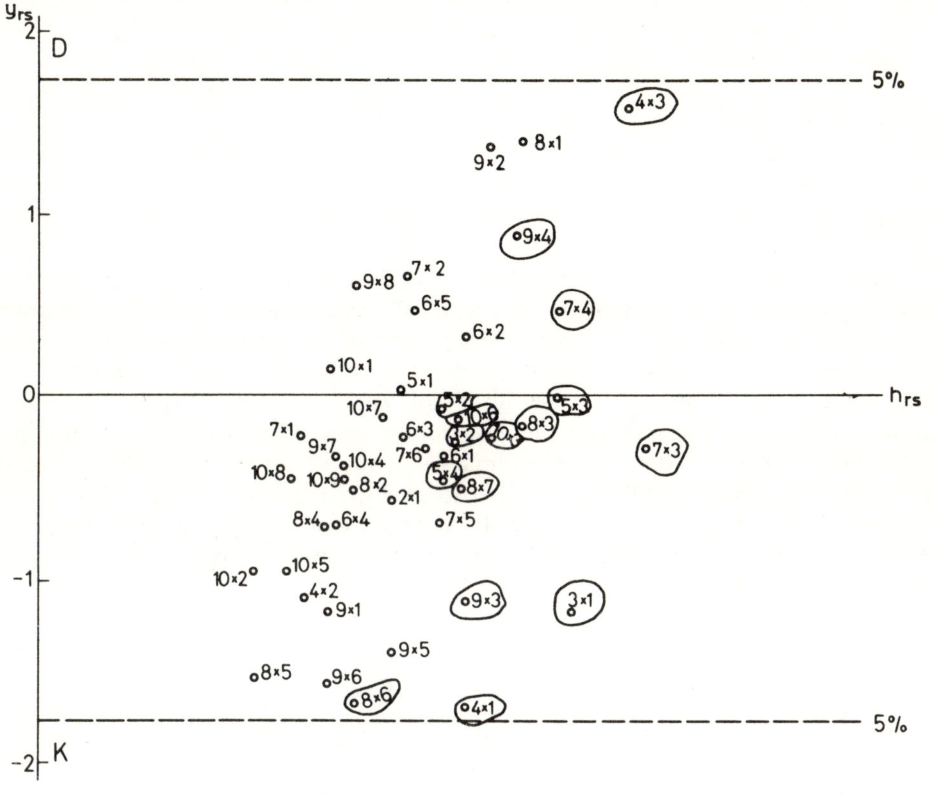

Fig. 9. Grain weight/plant (Moscow).

(106·7% of the better parent) and Selkirk × Lutescens 758 (104·9%), heterosis is caused by duplicate genes. When duplicate epistasis takes place the level of heterosis on the average is not high (105·8% of the better parent); when complementary epistasis acted, the level of heterosis was 133·6%.

When complementary or duplicate epistasis take place (at the significance level of $P < 0.05$ and low dominance), as in the hybrids Selkirk × Lutescens 758, ISWRN N 225 × Kzil-Bass, ISWRN-N 225 × Minskaya and Minskaya × Lutescens 758, the degree of heterosis is low and cannot meet the requirements of heterotic breeding (Fig. 8). It has also been shown that in ISWRN-N 225 and CB 6009, when grown in Derbent, epistasis takes place together with high expressiveness of additive genes. This is explained by positive reaction of genotypes to irrigation combined with fertilizer application. We believe that non-allelic gene interaction is the basis of breeding intensive type varieties, that is varieties that are responsive to high rates of fertilizer and irrigation. Thus the phenotypic manifestation of the effect "additive × additive" in the form of high productivity requires high soil fertility conditions.

Under Moscow conditions heterosis in a number of intervarietal F_1 wheat hybrids has been ensured by the effect of dominance, because statistically significant epistasis has not been observed (Fig. 9). At the same time dominant gene interaction has ensured a fairly high level of heterosis in a number of F_1 hybrids, namely: Selkirk × CB 151-a (142% of the better parent), Krasnozernaya × ISWRN-N 225 C (135%), Minskaya × CB 6009 (128%), Selkirk × Saratovskaya 29 (140%), etc.

We were pleased to note that in some combinations heterosis of productivity was accompanied by high grain quality.

CONCLUSION

The analysis of the data obtained from the studies of intervarietal F_1 wheat hybrids and their parental forms enables us to state that dominant gene action is the source of heterosis of the character "Height of plants"; dominance and epistasis are the source of heterosis of the character "Number of tillers"; dominance causes heterosis of the characters "Kernel number and grain weight per ear" and "Protein content of grain"; dominance and pleiotropic correlation causes heterosis of the character "Number of spikelets per ear"; dominance or dominant-epistatic effects are the source of heterosis of the character "Kernel weight per plant". In a number of cases interaction between nucleus and cytoplasm causes a stimulating or inhibiting effect on the character value. Dominant gene action is more stable to environmental changes than is epistasis.

REFERENCES

ALLARD, R.W. (1956). The analysis of genetic-environmental interactions by means of diallel crosses. *Genetics* **41,** 3:305–318

GRIFFING, B. (1956). Concept of general and specific combining ability in relation to diallel crossing systems. *Aust. J. Biol. Sci.* **9,** 4:463–493

HAYMAN, B.I. (1954a). The theory and analysis of diallel crosses. *Genetics* **39,** 8:789–809

HAYMAN, B.I. (1954b). The analysis of variance of diallel tables. *Biometrics* **10,** 9:235–244

HAYMAN, B.I. & MATHER, K. (1955). The description of genic interaction in continuous variation. *Biometrics* **11,** 69–82

HAYMAN, B.I. (1957). Interaction heterosis and diallel crosses. *Genetics* **42,** 3:338–355

KEEBLE, F. & PELLEW, C. (1910). The mode of inheritance of stature and of time of flowering in peas *(Pisum sativum). J. Genetics* **1,** 47–56

MATHER, K. & JINKS, J.L. (1971) *Biometrical genetics.* Chapman and Hall, London.

New trends and results of the use of heterosis in tomato

CHR. DASKALOV AND CHR. GEORGIEV

Institute of Genetics and Plant Breeding, Sofia, Bulgaria

SUMMARY

Concentration and specialization in agriculture in Bulgaria, and especially in vegetable growing, the formation of large scale farms for industrial vegetable production including tomato growing on large areas, and the shortage of manpower in the village posed the problem of new trends in tomato breeding. The basic trend comprises development of determinant, simultaneously ripening, high yielding with resistance to diseases, varieties of high qualities for early and medium early production, suitable for mechanized cultivation, including mechanized harvest.

An extensive programme has been developed and is being carried now in the following directions: development on a sterile basis of early, high yielding, determinant F_1 hybrids of high quality, resistant to tobacco mosaic virus, *Verticilllium, Fusarium,* blotchy ripening, etc. and adapted for early field production; development on a sterile basis of determinant F_1 hybrids suitable for the canning industry and for industrial production.

The results of the tests of the determinant F_1 hybrids, obtained on the basis of functional sterility, show new and wider possibilities for the use of heterosis. The introduction of determinant F_1 hybrids, adapted for industrial technology, for early and medium early production represents a new stage in tomato growing.

The use of heterosis in tomato breeding in Bulgaria is of 40 years standing. On the basis of the combined use of interspecific hybridization and heterosis, a number of valuable indeterminate hybrid varieties, such as Zarya × Komet, No. 10 × Bison, Triumph, Ogosta, etc. have been obtained. They are widely distributed and mark a significant stage in tomato production.

The increasing concentration in tomato growing during the recent years and the shortage of manpower pose the problem of new trends in tomato breeding. It is necessary to create determinate tomato varieties which ripen simultaneously, are resistant to deseases, of high quality and suitable for mechanized cultivation, including mechanized harvest. The combination of so many traits in a variety is a problem which can be solved most completely and rapidly through the heterosis method. Moreover, we start from the concept that the use of determinate F_1 hybrids is completely realizable and economically justifiable not only for early but also for medium early production to meet the requirements of the canning industry. This concept is based on two conditions: first, the possibility of large scale production of cheap hy-

Table 1. Comparative test of determinate sterile basis F_1 tomato hybrids for processing

No. F_1 hybrids	Yield of the first four harvests		Total yield		Mean fruit weight *G*	Dry matter content %
	kg/1000 m²	%	kg/1000 m²	%		
1. Hebros — ST	174	100	6·546	100	80	4·8
2. Start × Družba	1·172	674·0	6·288	96·0	90	5·3
3. Start × Heinz–1370	549	315·0	8·179	124·9	85	5·1
4. Start × Standard	1·766	1015·0	6·182	94·4	60	5·3
5. Start × Ači	533	306·0	7·955	121·5	96	5·5
6. Stahir × Antej	400	230·0	7·256	110·8	68	6·1
7. Stahir × Standard	282	162·0	8·455	129·1	60	5·8
8. Stahir × Družba	164	94·0	8·022	122·5	80	5·8

brid seeds through the use of male sterility; second, the high potentialities of hybrid varieties for ensuring a considerable decrease in the number of plants per unit area and hence a decrease in planting rate.

For realization of this concept a concrete program has been set up and work on it is in progress.

First, extensive studies have been conducted on the different types of male sterility *(ms, sl, ps* and *ex)* in regard to their suitability to industrial hybrid seed production. The most suitable types, at least for the present stage, proved to be the positional sterility of the Vrbičansky niske type and we designated this type as ps^2 in contrast to the John Bear type designated as *ps*. In the positional sterile forms ps^2, emasculation and pollination were made at the same time and thus labour productivity was increased twofold. By combining positional sterility (ps^2) and exertion of styles *(ex)*, labour productivity in the process of crossing was increased even more due to elimination of emasculation. For the purpose of facilitating and accelerating the control of authenticity of hybrid seeds, the genetic markers *a* (absence of anthocyanin) and *c* (potato leaf), which appear at the earliest growing stage, were transferred to the maternal components.

Through the method of combination breeding and back-crossing maternal lines with positional sterility, exerted styles and genetic markers were obtained for the different trends in heterosis breeding, namely the lines Start, Stahir and Hera. Work is being carried out on the development of analogues of sterile lines resistant to tobacco mosaic virus (Tm_2^2) and with jointless fruit stalks (j^2).

The following Bulgarian and introduced varieties were used as paternal components: Druzhba, Hebors, Ace, VF 198, Ventura, Roma VF, etc., as well as lines with valuable characters obtained through interspecific hybridization with *L. pimpinneli-folium,* namely Vranja 123, Standard, Antej, Kompakta, etc. Transferring of resistance to tobacco mosaic virus, *Verticillium* and *Fusarium* wilt, and jointlessness of fruit stalks is being performed in the lines, showing good combining ability.

Many of the hybrid combinations obtained on a sterile basis proved to be highly promising. As is evident from the data in Table 1, the hybrids are much superior to the direct variety Hebros, commercially grown in our country, in earliness, yield and fruit quality.

The introduction of determinate hybrid tomato varieties into large scale production is an enormous reserve for enhancement of yield, fruit quality and especially earliness, and it will provide possibilities for a continuous and more effective use of the capacity of the canning factories.

Genetics of rice hybrid heterozygosis

V. A. DZUBA

Genetics and Cytology Laboratory of the All-Union Rice Research Institute, Krasnodar, USSR

SUMMARY

The genetic background of heterosis in rice has been studied in F_1 and F_2 generations of four hybrids.

The following characters were considered in the analysis of heterosis effects: plant height, length of panicle, number of spikelets per panicle, and grain weight per panicle.

In the world scientific literature there is little information on rice hybrid heterozygosis. Some available reports link heterozygosis of F_1 hybrids with high male pollen sterility.

We studied F_1 and F_2 generations in comparison with parental forms to determine the effect of heterosis on some characters and the dominance coefficient in the following hybrids: Redi × Kros 4428, Krasnodarski 424 × Anao, Mutant Shilovski × CNIIR 128 and Korbetta × Soyuzni 244.

Such samples as Redi, Kros 4428, VNIIR 128 and Korbetta carry dominant genes for tallness, and lines of Anao and Mutant Shilovski – recessive genes for dwarfness.

Hybrid heterosis value (expressed as a percentage) has been determined in comparison with the female and male forms and the mean value of both parental forms.

Heterosis for plant height was noted in all hybrids (Table 1).

The decrease of heterosis pointed out in F_2 generation of Krasnodarski 424 × Anao hybrid has happened due to the presence of homozygous forms with a recessive gene of dwarfness.

As far as panicle length is concerned heterosis manifested itself only in two hybrids, Redi × Kros 4428 and Korbetta × Soyuzni 244. In the hybrid Krasnodarski 424 × Anao, heterosis was not discovered in comparison with maternal form. The hybrid mutant Shilovski × VNIIR 128 showed negative heterosis to parental form in F_1 and F_2 (Table 2).

All hybrids have a very high hereditability character of spikelet number per panicle. The heterosis value in all hybrids was high (Table 3). It points to the fact, that the number of spikelets per panicle (as a polygenic character) in all cases manifests the effect of complete dominance and even over-dominance.

All hybrids show a high heterosis value for the important character weight of grains per panicle, though the F_1 hybrid Korbetta × Soyuzni 244 showed negative heterosis to its maternal form due to the low combining ability of the male form of Sojuzni 244 (Table 4).

Table 1. Height of hybrids and parental forms

Hybrids and parental forms	Symbols of pedigree	$s_{\bar{x}}$ (cm)	s%	$s_{\bar{x}}$%	Heterosis value in % to: maternal form	Heterosis value in % to: paternal form	Heterosis value in % to: mean value of both parents
Redi	♀	$105 \pm 1{\cdot}3$	4·1	1·3	—	—	—
Redi × Kros 4428	F_1	$119 \pm 1{\cdot}3$	3·5	1·1	113	168	130
Redi × Kros 4428	F_2	$114 \pm 3{\cdot}5$	9·7	3·0	109	160	129
Kros 4428	♂	$74 \pm 1{\cdot}4$	6·5	2·0	—	—	—
Krasnodarski 424	♀	$108 \pm 3{\cdot}3$	9·6	3·1	—	—	—
Krasnodarski 424 × Anao	F_1	$100 \pm 2{\cdot}1$	6·7	2·1	105	164	124
Anao	F_2	$86 \pm 6{\cdot}0$	31·5	7·0	89	141	102
Anao	♂	$61 \pm 0{\cdot}4$	2·1	0·6	—	—	—
Mutant Shilovski	♀	$50 \pm 0{\cdot}6$	4·0	1·2	—	—	—
Mutant Shilovski × VNIIR 128	F_1	$97 \pm 3{\cdot}3$	11·0	3·4	197	116	145
Mutant Shilovski × VNIIR 128	F_2	$85 \pm 4{\cdot}0$	23·3	4·7	170	101	127
VNIIR 128	♂	$84 \pm 2{\cdot}4$	9·0	2·8	—	—	—
Korbetta	♀	$98 \pm 2{\cdot}2$	7·1	2·2	—	—	—
Korbetta × Soyuzni 244	F_1	$122 \pm 2{\cdot}4$	6·2	1·9	124	157	139
Korbetta × Soyuzni 244	F_2	$121 \pm 4{\cdot}2$	11·7	3·5	124	155	139
Soyuzni 244	♂	$78 \pm 1{\cdot}2$	5·0	1·6	—	—	—

Table 2. Panicle length of hybrids and parental forms

Hybrids and parental forms	Symbols of pedigree	$s_{\bar{x}}$ (cm)	s%	$s_{\bar{x}}$%	Heterosis value in % to: maternal form	Heterosis value in % to: paternal form	Heterosis value in % to: mean value of both parents
Redi	♀	$17{\cdot}4 \pm 0{\cdot}37$	6·8	2·1	—	—	—
Redi × Kros 4428	F_1	$19{\cdot}0 \pm 0{\cdot}64$	10·8	3·4	109	154	128
Redi × Kros 4428	F_2	$17{\cdot}8 \pm 0{\cdot}34$	6·1	1·9	102	145	120
Kros 4428	♂	$12{\cdot}3 \pm 0{\cdot}51$	13·2	4·2	—	—	—
Krasnodarski 424	♀	$22{\cdot}8 \pm 0{\cdot}91$	12·7	4·0	—	—	—
Krasnodarski 424 × Anao	F_1	$17{\cdot}8 \pm 0{\cdot}59$	10·7	3·3	78	152	103
Krasnodarski 424 × Anao	F_2	$16{\cdot}6 \pm 0{\cdot}81$	21·8	4·9	73	142	96
Anao	♂	$11{\cdot}7 \pm 0{\cdot}36$	9·7	3·1	—	—	—
Mutant Shilovski	♀	$13{\cdot}1 \pm 0{\cdot}45$	10·8	3·4	—	—	—
Mutant Shilovski × VNIIR 128	F_1	$16{\cdot}9 \pm 0{\cdot}55$	10·4	3·3	129	97	110
Mutant Shilovski × VNIIR 128	F_2	$16{\cdot}6 \pm 0{\cdot}55$	16·8	3.3	127	95	109
VNIIR 128	♂	$17{\cdot}4 \pm 0{\cdot}66$	12·1	3·8	—	—	—
Korbetta	♀	$19{\cdot}4 \pm 0{\cdot}59$	9·5	3·0	—	—	—
Korbetta × Soyuzni 244	F_1	$21{\cdot}1 \pm 0{\cdot}43$	6·5	2·0	109	145	124
Korbetta × Soyuzni 244	F_2	$21{\cdot}3 \pm 0{\cdot}89$	13·9	4·2	110	147	126
Soyuzni 244	♂	$14{\cdot}5 \pm 0{\cdot}49$	10·7	3·4	—	—	—

Table 3. Number of spikelets per panicle of hybrids and parental forms

Hybrids and parental forms	Symbols of pedigree	$s_{\bar{x}}$ (number)	s%	$s_{\bar{x}}$%	Heterosis value in % to: maternal form	paternal form	mean value of both parents
Redi	♀	97 ± 2·9	9·6	3·0	—	—	—
Redi × Kros 4428	F_1	111 ± 3·3	9·6	2·9	114	308	168
Redi × Kros 4428	F_2	103 ± 6·3	19·5	6·2	106	286	155
Kros 4428	♂	36 ± 3·3	28·9	9·1	—	—	—
Krasnodarski 424	♀	85 ± 7·7	29·2	9·1	—	—	—
Krasnodarski 424 × Anao	F_1	113 ± 3·9	11·1	3·5	133	122	103
Krasnodarski 424 × Anao	F_2	113 ± 8·0	32·5	7·2	130	182	152
Anao	♂	61 ± 4·7	24·9	7·8	—	—	—
Mutant Shilovski	♀	76 ± 8·5	35·3	11·2	—	—	—
Mutant Shilovski × VNIIR 128	F_1	75 ± 3·3	13·9	4·4	149	191	167
Mutant Shilovski × VNIIR 128	F_2	76 ± 6.5	41·7	8·5	100	129	113
VNIIR 128	♂	59 ± 7·5	40·4	12·8	—	—	—
Korbetta	♀	149 ± 11·9	25·1	7·9	—	—	—
Korbetta × Soyuzni 244	F_1	193 ± 16·4	27·2	8·5	129	254	171
Korbetta × Soyuzni 244	F_2	205 ± 21·9	35·5	10·4	137	260	182
Soyuzni 244	♂	76 ± 3·6	14·9	4·7	—	—	—

Table 4. Grain weight per panicle of hybrids and parental forms

Hybrids and parental forms	Symbols of pedigree	$s_{\bar{x}}$ (g)	s%	$s_{\bar{x}}$%	Heterosis value in % to: maternal form	paternal form	mean value of both parents
Redi	♀	3·1 ± 0·57	17·1	5·5	—	—	—
Redi × Kros 4428	F_1	3·2 ± 0·51	51·2	15·9	103	320	156
Redi × Kros 4428	F_2	3·1 ± 0·32	32·3	10·3	100	310	151
Kros 4428	♂	1·0 ± 0·17	53·0	17·0	—	—	—
Krasnodarski 424	♀	2·9 ± 0·26	31·1	9·6	—	—	—
Krasnodarski 424 × Anao	F_1	3·2 ± 0·37	36·9	11·6	110	139	123
Krasnodarski 424 × Anao	F_2	3·4 ± 0·34	44·7	10·0	117	148	131
Anao	♂	2·3 ± 0·15	21·3	6·5	—	—	—
Mutant Shilovski	♀	1·5 ± 0·23	48·0	15·3	—	—	—
Mutant Shilovski × VNIIR 128	F_1	2·9 ± 0·37	41·4	12·8	193	181	187
Mutant Shilovski × VNIIR 128	F_2	2·3 ± 0·22	47·8	9·6	153	144	148
VNIIR 128	♂	1·6 ± 0·17	32·5	10·6	—	—	—
Korbetta	♀	5·1 ± 0·41	25·5	8·0	—	—	—
Korbetta × Soyuzni 244	F_1	4·8 ± 0·62	41·3	12·9	94	228	133
Korbetta × Soyuzni 244	F_2	5·8 ± 0·60	34·5	10·3	114	276	161
Soyuzni 244	♂	2·1 ± 0·09	14·3	4·3	—	—	—

Table 5. Coefficient of dominance of hybrids

Hybrids	Generation of hybrid	Coefficient of dominance			
		plant height	panicle length	spikelet number per panicle	grain weight per panicle
Redi × Kros 4428	F_1	1·8	1·6	1·5	1·1
Redi × Kros 4428	F_2	1·5	1·2	1·2	1·0
Krasnodarski 424 × Anao	F_1	0·66	0·09	3·3	2·0
Krasnodarski 424 × Anao	F_2	0·06	0·13	3·3	2·7
Mutant Shilovski × VNIIR 128	F_1	1·8	0·8	0·9	27·0
Mutant Shilovski × VNIIR 128	F_2	1·1	0·6	1·0	15·0
Korbetta × Soyuzni 244	F_1	3·4	1·7	2·2	0·8
Korbetta × Soyuzni 244	F_2	3·3	1·8	2·5	1·5

The coefficient of dominance determined on every hybrid points out the heredity of the studied characters and their variability in the F_2 generation (Table 5). The hybrid heterosis value can serve as a practical test to select parental forms for hybridization.

The high positive correlation between grain weight per panicle and number of panicles ($r = 0{\cdot}92 \pm 0{\cdot}07$) and panicle density ($r = 0{\cdot}77 \pm 0{\cdot}12$) has been obtained. These data can be taken as tests for selection of desirable parental pairs among hybrid populations.

ABSTRACTS OF PAPERS SUBMITTED TO THE CONGRESS

Heterosis in a partially allogamous species

J. I. CUBERO-SALMERÓN

Instituto Nacional de Investigaciones Agrarias, Centro Regional de Andalucía, Córdoba, Spain

Many species of economical value are partially allogamous, and studies on heterosis in this kind of material can give useful indications both on breeding programs and on the evolution of these species.

I have studied the problem in *Vicia faba* L., known as partially allogamous from the time of Darwin, who observed (Darwin, 1900) lack of full fertilization in plants grown under isolation from insect visits. Many authors (Sirks, 1931; Holden & Bond, 1960; Picard, 1960 *inter alia*) have calculated the rate of allogamy using different methods. The rate depends on genetical and environmental factors (Bond, Picard, *loc. cit.*) and seems to vary from full autogamy to full allogamy. The varieties from the Mediterranean countries seem more self-fertile than those originating from North-Western and Central Europe, at least under our experimental conditions. These points have been discussed recently (Hanelt, 1972; Cubero, 1973, 1974) in connection with evolution of *V. faba.*

MATERIAL AND METHODS

The subject has been studied in a 7 × 7 diallel cross including two *major* lines (accession numbers 164 and 165) of Spanish origin, one *equina* (acc. N° 169) and two *minor* (acc. N°s 166, 168) of Central-European origin, and two *pauci juga* (acc. N°s 171, 172) of Indian origin. All these lines showed at least five self-pollinations. I have used the terminology of Muratova (1931) though not exactly; following the cited works of Hanelt and Cubero the latin terms mean here "groups", not subspecies *(pauci juga)* or botanical varieties.

For analysis the method of Hayman (1954 a, b) has been applied. Five plants (real number: 4.1) per cross or parental line have been studied, in two replications each.

The studied characters were: height, rachis length, number of leaflets per leaf, leaflet width, width/length of leaflet, grain weight per plant, number of seeds per plant, number of flowers per raceme, number of pods per node, number of ovules per ovary, number of seeds per pod, pod width, seed length, thickness/length of seed, seed weight, time of blossom and time of maturity.

The lines have been chosen as being very different, with the object to find the maximum reciprocal differences and expecting the maximum of heterosis.

DISCUSSION

The results indicated the following systems:

Overdominance: Rachis length, number of leaflets per leaf, weight per plant, number of seeds per plant and number of pods per node. Strong tendency towards overdominance: leaflet width and number of seeds per pod. Thus, these characters show to be directly related with fitness.

Dominance: Number of ovules per ovary and time of maturity. Perhaps also time of blossom and width/length of leaflet.

Table 1. Summary of results (*)

Character	Dominance	Direction of dominance	Genic symmetry	Dominants recessives	Reciprocal differences
Height	partial	positive	yes	1·0	(yes)
Rachis length	overdom.	positive	not	0·4	not
Leaflets/leaf	overdom.	positive	yes	1·5	not
Leaflet width	(overdom.)	positive	not	0·3	not
Width/length	(total)	negative	(not)	0·4	not
Weight/plant	overdom.	positive	not	2·1	(yes)
Seeds/plant	overdom.	positive	not	1·4	(yes)
Flowers/raceme	partial	positive	not	1·4	yes
Pods/node	overdom.	positive	not	1·1	not
Ovules/ovary	total	positive	yes	1·8	(yes)
Seeds/pod	(overdom.)	positive	not	2·6	not
Pod width	–	–	yes	1·5	not
Seed length	partial	negative	not	2·7	not
Thickness/length	partial	negative	yes	0·9	(not)
Seed weight	partial	negative	not	2·0	not
Time of blossom	(total)	(negative)	not	0·4	not
Time of maturity	total	negative	not	1·6	not

(*) Parentheses indicate tendencies.

Partial dominance: Plant height, number of flowers per raceme, pod width, seed length, thickness/length of seed and seed weight. For four characters (plant height, number of flowers per raceme, seed length and seed weight), partial dominance represented in this experiment a "compound result or system". This fact has been recognized comparing this diallel cross with a previous one (Cubero, 1970) which was only 4 × 4 and the four lines were of Central-European origin. The very small subdiallel table including only lines 166, 168 and 169 (Table 2) shows that different lines (*i.e.* populations) seem to display different genetic systems. Two of the characters analyzed show changes also in the direction of dominance. I have discussed recently (Cubero, 1974) the implications of this finding on the evolution of *V. faba*. Here I retain only the fact that some characters may or may not present overdominance depending on environmental conditions. As a summary it can be said, that the results are a consequence of the not very old partial allogamy of this species and the border mechanisms reinforcing the basic random drift due to the obligatory consanguinity of a partially allogamous species. This second point agrees with the thought of Bond (pers. com.) that "Spanish populations (of *V. faba*) are better adapted to Spanish conditions than North-European varieties are to North-European conditions".

Overdominance is shown by seven (eleven in some populations or environmental conditions) out of the seventeen characteristics. This proportion is rather heigh. In connection with breeding programmes, where the environmental conditions permit the full manifestation of heterosis (for this a high level of pollination is necessary), the best thing to do is to produce hybrids, if it is possible, or synthetic varieties. But where the favorable conditions are not constant, and this is the case in the South of Spain (great variability in climatic conditions at time of blossom causing great differences between years in the coefficient of allogamy), the use of overdominance to obtain full heterosis manifestation seems problematic.

Table 2. Characteristics that present discrepancy in the genetical systems (from Cubero, 1974)

Characteristics	7 × 7 diallel cross			4 × 4 diallel cross		
	All lines	Dominance direction	Central-Eur. lines	Dominance direction	System	Dominance direction
Plant height	partial dom.	positive	overdom.	positive	overdom.	positive
Flowers/raceme	partial dom.	positive	overdom.	positive	overdom.	positive
Seed length	partial dom.	negative	overdom.	positive	overdom.	positive
Seed weight	partial dom.	negative	doubtful	doubtful	overdom.	positive

This problem can be studied using the Wright formulae of equilibrium in a population with inbreeding (Wright, 1969). If the inbreeding coefficient is F, then the proportion of homozygotes and heterozygotes for a single pair of alleles A and a, of frequencies p and q are:.

$$AA\colon p^2 + pqF \qquad Aa\colon 2pq - 2pqF \qquad aa\colon q^2 + pqF \tag{1}$$

The relationship between F and the allogamy coefficient, α is known to be

$$F = \frac{1-\alpha}{1+\alpha}$$

If we suppose that the fitness of AA, Aa and aa are $1-S_1$, 1 and $1-S_2$, respectively, the change of q in the following generation is (Li, 1968)

$$\Delta q = pq\,\frac{(S_{1p} - S_{2q}) + F(S_{1q} - S_{2p})}{\overline{w}} \tag{2}$$

where

$$\bar{w} = 1 - S_1\,(p^2 + pqF) - S_2\,(q^2 + pqF)$$

Rearranging and remembering that $\dfrac{S_2}{S_1+S_2}$ and $\dfrac{S_1}{S_1+S_2}$ are the equilibrium values of p and q, respectively, in panmictic conditions (p_{ep} and q_{ep}):

where
$$\Delta q = \frac{pq}{W} (S_1 + S_2) \tag{3}$$

K is essentially positive.

An equilibrium situation is possible, when $q = 0$:

$$F_e = \frac{q_{ep} - q}{p_{ep} - q} \tag{4}$$

Logically, q cannot take values in the interval (p_{ep}, q_{ep}). Expression (4) shows that there are infinite positions of equilibrium. It is logical, because of the introduction of a new variable *(F)* in the description of the genetical properties of the population.

Expression (3) shows that if F falls from its equilibrium point (that is to say, if the allogamy rate increases because of, for example, an increase in insect visits) Δq will be positive. In the next generation the sign of Δq will not necessarily change (as in the case of a panmictic situation), depending on the new values q' and F'. The changes in genic frequencies are not, at least not strongly, buffered but can largely change.

Using the diffusion equations, Crow and Kimura (1970) have shown the action of overdominance on the fixation of an allele. Overdominance acts as a "retardation factor" on fixation, but only if the equilibrium value q_e is maintained between certain limits (roughly 0·2 and 0·8). Outside of these limits, overdominance acts in favour of the fixation process. The same result was obtained by Robertson (1962) with other methods of analysis. It is important, anyhow, to know the q_e value of the population we are concerned with.

To cut short a long discussion, in the case of partial allogamous species, overdominance produces a situation of equilibrium. This equilibrium is not as stable as in the case of panmictic conditions producing changes in allelic frequencies, which are variable and depending on variety location, culture methods intensity of gene migration, climatic conditions etc. The genetic array of a population (and even of a hybrid), can change from place to place and year to year. This can easily explain the differences between populations which have the same original name (for example, *Aguadulce* and *Muchamiel* in Spain).

Therefore, partially allogamous populations must be carefully handled by breeders and even by farmers. Otherwise the results of breeding programmes or culture may also be "random drift".

REFERENCES

CROW, J.F. & KIMURA, M. (1970): *An Introduction to Population Genetics Theory*. Harper & Row, 409–414.

CUBERO, J.I. (1970). Herencia de quince caracteres cuantitativos en *Vicia faba*. *Anales I.N.I.A.* **19** (1), 11–50.

CUBERO, J.I. (1973). Evolutionary trends in *Vicia faba* L. *Theor. Appl. Genetics* **43** (2), 59–65.

CUBERO, J.I. (1974). On the evolution of *Vicia faba* L. *Theor. Appl. Genetics* (in press).

DARWIN, CH. (1900). In "*Cross and self-fertilization of plants*". 2nd ed., London, 487.

HANELT, P. (1972). Die infraspezifische Variabilität von *Vicia faba* L. und ihrer Gliederung. *Die Kulturpflanze* **20,** 75–128.
HAYMAN, B.J. (1954*a*). The theory and analysis of diallel crosses. *Genetics* **39,** 789–809.
HAYMAN, B.J. (1954*b*). The analysis of variance of diallel tables. *Biometrics* **10,** 235–244.
HOLDEN, J.H. & BOND, D.A. (1960). Studies on the breeding system of the field bean, *Vicia faba* L. *Heredity* **15,** 175–192.
LI, CH.CH. (1968). *Population Genetics.* The University of Chicago Press. 6th ed.
MURATOVA, V. (1931). Common beans (*Vicia faba* L.) Suppl. **50.** *Bull. appl. Bot. Genet. Plant Breed.* 285.
PICARD, J. (1960). Données sur l'amélioration de la Féverole de printemps. *Ann. Amélior. Plantes* **10,** 121–153.
ROBERTSON, A. (1962). Selection for heterozygotes in small populations. *Genetics* **47,** 1291–1300.
SIRKS, M.J. (1931). Beitrag zu einer genotypischen Analyse der Ackerbohne, *Vicia faba* L. *Genética* **13,** 210–631.
WRIGHT, S. (1969). *Evolution and the Genetics of Populations.* Vol. **2:** *The Theory of Gene Frequencies.* The University of Chicago Press. 173–174, 243.

Prediction of vigour and choice of parents for a synthetic variety; application to cocksfoot (*Dactylis glomerata* L.)

A. GALLAIS

Station d'Amélioration des Plantes Fourragères, Institut National de la Recherche Agronomique, Lusignan, France

In order to efficiently select the parents of a synthetic variety, it is essential to be able to predict the vigour of its commercial generation from characteristics of the parents. Until now, the most used formula of prediction has been the so-called "Wright's formula". In its generalized form it predicts the value of a diploid synthetic variety in equilibrium (Syn *e*) knowing the number of parents *(k)*, the mean of their progenies by self fertilization *($\bar{P}$)* and the mean of all their possible crosses *($\bar{C}$)*.

$$\text{Syn}\, e = 1/k \qquad \bar{P} + (1 - 1/k)\, \bar{C} \tag{1}$$

It is strictly valid only in the absence of epistasis. To make the prediction more efficient and to know its limits a new approach of prediction formulae has been developed, first from a fixed standpoint and second from a random one.

To predict the vigour of a synthetic variety, it is possible to use the relationship between vigour and genetic structure *(S)* according to the identity by descent relationship at one locus or several loci considered simultaneously.

$$\text{Syn}\, e = a + bS \tag{2}$$

where S is a vector of the genetic structure and b a vector of weighting coefficients.

It has been shown that, using one locus parameters, the prediction will be unbiased for all ploidy levels only in the absence of epistasis and with completely inbred parents. However if these conditions are not satisfied, such a relationship may be useful from a statistical point of view and will be sufficiently efficient with a relatively great number of parents, not highly selected and with many randomly distributed loci of weak effect. The weighting coefficients in (2) *(b)* can be determined by using several generations derived from the parents of the synthetics. For autotetraploids, two new useful formulae were derived, in the absence of epistasis and restricting the interactions between alleles to the first order:

$$\text{Syn}\, e = \bar{C} - \frac{3}{2k}\,(\bar{C} - \bar{P}) \quad \text{for inbred parents} \tag{3}$$

$$\text{Syn}\, e = \bar{C} + \frac{1}{2} - \frac{3}{2k}\,(\bar{C} - \bar{P}) \quad \text{for homozygous parents} \tag{4}$$

Using the estimation of genetic effects which contribute to the value of a synthetic at equilibrium from the mean of a number of generations issued from the parents, new formulae have also been developed taking into account different types of epi - stasis for diploids, and interactions between three and four alleles in autotetraploids. For example, in diploids an unbiased predictor, with epistasis restricted to additive × additive type, will be: $\text{Syn}\ e = 2\left(\frac{k-1}{k}\right)\overline{2\ \text{Syn}\ e} - \frac{k-2}{k}\bar{P}$, $\overline{2\ \text{Syn}\ e}$ representing the mean at equilibrium of all two parent synthetics which can be derived from the parents. Three way, double crosses, etc. can also be used for the prediction.

From equations of prediction using $\bar{P}$ and $\bar{C}$ it is possible to replace $\bar{C}$ by $\bar{G}$, twice the mean of the general combining abilities of the parents, if the specific combining ability is not important. So applying the equation to each parent, a general synthesizing ability – i.e. the ability to give a good synthetic – can be defined for each parent as a linear combination of its value under self fertilization (for one generation) and twice its general combining ability.

The theory has been applied to two experiments with cocksfoot, an autotetraploid forage grass, evaluated at high density. First, it was verified that due to the autosterility phenomenon the Syn 1 can be easily predicted by the mean $\bar{C}$ of all crosses among the parents and also by $\bar{G}$ (the mean of the general combining abilities of the parents). When synthetics with different numbers of parents are considered, predictors including a parameter of the genetic structure (P_4 = frequency of tetragenic loci) with $\bar{G}$ or with the Syn 1 values, give a more accurate prediction (Table 1). Using the formula of prediction with $\bar{C}$ (or $\bar{G}$) and $\bar{P}$, it was found that formula (4) for

Table 1. Correlations between observed and predicted values showing the interest of the genetic structure (P_4 frequency of tetragenic loci). $\bar{C}$: mean of all possible crosses among the parents, $\bar{G}$: mean of the general combining abilities of the parents, Syn 1, Syn 2, Syn 3, (1st, 2nd and 3rd generation of the synthetics)

Variables in the predictors		1st experiment		2nd experiment		
	$\bar{C}$	$\bar{C}, P_4$	$\bar{G}$	$\bar{G}, P_4$	Syn 1	Syn 1, P_4
Syn 2	0·73**	0·85**	0·64**	0·78**	0·71**	0·84**
Syn 3	X	X	X	X	0·51**	0·66**

*, ** significance respectively at 0·05 and 0·01
1st experiment : 21 synthetics (3 with 8 parents, 6 with 4, 12 with 2)
2nd experiment : 25 synthetics (7 with 2 parents, 7 with 3, 7 with 4) (2 with 5 parents, 1 with 6, 1 with 7)

Table 2. Efficiency of some predictors showing, for the 1st experiment, the interest of adding the value of the parents under self fertilization to their general combining ability (measured by the correlation between observed and expected values). (1) predictor of equation 4. (2) predictor derived from equation 4 in which $\bar{G}$ replaces $\bar{C}$

Variables in the predictors	$\bar{P}$	$\bar{G}$	$\bar{C}$	k, $\bar{P}$, $\bar{C}$ (1)	k, $\bar{P}$, $\bar{G}$ (2)
Syn 2	0·58**	0·62**	0·73**	0·82**	0·79**

homozygous parents was better than formula (3) for non inbred parents, in spite of the fact that the parents were non or little inbred. Some results are given in Table 2, which indicates the interest of the simultaneous consideration of $\bar{C}$ (or $\bar{G}$) and $\bar{P}$. The specific combining abilities of the parents do not disturb the prediction, except with the more inbred parents. Both experiments revealed the great importance of the origin of the parents. The predictor using the value under one generation of self fertilization and the general combining ability of the parents appears very useful.

In conclusion, it appears that it is possible, theoretically and practically, to adapt new formulae enabling a better prediction of the commercial generation of k parent synthetics. So, for the breeder, it will not be necessary to test all combinations of parents in synthetics but only those expected to be the best.

EUCARPIA VII (1976)

Correlations between ability of pollen fertility restoration and some floral characters of wheat

M. JOŠT and J. MILOHNIĆ

Institute for Breeding and Production of Field Crops of the Agricultural Faculty, University of Zagreb, Zagreb, Yugoslavia

INTRODUCTION

The backcross method is used for development of R-lines, i.e. for carrying Rf-genes in the desired genotypes. In segregating progenies plants with Rf-genes are to be identified. For the present, this can be done only by test-crossing on an *ms*-tester line.

Relying on our previous investigations (Milohnić & Jošt, 1970), a supposition was made that Rf-genes, in addition to restoring fertility in F_1 hybrids with "sterile" cytoplasm, might also influence some other characters of the plant. It is most likely that their pleiotropic effect is displayed on floral parts. If this pleiotropic effect of Rf-genes could be easily measured, it might be used as a direct criterion for determining the presence of Rf-genes in progenies after back-crossing.

MATERIALS AND METHODS

18 various Rf-sources were tested during 3 years on an experimental field, at Zagreb. Two Rf-sources Nebraska (R-5) and Rw Bezostaya 1^4S^4(R-8) tested in the first year, were replaced by new Rf-sources (PI 277013 and PI 277016) in the subsequent seasons. The following studies were performed on the floral parts of Rf-sources:

1. Size of fresh anthers: on each Rf-source, in each year, length and width of 48 anthers (primary and secondary floret from the medium part of 12 randomly taken heads) was measured;
2. Number of pollen grains per anther was determined by the method of Pederson Johansen-Jorgensen in 1961, modified by Joppa *et al.* (1968). 6 samples of 20 anthers (in 2 ml of 1% aceto carmine) were taken from each Rf-source;
3. Diameter of pollen grains was measured on 60, with aceto carmine stained grains.

RESULTS

Ability for restoring pollen fertility, determined at 3 *cms*-testers for all the Rf-sources examined during three years, was previously reported (Milohnić & Jošt, 1974).

Anther size. Significant differences in length and width of anthers were obtained for the Rf-sources tested. The smallest anthers were recorded for line IBO 460/616

(3.29 × 0.829 mm) and this line hardly displayed any ability for restoring fertility under the climatic conditions of Zagreb. The largest size of anthers was determined for line CH 4-II-2 (5.07 × 0.968 mm), a bad restorer, as well. The best restorers Primépi and CH1 × 9E1[3] were among the Rf-sources ranking first in anther size. No significant difference was found in anther width among years of testing (average of 16 Rf-sources tested in all the 3 years), while differences in anther length were highly significant.

Number of pollen grains/anther. Rf-sources tested significantly differ in number of pollen grains/anther. The lowest values were obtained for line IBO 519/616 (2188), the highest for variety Primépi (3983), which was also found to be the most efficient restorer. Also in previous studies of Milohnić and Jošt (1970), variety Primépi exceeded in number of pollen grains/anther 26 wheats tested. For the average number of pollen grains/anther highly significant differences were determined between years of testing.

Diameter of pollen grain. The pollen grain size in wheat increases with the level of polyploidy. However, genoms differ in their influence on the pollen size. According to Cetl (1960) "genoms form a degressive sequence A>B>D>G from the point of view of their influence on the pollen grain volume". Although all the Rf-sources tested belong to hexaploid wheats, significant differences were recorded in the pollen grain size. The smallest pollen diameter was shown by the variety Palmaress (57.4 μ), and the largest by line IBO 460/616 (64.7 μ). Differences between years of testing are significant also for this character.

Correlations of pollen restoration ability and floral characters. Simple correlation coefficients were calculated for restoration of pollen fertility in F_1 hybrids with *timopheevi* cytoplasm (seed set on bagged and unbagged heads) and for all four floral characters of the Rf-sources tested. Correlation coefficients (calculated per year of testing, regardless to tester linen, and according to tester lines with no regard to year), showed highly significant positive correlation between ability for restoring fertility and number of pollen grains/anther for all the years of testing and for all the tester lines correlated. Simple correlation coefficients for the mean of three years and three *ms*-tester lines were: $r = 0.885^{**}$ for bagged heads, and $r = 0.814^{**}$ for unbagged heads. Positive correlation, although not always significant, was also determined for the characters: anther size and restoration of pollen fertility. Pollen diameter and fertility restoration are in a negative correlation, being significant only in year 1971 and on tester line ms^t-Mirna.

On the basis of correlations determined the supposition seems justified that in the process of breeding R-lines the characters: anther length and number of pollen grains/anther might be used for identifying those plants in the segregating progenies, which possess Rf-genes. By this way the number of test-crosses would be reduced and, obviously, the process of developing R-lines would be accelerated. The efficiency of the proposed method should be specially tested.

REFERENCES

CETL, I. (1960). The size of pollen grain of the genus *Triticum* L. *Biologia Plantarum (Praha)* **2**/4, 287–291.

JOŠT, M. (1973). Phenotypical expression of genes for pollen fertility restoration (Rf-genes) in fertile wheat genotypes *Triticum aestivum* ssp. *vulgare*. Ph. D. Thesis., The Agricultural Faculty, University of Zagreb, 59 pp.

MILOHNIĆ, J. & JOŠT, M. (1970). Pollen production and anther extrusion of wheat (*Triticum aestivum* L. em. Thell.). *Acta Agron. Acad. Sci. Hung.* **19**(1–2): 17–23.

MILOHNIĆ, J. & JOŠT, M. (1974). Fertility restoration capacity of different Rf-sources in *Tr. timopheevi* cytoplasm. *VIIth Congress of Eucarpia, Budapest, 24–29 June.*

VRIES, A. PH. DE (1971). Flowering biology of wheat, particularly in view of hybrid seed production – A review. *Euphytica* **20,** 152–170.

JOPPA, L. R., MCNEAL, F. H. & BERG, M. A. (1968). Pollen production and pollen shedding of hard red spring (*Triticum aestivum* L. em Thell.) and durum (*T. durum* Desf.) wheats. *Crop Sci.* **8,** 487–490.

Expression of heterosis and combining ability in a diallel wheat cross

I. MIHALJEV

Faculty of Agriculture, Novi Sad, Yugoslavia

Many studies have demonstrated heterosis for grain yield and yield components in intervarietal wheat crosses, both with space-planted stands, based on the single plan yield (Briggle, 1963; Knott, 1965; Brown *et al.,* 1966; Gyawali *et al.,* 1968) and close-planted stands, based on the yield per plot unit (Fonseca & Patterson, 1968; Wells and Lay, 1970; Keydel, 1973). According to Zeven (1972) and Fonseca & Patterson (1968) the plant density has no effect on the degree of heterosis, while according to Griffing & Zsíros (1971) it has. However, for successful hybrid wheat breeding it is necessary that F_1 hybrids express heterosis at normal seeding rates used in practice and therefore F_1 hybrids should be evaluated with such seeding rates.

This study was conducted to evaluate parents and F_1 hybrid combinations under conditions of normal seeding rate, and to estimate combining ability of the parents in order to obtain information for further testings and development of wheat hybrids.

MATERIALS AND METHODS

The winter wheat varieties Sava, NS-732, Bezostaya-1, Kavkaz, Libellula, N. lot-1 and spring wheat varieties Cajeme 71 and Siete Ceross were crossed by hand in a diallel system excluding reciprocals, in 1972. Parents and F_1 hybrids were planted in the field in a randomized complete block design with three replications and 375 viable seedsj/m^2, in 1973. The plot unit was 1 row 1 m long with 0.2 m between rows. Four characters were measured (see tables).

F_1 hybrids were compared with the mid-parent value and with the better parent of the cross to determine percentage of heterosis. Estimates of general combining ability (GCA) and specific combining ability (SCA) were obtained using Griffing's (1956) combining ability analysis, method 2, model I.

RESULTS AND DISCUSSION

Heterosis in F_1. Five of the 28 F_1 hybrids exceeded in grain yield per plot the better parent of the cross, ranging from 3–21%, and 12 of the 28 surpassed the mid-parent value, ranging from 3–25%. The yield components measured also showed heterosis in F_1 and among them the 1000 kernels weight showed heterosis in most F_1 hybrids. Compared to the better parent, heterosis for 1000 kernel weight was expressed in 12

Table 1. Estimates of GCA effects for yield components and grain yield (1973)

Variety	Kernels per spike	Kernel weight per spike	1000 kernel weight	Grain yield per plot
Sava	+ 2·15**	+ 0·03	− 0·74	+ 4·07
Bezostaya-1	− 0·42	+ 0·05	+ 2·73**	+ 6·06
Kavkaz	+ 1·29*	+ 0·14**	+ 3·49**	+ 11·40**
Libellula	+ 0·78	− 0·08	+ 0·93	− 3·80
Cajeme 71	− 1·95	− 0·05	− 1·81	− 6·66
Siete Ceross	+ 4·76**	+ 0·06	− 0·71	+ 5·38
NS-732	− 4·41	− 0·12	− 3·84	− 20·64
N. lot-1	− 2·66	− 0·08	− 0·06	+ 4·18
L.S.D.: 0·05	1·25	0·06	1·07	6·71
0·01	1·67	0·08	1·43	8·93

*, ** Significant at the 0·05 and 0·01 level, respectively.

Table 2. Estimates of SCA effects for yield components and grain yield (1973)

Characters and combinations		Estimates
Bezostaya-1 × Cajeme 71	*Kernels per spike*	+ 4·51*
Libellula × Cajeme 71		+ 4·75**
Sava × Siete Ceross	*Kernel weight per spike*	+ 0·19*
Bezostaya-1 × Cajeme 71		+ 0·18*
Bezostaya-1 × Cajeme 71	*1000 kernel weight*	+ 3·46*
Libellula × NS-732		+ 3·36*
Cajeme 71 × N.lot-1		+ 4·07**
NS-732 × N.lot-1		+ 4·02*
Kavkaz × N.lot-1	*Grain yield per plot*	+23·25*

*, ** Significant at the 0·05 and 0·01 levels, respectively
Note: Only combinations are shown which were significantly better

of the 28 F_1 hybrids ranging from 1–21%, and compared to the mid-parent value in 25 of the 28 F_1 hybrids ranging from 3–25%(Tabl. 1). The results obtained suggest that a considerable amount of heterosis for grain yield and yield components may be found only in certain F_1 hybrids under normal plant density. Since the production of hybrid wheat will depend on the use of male sterility, in the case of cytoplasmic male sterility the expression of heterosis may be affected by male sterile cytoplasm. Thus, the obtained results should be considered from this point of view.

Combining ability. There was a significant difference between genotypes (F_1 hybrids and parents) for all characters studied and the major component of variation was associated with highly significant GCA. Only for 1000 kernel weight and kernel weight per spike were the SCA mean squares significant.

Comparison of GCA performances of individual varieties for all characters studied shows that Kavkaz has the highest positive GCA, while NS-732 has the highest negative GCA. Siete Ceross gave the highest positive GCA for kernels per spike and Kavkaz for the other three characters (Table 1).

Highly significant GCA indicates that additive gene action was responsible for an important portion of the genetic variance associated with yield components and grain yield. Much of the positive GCA encountered was contributed by the variety Kavkaz, suggesting that this variety might be quite useful in a hybrid wheat program. The opposite situation is seen with the variety NS-732 and most others. The statistical significance and the magnitude of SCA mean squares for 1000 kernel weight and kernel weight per spike indicates that non-additive gene action controlled a portion of genetic variance associated with these characters.

Estimates of SCA effects for yield components and grain yield (Table 2) show that the SCA was significantly higher in only a few combinations. The highest positive SCA for grain yield was shown by the combination Kavkaz × N. lot-1, for 1000 kernels weight by Cajeme 71 × N.lot-1, for kernel weight by Sava × Siete Ceross and for kernels per spike by Libellula × Cajeme 71.

Failure to detect more combinations with significant positive SCA for the characters studied does not necessarily mean that SCA does not exist more frequently in wheat. It might have happened because SCA is more dependent on the set of parents chosen than GCA.

REFERENCES

BRIGGLE, L.W. (1963). Heterosis in wheat – A review. *Crop Sci.* **3,** 407–412.

BROWN, C.M., WIEBEL, R.O. & SEIF, R.D. (1966). Heterosis and combining ability in common winter wheat. *Crop Sci.* **6,** 382–383.

FONSECA, A. & PATTERSON, F.L. (1968). Hybrid vigor in a seven-parent diallel cross in common winter wheat (*Triticum aestivum* L.). *Crop Sci.* **8,** 85–88.

GRIFFING, B. (1956). Concept of general and specific combining ability in relation to diallel crossing systems. *Aust. J. Biol. Sci.* **9,** 463–493.

GRIFFING, B. & ZSÍROS, E. (1971). Heterosis associated with genotype-environment interactions. *Genetics* **68,** 443–445.

GYAWALI, K.K., QUALSET, C.O. & YAMAZAKY, W.T. (1968). Estimates of heterosis and combining ability in winter wheat. *Crop Sci.* **8,** 3, 322–324.

KEYDEL, F. (1973). Untersuchungen über die Leistungseigenschaften von F_1-Hybriden von Winterweizen. *Z. f. Pflanzenzüchtung* **69,** 3, 239–255.

KNOTT, D.R. (1965). Heterosis in seven wheat hybrids. *Canad. J. Pl. Sci.* **45,** 499–501.

WELLS, D.G. & LAY, C.L. (1970). Hybrid vigor in hard red spring wheat crosses. *Crop Sci.* **10,** 220–223.

ZEVEN, A.C. (1972). Plant density effect on expression of heterosis for yield and its components in wheat and F_1 versus F_3 yields. *Euphytica* **21,** 397–563.

Study of heterosis effect in some intervarietal hybrids of tobacco

P. P. NOSOVA

All-Union Research Institute of Tobacco and Makhorka, Krasnodar, USSR

Heterosis effect was studied in a five-year experiment in F_1 and F_2 tobacco hybrids on the characters: plant height, number of leaves, length and width of middle leaf. The experiments were carried out in our institute at Krasnodar, under field conditions, using hybrids between the following native tobacco varieties:

"Immunny 580" sessile-leaved, with long term of maturity. Height of plants 130–150 cm. Number of leaves 32. Length of middle leaf 40–42 cm. Width of middle leaf 25–30 cm;

"Dubek 84" (spontaneous mutant) sessile-leaved, with middle term of maturity. Height of plants 2–2.5 m. Leaf number 80–85. Length of middle leaf 25–30 cm. Width of middle leaf 10–15 cm;

"Ostrolist 1519" sessile-leaved, with middle term of maturity. Height of plants 140–150 cm. Number of leaves 30–32. Length of middle leaf 40–42 cm. Width of middle leaf 20–25 cm;

"American 273" sessile-leaved, with long term of maturity. Plant height 145–150 cm. Number of leaves 30–32. Length of middle leaf 35–40 cm. Width of middle leaf 20–25 cm;

"Ostrolist 518" sessile-leaved, with middle term of maturity. Plant height 150–160 cm. Leaf number 30–32. Length of middle leaf 45 cm. Width of middle leaf 25–30 cm;

"Samsun 959" petiol-leaved, with middle term of maturity. Plant height 150–170 cm. Leaf number 55–60. Width of middle leaf 12–13 cm;

"Ostrolist 125" sessile-leaved, with middle term of maturity. Plant height 110–115 cm. Number of leaves 30. Length of middle leaf 35–40 cm. Width of middle leaf 20–25 cm;

"Sobolchsky 174" sessile-leaved with middle term of maturity. Plant height 140–150 cm. Number of leaves 30. Length of middle leaf 40 cm. Width of middle leaf 20–24 cm;

"Trapezond 219" petiol-leaved, with middle term of maturity. Plant height 120–130 cm. Number of leaves 30–32. Length of middle leaf 30–35 cm. Width of middle leaf 20–25 cm.

The seedlings were transplanted in the field, according to the scheme: *mother's form-hybrid-pollinizer* on two, ten-meter rows with nutrition area of 60×25 cm.

The measurement of plant height and estimation of leaf number was carried out at the period of total flowering; the estimation of leaf size was carried out at the period of maturity of middle leaves.

Table 1. Inheritance of quantitative characters in F_1 hybrids (summarized data for 1968—1972)

Year	Number of combi-nations	Frequency	Plant height				Leaf number				Leaf length				Leaf width			
			heter-osis	domi-nance	inter-medi-ate	reces-sive	heter-osis	domi-nance	inter-medi-ate	reces-sive	heter-osis	domi-nance	inter-medi-ate	reces-sive	heter-osis	domi-nance	inter-medi-ate	reces-sive
1968	16	number	6	3	4	3	—	2	5	9	9	3	1	3	7	6	—	3
		%	37·6	18·7	25·0	18·7	—	12·5	31·2	56·3	56·3	18·7	6·3	18·7	43·7	37·6	—	18·7
1969	16	number	5	4	4	3	2	2	4	8	8	3	3	2	6	4	4	2
		%	31·3	25·0	25·0	18·7	12·5	12·5	25·0	50·0	50·0	18·7	18·7	12·5	37·5	25·0	25·0	12·6
1970	16	number	3	4	3	6	1	1	4	10	4	7	2	3	3	8	2	3
		%	18·7	25·0	18·7	37·5	6·2	6·2	25·0	62·5	25·0	43·7	12·6	18·7	18·7	50·0	12·6	18·7
1971	28	number	7	9	8	4	2	5	5	16	3	8	8	9	6	7	11	4
		%	25·0	32·1	28·6	14·3	7·1	17·9	17·9	57·1	10·3	28·7	28·7	32·0	21·4	25·0	39·3	14·3
1972	42	number	21	7	2	12	12	4	2	24	18	7	6	11	19	11	7	5
		%	50·0	16·8	4·5	26·7	28·7	9·5	4·4	57.1	42·8	16·7	14·2	26·6	45·2	26·6	16·7	11·9

Heterosis and transgressive variation was only shown for certain characters in crossing varieties combining the dominance of analyzed characters.

Heterosis and positive transgressions for plant height over both parents were noted in seven out of 10 studied F_1 hybrids.

Two F_1 hybrid combinations (Dubek 84 × Ostrolist 1519, Dubek 84 × Trapezond 219) displayed non-significant heterosis only with regard to pollenizer (male parent) – 11·3 and 8·6 per cent.

F_1 hybrid (Samsun 959 × American 273) displayed heterosis only with regard to the mother form – 9.0 per cent.

In the second generation 6 hybrid combinations revealed heterosis foı plant height, which equals to 60 per cent of the total number of analyzed hybrids.

The hybrid plant combinations (Ostrolist 518 × Immunny 580, Ostrolist 518·× Dubek 84, American 273 × Sobolchsky 174) displaying heterosis both in the 1st and the 2nd generations, showing a high degree and frequency of transgressions.

The F_2 hybrid Ostrolist 125 × Immunny 580 displayed a negative degree of transgression for height of plant, but with non-significant frequency.

The hybrids Ostrolist 1519 × Immunny 580, Ostrolist 125 × Immunny 580, Ostrolist 518 × Immunny 580 revealed heterosis for leaf number both in the 1st and the 2nd generations. These hybrids had a high degree and frequency of transgression.

The hybrid Ostrolist 518 × Immunny 580 both in the 1st and the 2nd generations exceeded the parents in plant height and number of leaves.

The presence of heterosis in the studied hybrid combinations obviously may be explained by heterozygosis of the new varieties for certain characters, in connection with combining ability.

In the cross of the multileaved varieties Samsun 959 × Dubek 84, the F_1 hybrids take an intermediate position. In all cases in the F_2's negative transgression with regard to the recessive multileaved parent was revealed.

Thus in the hybrid combinations studied heterosis was considered to depend not only on the efficiency (productivity) of initial forms, but also on the biological differences between them.

In crossing large-leaved varieties with small-leaved ones the type of inheritance in F_1 and F_2 hybrids was intermediate.

However, F_2 hybrid American 273 × Sobolchsky 174 displayed non-significant heterosis for leaf length.

On the parent level (at the error limits) inheritance also took place in F_2 hybrids (Ostrolist 1519 × Immunny 580, Ostrolist 518 × Immunny 580).

Only few of all studied combinations showed heterosis for leaf width and length.

It is worthy to note that in F_2's greater heterosis for the mother form of leaf width was displayed than in F_1's. The highest degree and frequency of transgression occured in F_2 hybrids (Ostrolist 1519 × Immunny 580, Ostrolist 125 × Immunny 580, Dubek 84 × Ostrolist 1519).

Summarized Table 1 illustrates the data of F_1 hybrids in 1968–1972. The data show that in the hybrid combinations studied in a 5 year experiment heterosis was not constantly displayed for all characters and, being the summarized effect of dominance of intermediate inheritance differed depending on the year of growing.

Therefore, in tobacco the heterosis effect is of discrete a character. That is why for obtaining heterotic hybrid combinations it was necessary to choose pairs using the complementary dominant genes principle.

CONCLUSIONS

1. Heterosis may be estimated either by certain studied characters or by their complex, depending on the hybrid combination.

2. The degree and frequency of transgression in tobacco hybrids was unequal for certain characters and was estimated to depend not only on the efficiency of initial forms, but also on the biological differences between them.

3. It is possible to obtain hybrids exceeding the level of parental forms by carefully selecting parental pairs, which differ in the analyzed characters.

Quantitative studies on heterosis in *Arabidopsis thaliana* (L.) Heynh.

H. EL ASHMI and N. CHALBI

Department of Genetics and Biometrics, Faculty of Sciences, University of Tunis, Tunis, Tunesia

Studies on heterosis have been realized with *Arabidopsis thaliana* for many traits such as earliness of flowering (Van der Veen, 1965) or heterosis expression under effects of different temperature levels (Langridge, 1962), or in various conditions of environment (Griffing & Zsíros, 1971) and many others.

In relation to the quantitative point of view, we have made an analysis of the rosette diameter length at different stages of the growth.

Five inbred lines of *Arabidopsis thaliana* were chosen: A_5 (Dijon, France), A_9 (Eastland, Baltique), A_{64} (Edinburgh, Scotland), A_{51} (Bologna, Italy) A_{80} (Turin, Italy). Observations on flowering time of these lines show that A_{51} and A_{80} are the latest. The experiment was designed as an incomplete diallel cross, described by Griffing (1956) method 2, including the five mentioned lines taken as parents and their ten direct combinations. Reciprocal crosses were not used.

The F_1 generations of all these crosses and parents were arranged according to a sixteen complete randomized block design (finally 16 repetitions for each parent or F_1), disposed under a light varying from 6000 to 8000 lux, 18 hours a day. Plants were grown in "giffy-pots" (6 cm of diameter) with a mixture of vegetable mould and sand.

Many characteristics were examined and will be presented elsewhere.

This paper presents only the results of analysis made on the rosette diameter length at an early stage, 19 days after sowing (Table 1).

The data show a real superiority for a large number of the hybrids relatively to their parents. These differences were analysed using a measure of heterosis expressed according to Falconer (1960) which is as follows:

If $\frac{1}{2}(m_1 + m_2)$ = mean value of the parents P_1 and P_2 and m_{F_1} = mean value of the F_1 between P_1 and P_2, then the heterosis for rosette diameters expressed on phenotype (H_{F_1}) is measured by the difference:

$$m_{F_1} - \frac{1}{2}(m_1 + m_2)$$

The significance of this difference was calculated at the 5% level.

Table 2 summarizes the H_{F_1} values obtained with this material (disposed according to an increasing arrangement) as well as the gain of the hybrid (in % relative to the parental mean).

Table 1. Rosette diameter (in cm) (mean value 19 days after sowing)

♀ Lines ♂	A_5	A_9	A_{51}	A_{64}	A_{80}
A_5	*1·96*	2·47	2·99	2·94	2·43
A_9		*2·88*	3·51	3·02	3·70
A_{51}			*2·39*	3·40	3·44
A_{64}				*2·45*	3·44
A_{80}					*2·38*

Table 2. Analysis of the measure of heterosis (NS = not significant, S = significant at the 5%, TS at the 1% and HS at the 0·1% level)

Hybrids	$A_5 \times A_9$	$A_5 \times A_{80}$	$A_9 \times A_{64}$	$A_5 \times A_{64}$	$A_{51} \times A_{80}$	$A_5 \times A_{51}$	$A_9 \times A_{51}$	$A_{51} \times A_{64}$	$A_{64} \times A_{80}$	$A_9 \times A_{80}$
H_{F_1}	0·04	0·26	0·36	0·73	0·80	0·82	0·88	0·98	1·04	1·16
%	1·53	11·80	13·37	33·12	33·67	37·40	33·32	40·38	43·22	44·30
t	NS	NS	S	S	TS	HS	HS	HS	HS	HS

According to these results we may conclude that there is evidence for heterosis in rosette diameter at the early stages of the growth, the magnitude varying in relation with the parents crossed. Further observations on this material indicated a persistence of this effect for the rosette diameter until the end of its growth (about 50 days after sowing).

The hypothesis of a genetic cause in the quantitative differences observed seems to be plausible. In respect with this hypothesis, the analysis of data by the diallel cross method (Griffing, 1956) gives a clear indication of highly specific combining ability. Similar results were obtained for some other characteristics, such as spike length.

REFERENCES

FALCONER, D.S. (1960): *Introduction to quantitative genetics.* Olliver & Boyd, Edinburgh.

GRIFFING, B. (1956): Concept of general and specific combining ability in relation to diallel crossing systems. *Austral. J. Biol. Sci.,* **9,** 469–493.

GRIFFING, B. & ZSÍROS, E. (1971): Heterosis associated with genotype environment interactions. *Genetics,* **68** (3) 443–455.

LANDGRIDGE, J. (1962): A genetical and molecular basis for heterosis in *Arabidopsis* and *Drosophila. Am. Naturalist* **96,** 5–27.

VAN DER VEEN, J.H. (1965): Genes for late flowering in *Arabidopsis* research. *Rep. Int. Symp. Göttingen* (Ed. G. Röbbelen).

GENERAL ASSEMBLY

Report of the General Assembly

The Seventh General Assembly of EUCARPIA was held at the University of Horticulture, Budapest, on June 26th, 1974. The President, Dr. F.G.H. Lupton was in the chair.

The President read the report of the Board for 1971–74. The Assembly approved the report.

The Financial Report was presented by the Treasurer, Dr. Lamberts, with audited statements for the accounts for 1972 and 1973. The report was accepted.

The President outlined the proposed amendments to the Constitution pointing out that the changes were mainly concerned with clarification of procedures which had been adopted for many years. They also included provision for formal auditing of the accounts, and a simplified procedure for future amendments to the statutes. Dr. Odenbach considered it desirable that it should be possible for members to be represented at General Assemblies and proposed that Article 6 C 2 be amended to read "Members can be represented". The President checked that a sufficient proportion of the membership of the Association was present or represented at the meeting to enable it to consider changes to the Statutes. This being the case, Dr. Odenbach's amendment was approved, and the new Statutes, including this amendment, were accepted.

The Internal Regulations as recommended by the Board were accepted.

Professor A. Jánossy was elected President of EUCARPIA for the period 1974–77. He thanked the Assembly and asked Dr. Lupton to continue as chairman until the end of the General Assembly.

Professor E. Sánchez-Monge was elected 1st Vice President and agreed to organize the Eighth Congress in Spain in 1977.

The following new members were elected to the Board for the next six years:

Representatives of countries:

Steineck, O.	Austria
Laeremans, R.A.F.	Belgium
Munk, A.	Denmark
Tigerstedt, P.	Finland
Cauderon, A.	France
Plarre, W.K.F.	Federal Republic of Germany
Kehoe, H.W.	Ireland

Scarascia-Mugnozza, G.T.	Italy
Bojanowski, J.S.	Poland
Popow, G.	Switzerland
Holland, H.	UK
Budin, K.Z.	USSR

Chairmen of Sections:

Wricke, G.	Biometrics in Plant Breeding
Palaversić, D.	Maize and Sorghum
Dorsman, C.	Fruit

Other members:

Mota, M.	Portugal
Mesken, M.	Representative of the Secretariat

Professor Fedin proposed that the Ninth Congress in 1980 should take place in the USSR. This proposal will be considered in due course by the Board.

On behalf of the Swedish members, Mr. Ewertson presented the following proposal: "We would like to propose that Eucarpia as soon as possible gets a plant breeder as representative in the Plant Breeders' Rights Organization UPOV. UPOV are making decisions vital for plant breeders. We think that we must participate in the UPOV council in order to arrive at rules and regulations, which are realistic as far as practical breeding is concerned. The UPOV council has a meeting in October or November this year. We think that we ought to make our first approach before then."

Dr. Lupton informed the Assembly that a request in this matter had already been sent to UPOV. This request will be discussed by the council of UPOV at its next meeting in November, 1974. A representation of Eucarpia at the technical level should also be considered.

Before closing the Assembly, Dr. Lupton thanked Professor Jánossy and his colleagues for their hospitality during the Congress, and for the excellent arrangements made to ensure its success.

Report of the Board of Eucarpia for the period 1971–1974

F.G.H. LUPTON,

President

The greater part of the activity of our Association since the Sixth Congress at Cambridge three years ago has been organised by the Sections, and the principal duty of the Board has been to co-ordinate and encourage activities and to allocate financial support where necessary.

The *Cereals Section* has held four meetings since the last Congress. Their first meeting, held at Munkebjerg, Denmark, considered non-rust diseases of cereals, and led to the setting up of a co-operative project with the International Organisation for Biological Control (IOBC), who are meeting the greater part of the financial costs involved. This project is designed to investigate yield losses caused by powdery mildew in barley and the effects of varied rates of fungicide application in controlling these losses. It represents an important advance in two respects: first, it shows the need for plant breeders to co-operate with manufacturers of fungicides in obtaining control of plant diseases; and secondly it establishes the principle that control of pests or diseases by plant breeding is as much a form of biological control as that of insect pests by the introduction of predators. The second meeting, on *durum* wheat breeding, took place at Bari, Italy, and enabled specialists in this crop to consider their problems at one of the world centres for *durum* wheat breeding. The third meeting of the Cereal Section considered *Triticale* breeding and took place at Leningrad. This meeting, the first which Eucarpia has held in the Soviet Union, was attended by workers from all the main centres in the world where *Triticale* is being studied and recommended that Eucarpia should organise and maintain a world collection of *Triticale* varieties and arrange for internationally co-ordinated yield trials. A fourth meeting of the Section on rye breeding took place only last week at Poznan, Poland.

The Section also arranged for the publication in the Eucarpia Bulletin of a new decimalised code for describing the growth stages of cereals. It is hoped that this scheme, developed by an international team comprising J.C. Zadoks, T.T. Chang and C.F. Konzak, will be expanded to include a wider range of crops, and that it will be accepted by all concerned with describing the growth stages of crops.

The *Fodder Crops Section* held two plenary meetings: the first, on "The evaluation of herbage crop breeding material", was held at Dublin and included visits to breeding and experimental stations in Eire. The second, on "New methods of fodder crop breeding", was held at Wageningen and gave members an opportunity to visit the new Institute de Haaff (Foundation for Agricultural Plant Breeding). The Section also organised a number of active Working Groups. The Group considering *Medicago sativa* met three times, at Lodi (Italy), Aberystwyth (Great Britain) and at

Kompolt (Hungary) and considered a wide range of problems concerned with lucerne breeding. Other Working Groups considered competition in forage crop breeding and diseases of grasses. The Working Group on competition produced a most valuable glossary of terms and definitions used in their specialised field, copies of which may be obtained from the Secretariat.

The *Potato Section* held a single but important meeting jointly at Wageningen and Köln, considering the importance of haploids in potato breeding. Cytological, genetical and phytopathological aspects of the problem were considered as well as those concerned directly with the methodology of breeding, and it was generally agreed that haploids provide a stimulating new approach to the practical breeding of potatoes. Participants also had the chance to visit the new Institute de Haaff at Wageningen and the Max-Planck Institute for Research in Plant Breeding at Köln, and to see aspects of the work on haploid potato breeding being conducted at these two institutes.

The *Maize and Sorghum Section* held two meetings: the first at Freising-Weihenstephan (FRG) and the second at Zagreb (Yugoslavia). Both meetings considered a wide range of problems concerned with the maize and sorghum crops and visited experimental stations in the neighbourhoods of Weihenstephan and Zagreb. The second meeting included a session, organised jointly with the Physiology Section,on the whole plant physiology of maize. In addition to these meetings, the section organised extensive international exchange of breeding material and planned co-operative yield trials. These were organised in two groups, covering the northern and southern countries of Europe and have been most productive in enabling breeders to assess their material over a wide range of environments.

On account of the wide range of interest of its members, the *Vegetable Section* organised a range of specialist meetings, each dealing with problems concerned with specific crops. These comprised meetings on tomatoes at Plovdiv (Bulgaria); on *Capsicum* at Turin (Italy); on peas at Versailles, melons at Avignon and on *Asparagus* at Versailles (France) and on cucumber at Hannover (FRG). For the same reasons, the *Fruit Plants Section* also organised specialist meetings on strawberries at Dundee (Scotland), on *Ribes, Rubus & Vaccinium* at Belgrade (Yugoslavia) on pears at Anger (France) and on tree fruits at Canterbury (England); the *Ornamentals Section* on *Begonia* and *Ageratum* at Ahrensburg (FRG), on carnations at Frejus (France) and on the glasshouse environment at Norwich (England). The exchange of views and of breeding material amongst breeders is one of the most important functions of EUCARPIA, and the Board greatly welcomes this wide range of specialist meetings. They are particularly pleased to hear that the apple and pear breeders in the Fruit Section are planning to form a Breeding Co-operative to coordinate the exchange of material and experience.

The *Physiology Section* held a most successful meeting on "Aseptic culture methods in plant breeding" at Leeds, England. It was clear from the papers read that aseptic culture techniques have great possibilities as a tool for plant breeders. The Section therefore plans to hold a similar meeting on the same theme in a few years time. They also arranged, as already mentioned, a session on the whole plant physiology of maize at the meeting of the Maize & Sorghum Section at Zagreb.

The *Mutation and Polyploidy Section* held a meeting, jointly with the FAO/IAEA Division of Atomic Energy in Agriculture at Bari (Italy). It was felt at this meeting that there was a need for closer co-ordination between the Section and other European organisations concerned with the exploitation of mutations and polyploidy in plant breeding, and it is planned to integrate future meetings more closely.

The *Wild Species & Primitive Forms Section* held an important meeting at Izmir (Turkey) at which the present and future organisation of European gene banks were discussed and methods of data retrieval considered. On the recommendation of those attending this meeting, a specialist committee comprising the Directors of the European gene banks and representatives of EUCARPIA has been set up to co-ordinate activities and to standardise data retrieval.

The newly established *Biometrics in Plant Breeding Section* held its first meeting at Hannover (FRG) last year. It was clear from the papers read and the discussions following them that this new Section will form a most valuable link between biometricians and those engaged in practical breeding.

The co-ordination of this very wide and progressive range of meetings has made it necessary for the Board to lay down a few rules, so as to ensure that our limited financial resources are used to the best purpose. First, it has been suggested that section meetings shall only exceptionally be published; secondly, that sections must produce estimates for any meeting for which financial support is required and that this estimate must be agreed by the Board before any money is spent; and thirdly it is strongly recommended that funds shall only exceptionally be used for payment of travelling expenses of section officers.

The Board has maintained contact with other organisations concerned with agricultural research. We have been represented at meetings of the European and Mediterranean Plant Protection Organisation (EPPO) and of the Variety Certification Committee of OECD. We have maintained close contact with FAO/IAEA, IIRB, CSSA, SABRAO and, as earlier mentioned, with IOBC. We have also been in contact with the Secretary General of UPOV, as we feel that closer association with this organisation would be valuable. In addition, we have tried to extend our membership to countries where we had few or no members, and in this connection have been in correspondence with colleagues in the GDR, and are very pleased to have some of them with us here this week.

The Board has met four times since the last Congress, three times in Paris and once shortly before the present Congress, in Budapest. Reports of these meetings have been published in the Bulletin. Its membership has been:

Dr. F.G.H. Lupton
Professor A. Jánossy
Professor G. Haussman

Professor S. Borojević
Academician D.D. Brezhnev
Professor A. Cotte
Professor F. D'Amato
Dr.C. Daskalov

Professor F. Garcia-Olmedo
Professor A. Hagberg
Professor H. Hänsel
Dr. M. Ingold
Dr. B. Jacobsen
Dr. J. Kovács
Dr. H. Lamberts
Dr. A. Lein
Professor K. R. Manner
Professor A. Möes
Dr. C. North
Dr. T. J. R. O'Sullivan
Professor H. Ross
Professor J. Sneep
Professor E. Strand

Thirteen of these Board members have now served for six years and must, therefore, be replaced by new members to be elected by this Assembly. These comprise representatives of Austria, Belgium, Denmark, Finland, France, Italy, Switzerland and the USSR, three chairmen of Sections, and one new representative of countries with less than ten members. As there are now more than ten members from Ireland, it will also be necessary to elect a representative from that country. The Board has prepared a list of nominations which will shortly be presented for your consideration.

There have been several changes in the chairmanships of sections during the past three years. The present situation, excluding changes made during this Congress, is as follows: –

Cereals	Dr. H. Lamberts
Fodder Crops	Dr. G. Julen
Potatoes	Dr. J. G. T. Hermsen
Maize and Sorghum	Dr. D. Palaversić
Vegetables	Dr. C. North
Fruit Plants	Dr. C. Dorsman
Ornamentals	Dr. L. D. Sparnaay
Physiology	Dr. Z. Martinić
Mutation & Polyploidy	Mrs. Dr. K. Borojević
Wild Species & Primitive Forms	Professor J. G. Hawkes
Biometrics in Plant Breeding	Professor G. Wricke

The Secretariat of EUCARPIA was moved from Paris to Wageningen shortly after the Sixth Congress, and has subsequently operated under the guidance of our Treasurer, Dr. Lamberts, to whom we are all very grateful. During the last three years, however, it has become clear that the administrative duties involved in running our Association are greater than can be undertaken by the busy Director of a Research Institute, and, while the overall responsibility has been taken by Dr. Lamberts, much of the day to day management has been delegated to one of his colleagues, Dr. Lange, to whom we are also most grateful. I would also like to take this opportunity to thank Mrs. Meurs Heijnekamp the administrative secretary to the Secretariat, Mr. J. von der Brink, who has given much valuable financial advice, and, of course, our Honorary General Secretary, Dr. Ingold, for all their help during the past three years.

Following the move to Wageningen, the financial organisation of the Secretariat has been reviewed. EUCARPIA now holds banking accounts in Switzerland, where our reserve funds are held, and in Wageningen, where there is a current account for running expenses. Arrangements have been made for both of these accounts to be audited annually, and I will be asking the Treasurer to report on our finances shortly. We have also maintained a small account in Cambridge, which has been used to deal with the very considerable receipts we have obtained, and still continue to obtain, from sale of the proceedings of the Sixth Congress. This account has also been useful in paying my travelling expenses during my tour as President. Its balance is now being transferred to our main account.

I would like next to mention briefly the revised Statutes which have been circulated to all members and which we will be considering shortly. The Statutes of the Association were last amended in 1959. During the subsequent years a number of changes in our code of practice have been introduced, and our numbers have increased to such an extent that changes in representation seemed desirable. The Board have, therefore, gone to considerable trouble to bring the Statutes up to date, and to reword some parts which were ambiguous. There are, however, no changes of substance in these new Statutes, which will, I hope, be adopted when they are formally proposed in a few minutes.

The Board reports with regret the deaths of Dr. S. Broekhuisen, Professor W. Hoffman, Dr. R.L. Knight, Dr. J. Kostecki, Academician P. Lukyanenko and Professor K. von Rosenstiel, who have died during the past three years.

During this period our membership has increased from 650 ordinary and 52 corporate to 775 ordinary and 56 corporate members, representing 24 European and 17 non-European countries, and this I feel is the surest indication that our Association is providing a useful service to plant breeders throughout this continent.

In conclusion, I would like to close by thanking my colleagues on the retiring Board for their help and support and our Hungarian friends and colleagues for their generous hospitality.

FINANCIAL STATEMENT

Total balance in Sw. Fr. per 31-12-1973

debit		31-12-72	31-12-73	*credit*	31-12-72	31-12-73
Typewriter		2964·17	2964·17	Writing-off typewriter	400·00	800·00
Algemene Bank Nederland		581·80	445·22	Creditors	1542·61	
Swiss Bank Corporation		5735·00	4688·00	Capital	72338·36	76010·39
Account depot [1]		30000·00	26713·00			
Cash Debentures		10000·00	10000·00			
Cash Debentures		10000·00	10000·00			
Cash Debentures		5000·00	5000·00			
Debtors		10000·00	14000·00			
Gene Bank Proceedings			3000·00			
	Sw.Fr.	74280·97	76810·39	Sw.Fr.	74280·97	76810·39

[1] *Explanation*

account depot	Sw.Fr.	30000·–
	interest Sw.Fr.	1134·–
	Sw.Fr.	31134·–
transfer Wageningen		4421·–
	Sw.Fr.	26713·–

ANNEX

LIST OF MEMBERS IN EUCARPIA

	Individual	*Associate*
Netherlands	104	1
United Kingdom	97	1
Federal Republic of Germany	93	4
France	91	37
Italy	53	–
Sweden	41	1
Belgium	31	2
Denmark	31	1
USSR	30	1
Yugoslavia	30	2
Spain	29	2
Finland	20	–
Hungary	14	–
Ireland	14	–
Switzerland	14	–
Austria	10	1
Poland	10	–
Israel	9	
Norway	6	–
USA	6	–
Bulgaria	5	1
Czechoslovakia	5	–
Turkey	5	–
Brasil	3	–
Lebanon	3	–
Greece	2	–
Portugal	2	–
Tunis	2	–
Kenya	2	–
India	2	–
Australia	1	–
Canada	1	–
Japan	1	1
Iceland	1	–
Dominican Rep.	1	–
Madagascar	1	–
Marocco	1	–
Nigeria	1	–
Peru	1	1
Roumania	1	–
Uganda	1	–
	775	56

List of participants

AUSTRALIA

DRISCOLL, C.I.	The University of New South Wales School of Botany, P.P. Box 1. 2033 Kensington N.S.W.
LÖKÖS, A.	State Research Farm, Werribe, Vic.

AUSTRIA

BÁNLAKI, A.	Maribo Zuckerrübenzucht, Podlahagasse 1/3, A–1228 Wien
HÄNSEL, H.	Probstdorfer Saatzucht GmbH, Parkring 12, A–1010 Wien
HRON, R.	Bundesanstalt für Pflanzenbau und Samenprüfung, P.O. Box 64, A–1201 Wien
NEPPEL, K.	Saatzucht, A–8200 Gleisdorf
ROHRINGER, K.E.	Saatzucht, A–8200 Gleisdorf

BANGLADESH

MALEK, M.A.	(no address available)

BELGIUM

BOGAERT, G. van	Rijksstation voor Plantenveredeling, 9220 Merelbeke
LAEREMANS, D.A.F.	Katholieke University Leuven, B–3040 Korbeck–LO
NICLAES, J.	Institute for Plantbreeding of Heverlee, De Croylaan 6, B–3030 Heverlee

BULGARIA

Mrs. BAEVA, R.	(no address available)
GEORGIEV, H.	Institute of Genetics and Plant Breeding, Sofia 13
Mrs. PANCHEVA, J.	(no address available)

CZECHOSLOVAKIA

JAVOREK, E.	Maize Research Institute, Trstinska 13, Trnava
KOVÁČIK, A.	Prague 6 – Ruzyně 507
POLERECZKY, O.	Bratislavská, 2630 Piestany
ROD, J.	Vyzkumny Ustav Zakladni Agritechniky 664 62 Hrusovany u Brna
Mrs. TRONIČKOVÁ, E.	Institut of Genetics and Plant Breeding Praha 6 – Ruzyně 507

DENMARK

NIELSEN, H.M.	Department of Crop Husbandry and Plant Breeding, Höjbakkegaard, DK–2630 Taastrup
OVERGAARD, S.	The Breeding Station "Maribo" DK–4960 Holeby
SCHELBECK, E.V.	A/S Daehnfeldt, P.O. Box 15, DK–5100 Odense

EIRE

O'SULLIVAN, T.J.R.	Department of Agriculture and Fisheries Cereal Station, Ballinacura, Midleton, Co-cork.
PURCELL, J.	Department of Agriculture, Cereal Breeding Station, Backweston, Leixlip Co. Kildare

FINLAND

KIVI, E.	Hankkija Plant Breeding Institute, Tammisto, SF–01510 Hgin Pitaja
TIGERSTEDT, P.M.A.	University of Helsinki (Plant Breeding) Helsinki 71

FRANCE

ARNAULT, J.L.	Station de Recherches et Développement P.B. 5, Saint Martin le Beau, F–37270 Montlouis
Miss BERGOUNIOUX, Ch.	Université de Paris, XI. Faculté des Sciences, 91405 Orsay
BERVILLÉ, A.	Université de Paris, XI. Bat. 360, 91405 Orsay
BRICOUT, E.	Institut Technique des Cereales et des Fourrages, 8. Av. President Wilson, F–75116 Paris
DEMARLY, Y.	Université de Paris, XI. 91405 Orsay
DESPREZ, M.	Maison Florimond Desprez, 59–Cappelle par Templeuve

ETASSE, Cl.	I.R.A.T., 110 Rue de l'Université 75340 Paris VII.
FARNARIER, G.	111, Avenue de Lespinet, 31 Toulouse
FOARE, Ch.	Association générale de Production de Maïs, Place de Lestapis, 64 Pau
GALLAIS, A.	Station d'Amélioration des Plantes Fourragères, 86600 Lusignan
GAYRAUD, P.	1, Rue Hegesippe-Moreau, 77160 Provins
GUILLEMAN, Cl.	Société Cooperative Agricole, 60480 Froissy (Oise)
JACQUOT, M.	I.R.A.T., 110 Rue de l'Université 75340 Paris, VII.
JOSSET, J.P.	GRANAX S.A. – Station de Sélection, Ferme de la Grand-Maison Semonville, 28310 Janville
KISS, Ch.	S.A. Maïs Angevin Hodée, 49–Corné
MARCHAIS, L.	Centre National de Recherches Agronomique, Bambey (Sénégal)
PLANCHON, C.	École Nationale Supérieure Agr., 145 Avenue de Muret, 31076 Toulouse-Cedex
SIMIAND, J.P.	S.E.C.O.B.R.A., 78580 Maule – Ivelines

FRG

ALBER, D.	Kleinwanzlebener Saatzucht AG., Postfach 146, 3352 Einbeck
BAUER, R.	Max-Planck-Institut für Züchtungsforschung, 5000 Köln 30
DRESSLER, O.	Carl Sperling et Co., Hamburger-Strasse 27, 3140 Lüneburg
GRÜNEBERG, J.	Breeding Organization "MARIBO", von Hoheneggstrasse 1, 8961 Buchenberg
HEYLAND, K.V.	Institut für Pflanzenbau der Rheinischen Friedrich-Wilhelm-Universität, Katzenburgweg 5, 5300 Bonn
KERCHER, A.	Kartoffelzucht Böhm, 3042 Münster, Ortsteil Trauen
KEYDEL, F.	Hittostrasse 6, 8050 Freising
KNAPP, E.	Max-Planck-Institut für Pflanzengenetik, Rosenhof, 6802 Ladenburg
KOBABE, G.	Institut für Pflanzenbau und Pflanzenzüchtung, von-Siebold-Strasse 8 3400 Göttingen
KRAMER, J.	Nordsaat Saatzuchtges., 2322 Waterneversdorf

Krolow, K.D.	Institut für Vererbungs- und Züchtungsforschung, Albrecht Thaer Weg 6, 1000 Berlin 33
Kuckuck, H.	Institut für Angewandte Genetik, Herrenhäuserstrasse 2, 3000 Hannover
Lein, A.	F. von Lochow-Petkus GmbH, 3411 Stöckheim, Zuchtstation Wetze
Lichter, R.	Max-Planck-Institut für Pflanzengenetik, Rosenhof, 6802 Ladenburg,
Odenagh, W.	Institut für Angewandte Genetik der Technischen Universität Berlin, Albrecht Thaer Weg 6, 1000 Berlin 33
Plarre, W.	Freie Universität Berlin, Institut für Angewandte Genetik, Albrecht Thaer Weg 6, 1000 Berlin 33
Reimann-Philipp, R.	Bundesforschungsanstalt für Gartenbauliche Pflanzenzüchtung, Bornkampsweg, 2070 Ahrensburg
Röbbelein, G.	Institut für Pflanzenbau und Pflanzenzüchtung, von-Siebold-Strasse 8, 3400 Göttingen
Scheller, H.	Bayerische Landessaatzuchtanstalt, Vöttingerstrasse 38, 8050 Weihenstephan-Freising
Schnock, M.	ASGROW GmbH., 7806 Ebnet
Mrs. Schwertfeger, G.	Universität Hohenheim, Hohenheim
Sedlmayr, T.	Kleinwanzlebener Saatzucht AG, Grimsehlstrasse 29, 3352 Einbeck
Seidewitz, L.	Institut für Pflanzenbau und Saatgutforschung, Bundesallee 50, 3300 Braunschweig
Mrs. Snoy, M.L.	Universität Hohenheim, Hohenheim
Strube, H.	Saatzucht Strube GmbH., 3339 Soellingen
Vettel, F.	Saatzuchtwirtschaft F. Strube, Hauptstrasse 1, 3339 Soellingen
Weber, W.E.	Institut für Angewandte Genetik der Technischen Universität Hannover, Herrenhäuserstrasse 2, 3000 Hannover
Wricke, G.	Institut für Angewandte Genetik der Technischen Universität Hannover, Herrenhäuserstrasse 2, 3000 Hannover
Miss Ziegenbein, G.	Forschungsanstalt für Grünlandwirtschaft und Futterbau, 1430 Eichhof in Bad Hersfeld
Zscheischler, J.	Bayerische Landesanstalt für Bodenkultur und Pflanzenbau, Vöttingerstrasse 38, 8050 Freising

GDR

DAMISCH, W.	Institut für Getreideforschung, 4351 Bernburg
GOTTWALD, J.	V.V.B. Saat- und Pflanzgut, Quedlinburg
GRÖBER, K.	Zentralinstitut für Genetik und Kulturpflanzenforschung der AdW., 4325 Gatersleben
HAMMER, K.	Zentralinstitut für Genetik und Kulturpflanzenforschung der AdW., 4325 Gatersleben
KAPPEL, W.	Institut für Getreideforschung, 4351 Bernburg
METTIN, D.	Berliner Strasse 2, 4104 Hohenturm
SCHOLZ, F.	Zentralinstitut für Genetik und Kulturpflanzenforschung der AdW., 4325 Gatersleben
SCHRÖTER, W.	V.V.B. Saat- und Pflanzengut, Quedlinburg

HUNGARY

ALMÁSI, E.	University of Horticulture, H–1118 Villányi u. 35–43
ANDRÁSFALVY, A.	Research Institute of Horticulture, H–1223 Budapest, Park u. 2
ASZTALOS, K.	Head Centre of State Farms, H–1054 Budapest, Akadémia u. 1/3
Miss BABÓTH, É.	Research Institute of Horticulture, H–1223 Budapest, Park u. 2
BAGI, J.	Cereals Trust, H–1024 Budapest, Kisrókus u. 15/b
BALLA, L.	Agricultural Research Institute, H–2460 Martonvásár
BALOGH, S.	Trust of the Canning Industry, H–1097 Budapest, Földváry u. 4
BARABÁS, Z.	Research Institute for Cereal Crops, H–6726 Szeged, Alsókikötősor 5
BARNA, B.	University of Horticulture, H–1118 Budapest, Villányi u. 35–43
BÁCS, B.	Research Institute of Soil Management, H–5300 Karcag
BÁLINT, A.	University of Agricultural Sciences, H–2100 Gödöllő
BEDŐ, Z.	Company for Seed Production and Seed Trade, H–1077 Budapest, Rottenbiller u. 33

BELEA, A.	Biological Research Center of the Hungarian Academy of Sciences, H–6726 Szeged, Odesszai krt. 62
BIACSI, I.	Agricultural Research Institute, H–2462 Martonvásár
BÓCSA, I.	Research Institute of Crop Husbandry and Soil Conservation, H–3356 Kompolt
BÖJTÖS, Z.	Research Institute of Crop Husbandry and Soil Conservation, H–3356 Kompolt
BROKÉS, J.	Company for Seed Production and Seed Trade H–1077 Budapest, Rottenbiller u. 33
BRÓZIK, G.	Research Institute of Horticulture, H–1223 Budapest, Park u. 2
BUGLOS, J.	Research Institute of Crop Husbandry and Soil Conservation, H–3356 Kompolt
Miss BUJDOSÓ G.	Research Institute of Vegetable Crops, H–6000 Kecskemét
BURGERT, R.	Head Centre of State Farms, H–1054 Budapest, Akadémia u. 1/3
CSÁSZÁR, J.	Agricultural Research Institute, H–2462 Martonvásár
CSIZMADIA, J.	Research Institute of Viticulture and Oenology, H–1022 Budapest, Herman Ottó u. 15
CSÓKAY, I.	Head Centre of State Farms, H–1054 Budapest, Akadémia u. 1/3
DÁNIEL, L.	Research Institute of Vegetable Crops, H–6000 Kecskemét
DÉVAY, M.	Agricultural Research Institute, H–2462 Martonvásár
DUFFEK, T.	Company for Seed Production and Seed Trade, H–1077 Budapest, Rottenbiller u. 33
ÉLIÁS, Gy.	Trust of the Sugar Industry, H–1084 Budapest, Tolnai L. u. 25
FARKAS, J.	Research Institute of Vegetable Crops, H–6000 Kecskemét
FAZEKAS, J.	National Seed-Survey Inspectorate, H–1024 Budapest, Kisrókus u. 15/a
GÁSPÁR, L.	Agricultural Research Institute, H–2462 Martonvásár
GÁSPÁR, S.	National Seed-Survey Inspectorate, H–1024 Budapest, Kisrókus u. 15/a
GÁTI, Gy.	Trust of Tobacco Industry, H–1111 Budapest, Budafoki u. 59
GÄRTNER, K.	Trust of Brewery Industry, H–1106 Budapest, Maglódi u. 17

GRÁCZOL, G.	Research Institute for Cereal Crops, Experimental Station, H–9761 Táplánszentkereszt
Miss GRUSZ, E.	University of Horticulture, H–1118 Budapest, Villányi u. 35–43
GULYÁS, J.	Research Institute of Enterprise for Vegetable Oils and Detergents, H–1106 Budapest, Maglódi u. 6
GYÉMÁNT, Gy.	Head Centre of State Farms, H–1054 Budapest, Akadémia u. 1/3
GYULAVÁRI, O.	Research Institute for Cereal Crops, Exp. Stat., H–9761 Táplánszentkereszt
GYURÓ, F.	University of Horticulture, H–1118 Budapest, Villányi u. 35–43
HADNAGY, Á.	Company for Seed Production and Seed Trade, H–1077 Budapest, Rottenbiller u. 33
HAJDU, L.	Trust of Tobacco Industry, H–1111 Budapest, Budafoki ut 59
HALÁSZ, Gy.	Enterprise for Vegetable Oils and Detergents, H–1151 Budapest, Horváth M. u. 2
HERCZEG, M.	Agricultural Research Institute, H–2462 Martonvásár
HERPAY, B.	Head Centre of State Farms, H–1054 Budapest, Akadémia u. 1/3
HORVÁTH, A.	Company for Seed Production and Seed Trade, H–1077 Budapest, Rottenbiller u. 33
IZINGER, P.	Head Centre of State Farms, H–1054 Budapest, Akadémia u. 1/3
JÁNOSSY, A.	Institute of Agrobotany, H–2766 Tápiószele
JÁNOSSY, A. jr.	Biological Research Center of the Hungarian Academy of Sciences, H–6726 Szeged, Odesszai krt. 62
JÁNOSSY, G.	Company for Seed Production and Seed Trade, H–1077 Budapest, Rottenbiller u. 33
KANDRA, Gy.	Trust of Tobacco Industry, H–111 Budapest, Budafoki u. 59
KAPELLER, K.	Research Institute of Vegetable Crops, Experimental Station, H–6300 Kalocsa
KÉZDY, P.	Cereals Trust, H–1024 Budapest, Kisrókus u. 15/b
KLENCZNER, A.	Head Centre of State Farms, H–1054 Budapest, Akadémia u. 1/3
KISS, Á.	Research Institute of Vegetable Crops, H–6000 Kecskemét

Kiss, B.	Trust of Brewery Industry, H–1106 Budapest, Maglódi u. 17
Kiss, J.	Research Institute of Vegetable Crops, H–6000 Kecskemét
Konrád, Z.	Trust of Canning Industry, H–1097 Budapest, Földváry u. 4
Kopátsy, J.	National Institute of Agricultural Variety Testing, H–1024 Budapest, Keleti K. u. 24
Koródi, L.	University of Horticulture, H–1118 Budapest, Villányi u. 35–43
Mrs. Kőrös, L.	Research Institute of Vegetable Crops, H–6000 Kecskemét
Kovács, E.	Research Institute of Horticulture, H–1223 Budatétény, Park u. 2
Kovács, I.	Agricultural Research Institute, H–2462 Martonvásár
Kovács, K.	Agricultural Research Institute, H–2462 Martonvásár
Mrs. Kovács, K.	Research Institute of Irrigation, H–5540 Szarvas
Kovács, L.	Cereals Trust, H–1024 Budapest, Kis-rókus u. 15/b
Kováts, Z.	Research Institute of Horticulture, H–1223 Budatétény. Park u. 2
Kurnik, E.	Research Institute of Forage Crops, H–7095 Iregszemcse
Mrs. Kurucz, J.	Research Institute of Entérprise for Vegetable Oils and Detergents, H–1106 Budapest, Maglódi u. 6
Lengyel, I.	Trust of Brewery Industry, H–1106 Budapest, Maglódi u. 17
Mrs. Lukács, J.	Research Institute of Enterprise for Vegetable Oils and Detergents, H–1106 Budapest, Maglódi u. 6
Lukács, L.	Trust of Sugar Industry, H–1141 Budapest, Komócsy u. 41
Magassy, L.	Research Institute of Beet Crops, H–9463 Sopronhorpács
Mrs. Manninger, M.	Agricultural Research Institute, H–2462 Martonvásár
Manninger, I.	Agricultural Research Institute, H–2462 Martonvásár
Matuz, J.	Research Institute for Cereal Crops, H–6722 Szeged, Alsókikötősor 5

Márkus, F.	Research Institute of Vegetable Crops, Experimental Station, H–6300 Kalocsa
Mesch, J.	Institute of Agrobotany, H–2766 Tápiószele
Mesterházy, Á.	Research Institute for Cereal Crops, H–6726 Szeged, Alsókikötősor 5
Monostory, E.	Head Centre of State Farms, H–1054 Budapest, Akadémia u. 1/3
Mormer, M.	Head Centre of State Farms, H–1054 Budapest, Akadémia u. 1/3
Nagy, I.	Trust of Brewery Industry, H–1106 Budapest, Maglódi u. 17
Miss. Nagy, J.	Company for Seed Production and Seed Trade, H–1077 Budapest
Németh, J.	Research Institute for Cereal Crops, H–6726 Szeged, Alsókikötősor 5
Papp, I.	Head Centre of State Farms, H–1054 Budapest, Akadémia u. 5
Pálos, T.	Trust of Canning Industry, H–1097 Budapest, Földváry u. 4
Mrs. Rajki, E.	Agricultural Research Institute, H–2462 Martonvásár
Rajki, S.	Agricultural Research Institute, H–2462 Martonvásár
Sági, F.	Research Institute for Cereal Crops, H–6726 Szeged, Alsókikötősor 5
Sárvári, I.	University of Agricultural Sciences, H–8360 Keszthely
Schuszter, Z.	Company for Seed Production and Seed Trade, H–1077 Budapest, Rottenbiller u. 33
Simon, I.	University of Agricultural Sciences, H–8360, Keszthely
Mrs. Simon-Kiss, I.	Research Institute of Irrigation, H–5540, Szarvas
Skorka, J.	Research Institute of Irrigation, H–5540 Szarvas
Somos, A.	H–1118 University of Horticulture, Budapest, Villányi u. 35–43
Strahl, A.	Trust of Brewery Industry, H–1106 Budapest, Maglódi u. 17
Sutka, J.	Agricultural Research Institute, H–2462 Martonvásár
Sváb, J.	University of Agricultural Sciences, H–2100 Gödöllő

Szabó, G.	Trust of Sugar Industry, H–1141 Budapest, Komócsy u. 41
Szabó, J.	Company for Seed Production and Seed Trade, H–1077 Budapest, Rottenbiller u. 33
Szabó, L.	Research Institute of Forage Crops, H–7095 Iregszemcse
Szentiványi, P.	Research Institute of Horticulture, H–1223 Budapest, Park u. 2
Szilágyi, Gy.	Agricultural Research Institute, H–2462 Martonvásár
Takács, F.	Head Centre of State Farms, H–1054 Budapest, Akadémia u. 1/3
Takátsy, T.	Company for Seed Production and Seed Trade, H–1077 Budapest, Rottenbiller u. 33
Tamássy, I.	University of Horticulture, H–1118 Budapest, Villányi u. 33–45
Terpó, A.	University of Horticulture, H–1118 Budapest, Villányi u. 33–45
Terstyánszky, Gy.	Head Centre of State Farms, H–1054 Budapest, Akadémia u. 1/3
Toma, L.	Trust of Canning Industry, H–1097 Budapest, Földvári u. 4
Tóth, E.	Ministry of Agriculture and Food, H–1055 Budapest, Kossuth L. tér 11
Tóth, J.	Company for Seed Production and Seed Trade, H–1077 Budapest, Rottenbiller u. 33
Tóth, J.	Head Centre of State Farms, H–1054 Budapest, Akadémia u. 1/3
Török, Sz.	Trust of Canning Industry, H–1097 Budapest, Földváry u. 4
Ujj-Mészáros, K.	University of Agricultural Sciences, H–8360 Keszthely
Varga, S.	Enterprise for Vegetable Oils and Detergents, H–1106 Budapest, Maglódi ut 6
Virányi, S.	Institute of Agrobotany, H–2766 Tápiószele
Vukov, K.	Trust of Sugar Industry, H–1141 Budapest, Komócsy u. 41
Zimmermann, I.	Head Centre of State Farms, H–1054 Budapest, Akadémia u. 1/3

and members of the staff of the Institute of Agrobotany assisting in running the Congress

INDIA

DHAIF, M. A.	Agricultural Research Institute, H–2462 Martonvásár
GUPTA, D.	Agricultural Research Institute, H–2462 Martonvásár
JANA, S.	Crop Science Department, University of Saskatchewan, Saskatoon, Canada S7N OWO
TEWARI, V.P.	Division of Genetics, Iari, New Delhi

ITALY

ALESSANDRONI, A.	Istituto Sperimentale per la Cerealicoltura, Via Cassia Vecchia 176, 00191-Róma
BALDI, G.	Ente Nazonale Risi, Centro di Ricerche sul Riso, Piazza Martiri della Liberta 31, 27036–Mortara (Pavia)
BORGHI, B.	Istituto Sperimentale per la Cereali-coltura, 20079 St. Angelo Lodigiano, Milano
BOZZINI, A.	Centro Studi Nucleari Cassacia, S. Maria di Galeria, 00191–Róma
CHICCO, G.	Asgrow Italia S.P.A., Viale Sabotino 13, 20075 Lodi (Milano)
CUBADDA, R.	Istituto Nazionale Nutrizione (no address available)
LEONE, A.	Istituto Sperimentale per la Cereali-coltura, Via Cassia 176, 00191–Róma
MALAGONI, R.	Ente Nazionale Risi, 27036–Mortara (Pavia)
OTTAVIANO, E.	Istituto di Genetica, Universita di Milano, Milano
Miss PEZZALI, M.	Istituto Sperimentale per la Cereali-coltura, Via Cassia 176, 00191–Róma
PORCEDDU, E.	Laboratorio del Germoplasma, Via Amendola 165/A, 70126 Bari
PORCELLI, S.	Stazione Agricole Sperimentale, Bari
SILVETTI, E.	Centro Appenninico del Terminillo, Via Comunali 43, Rieti
UNCINI, L.	Istituto Sperimentale Orticoltura, Monsampolo, Tronto (AP)

JAPAN

HOSODA, T.	Faculty of Agriculture, University of Tokyo, Kyoiku, Komaba, Meguro-Ku, Tokyo
YAMAGUCHI, H.	Faculty of Agriculture, University of Tokyo, Bunkyo-ku, Tokyo 113

NETHERLANDS

AREND, W.T.M.v.d.	Nunhems Zaden B.V., Haelen (Lb)
BOS, I.	Institute for Plant Breeding, Law. Allee 166, Wageningen
DORSMAN, C.	Institute of Horticultural Plant Breeding, S.L. Mansholtlaan 15, Wageningen
FEEKES, W.	G. Geertsema-Groningen N.V., P.O. Box 405, Groningen
HAL, J.G. van	Unilever Research Duiven, P.O. Box 7, Zevenaar
HOGENSBOOM, N.G.	Institute for Horticultural Plant Breeding, S.L. Mansholtlaan 15, Wageningen
KENNEMA, C.	Breeding Station Zelder, Ottersum
LAMBERTS, H.	Institute de Haaff, Foundation for Agricultural Plant Breeding, P.O. Box 117, Wageningen
LANGE, W.	Institute de Haaff, Foundation for Agricultural Plant Breeding, P.O. Box 117, Wageningen
MASTENBROEK, C.	Plant Breeding Station "CEBECO", Lisdoddeweg 36, Lelystad
NIEUWHOF, M.	Institute for Horticultural Plant Breeding, S.L. Mansholtlaan 15, Wageningen
SNEEP, J.	Institute for Plant Breeding, Law. Allee 166, Wageningen
WEIJS, H.J.H.	Landgoed C.I.V. "ZELDER", Ottersum

POLAND

BACZYNSKI, Z.	63–743 Smolice/Krotoszyn
BOJANOWSKI, J.	Plant Breeding and Acclimatization Institute, Radzikow, Post 05–870 Blonie
Mrs. JAKUBIEC, A.	Novorusynowska Str., Warszawa
KROLIKOWSKI, Z.	63–743 Smolice
Miss NATECZYNSKA, A.	Koscelna 14/16, Warszawa
NALEPA, S.	Cereal Crops Department, ul. Zawila 4a 30–423 Krakow
STARZYCKI, S.	Marsalkowska 34/50, Warszawa
STASZEWSKI, Z.	Plant Breeding Institute, Radzików, Post 05–870 Blonie
SZORA, Z.	Bydgoszcz (no address available)
SOWA, W.	Plant Breeding Institute, Radzików, Post 05–870 Blonie
TRZEBISNKI, J.	Bydgoszcz (no address available)

PORTUGAL

RIBEIRO, M. do Valle	The Agricultural Institute, Carlow Eire

SPAIN

GARCÍA-OLMEDO, F.	Instituto Nacional de Investigaciones Agronomicas, Avda Puerta de Hierro, Madrid–3
SÁNCHEZ-MONGE, E.	Centro de Mejora del Maiz I.N.I.A., Avda Puerta de Hierro, Madrid–3

SUDAN

ABDALLA, A.M.	Khartoum University, Khartoum

SWEDEN

BINGEFORS, S.	Swedish Seed Association, 75105 Uppsala
ESKILSSON, L.	178, Scartho R.D., Grimsby, Lincs., UK
EWERSON, G.	Weibullsholm Plant Breeding Institute, S–26120 Landskrona
HAGBERG, P.	Swedish Seed Association, 26800 Svalöf
JOHANSSON, E.	Hilleshögs Frö AB, P.O. Box 86, 26122 Landskrona
JÖNSSON, I .Ö.	Weibullsholm Plant Breeding Institute, 26120 Landskrona
Mac KEY, J.	Department of Genetics and Plant Breeding, Agricultural College of Sweden, 75007 Uppsala
OLSSON, G.	Swedish Seed Association, 26800 Svalöf
TUVESSON, S.	J.E. Ohlsens Enke Seed Firm, Ny Munkegoard, Taastrup, Denmark

SWITZERLAND

INGOLD, M.	Station Fédérale de Recherche Agronomique, 1260 Nyon
NÜESCH, B.	Eidg. Forschungsanstalt für Landw. Pflanzenbau, 8046, Zürich-Reckenholz

TUNISIA

CHALBI, N.	Faculté des Sciences, Laboratoire de Génétique, 8, rue de Rome, Tunis

TURKEY

ELCI, S.	University of Ankara, College of Agriculture, Ankara

UK

ALDIS, P.R.	Seed Research Ltd. (part of Miln Marster Group) North End Farm, Docking, King's Lynn
DOIG, R.	Bigmore Farm Bungalow, Bigmore Lane, Stokenchurch, Bucks
EMECZ, T.I.	Soil Fertility Dunns Ltd. Hartham, Corsham, (Wiltshore)
ENGLAND, F.J.W.	Scottish Plant Breeding Station, Pentlandfield, Roslin (Midlothian)
FELLNER, C. de	R.W. Gunson (Seeds) Ltd., 20/21 St. Dunstan's Hill, London E.C. 3
GYMER, P.T.	Rothwell Plant Breeders, Rothwell, Lincoln
Miss HANSON, J.	Botany Department, University of Birmingham, Birmingham B15 2TT
HASWELL, D.R.	Suttons Seeds Ltd., London Road, Reading, Berks.
HAYES, J.D.	Welsh Plant Breeding Station, Plas Gogerddan, Aberystwyth, Wales
HOLDEN, J.H.W.	Scottish Plant Breeding Station, Pentlandfield, Roslin EH 25 9RF, Midlothian
HUGHES, W.G.	Plant Breeding Institute, Trumpington Maris Lane, Cambridge CB2 2LQ
INNES, N.L.	National Vegetable Research Station, Wellesbourne, Warwick
LAW, C.N.	Plant Breeding Institute, Trumpington, Maris Lane, Cambridge CB2 2LQ
LUPTON, F.G.H.	Plant Breeding Institute, Trumpington, Maris Lane, Cambridge CB2 2LQ
NORTH, C.	Scottish Horticultural Research Institute, Mylnefield, Invergowrie by Dundee
PAP, A.	Hurst, Gunson, Cooper, Taber Ltd., Witham, Essex
RUDDICK, D.P.	Miln Marster Group Ltd., Chester CH3 5AA
SAGE, G.C.M.	Plant Breeding Institute, Trumpington, Maris Lane, Cambridge CB2 2LQ
THOMAS, R.	Twyford Plant Breeders, Adderbury, Oxon
WILLIAMS, T.	Botany Department, University of Birmingham, Birmingham B15 2TT

USA

Galia, A.N.	20/21 St. Dunstan's Hill, London, E.C.3
Konzak, C.F.	Department of Agronomy and Soils, Washington State University, Pullman, 99163 Washington
Mrs. Thompson, M.	Department of Horticulture, Oregon State University, Corvallis, Oregon 97331

USSR

Brezhnev, D.	V.I.R. All-union Institute of Plant Industry. Herzenstreet 44, Leningrad
Dzuba, V.	Genetics and Cytology Laboratory, Rice Research Institute, Krasnodar
Fedin, M.A.	Laboratory of Heterosis of VASKNIL Institute, Moscow
Mrs. Gorina, E.D.	Minsk, Belorussia (no address available)
Konarev, V.G.	V.I.R. All-union Institute of Plant Industry, Herzenstreet 44, Leningrad
Mazur, K.	Moscow (no address available)
Mrs. Nosowa, P.P.	Research Institute of Tobacco and Mahorka, Krasnodar
Varenitca, E.T.	Agricultural Research Institute, Nemchinkovka, Moskovski obl.

YUGOSLAVIA

Mrs. Borojević, K.	Department of Genetics and Plant Breeding, Faculty of Agriculture, 21000 Novi Sad
Borojević, S.	Department of Genetics and Plant Breeding, Faculty of Agriculture, 21000 Novi Sad
Dokić, P.	Faculty of Agriculture, 21000 Novi Sad
Jošt, M.	Institute for Breeding and Production of Field Crops, P.O. Box 880, 41000 Zagreb
Jovanović, B.	Poljoprivredni Fakultet "Beograd", Nemanjina B.R. 6, 11080 Zemun
Jovanović, C.	Institut za Kukuruz, P.O. Box 89, Beograd
Mrs. Kraljević-Balačić, M.	Faculty of Agriculture, 2100 Novi Sad
Mihaljev, I.	Faculty of Agriculture, 21000 Novi Sad
Milohnić, J.	Faculty of Agriculture, Department of Plant Breeding and Genetics, Zagreb
Milutinović, M.	Faculty of Agriculture, P.O. Box 127, 11081 Beograd-Zemun
Misović, M.	Institute za Kukuruz, P.O. Box 89, Beograd-Zemun

PEJKIĆ, B.	Faculty of Agriculture, Nemanjina 6, P.O. Box 127. 11081 Beograd-Zemun
PETROVIĆ, D.	P.K. "BEOGRAD", Padinska Skela 11213, Beograd
PETROVIĆ, R.	Institute za Kukuruz, P.O. Box 89, Beograd-Zemun
POPOVIĆ, A.	Institute for Small Grains, Kragujevac
SATARIĆ, I.	P.K. "BEOGRAD", Padinska Skela 11213, Beograd
SAVIĆ, P.	Maksim Gorki 30, Novi Sad
TRIFUNOVIĆ, V.	Maize Research Institute, Zemun
ZONIĆ, I.	Faculty of Agriculture, University of Beograd, Zemun-Beograd

ZAIRE

MANYALA, B.	Cren, Kinshasa

F. A. O.

TAHIR, W.M.	Viale di Caracalla 11–13, 00153 Roma, Italy

I. A. E. A.

MICKE, A.	Kärtner Ring 11–13, 1010 Wien, Austria